Agronomics and Agro-Food Marketing

Agronomics and Agro-Food Marketing

Edited by **Alabaster Jenkins**

SYRAWOOD
PUBLISHING HOUSE
New York

Published by Syrawood Publishing House,
750 Third Avenue, 9th Floor,
New York, NY 10017, USA
www.syrawoodpublishinghouse.com

Agronomics and Agro-Food Marketing
Edited by Alabaster Jenkins

International Standard Book Number: 978-1-68286-246-9 (Hardback)

Printed in China.

Contents

Preface

Over the recent decade, advancements and applications have progressed exponentially. This has led to the increased interest in this field and projects are being conducted to enhance knowledge. The main objective of this book is to present some of the critical challenges and provide insights into possible solutions. This book will answer the varied questions that arise in the field and also provide an increased scope for furthering studies.

There has been a paradigm shift in the fields of agronomics and agri-food management because of several factors like globalization, rise in world population, climate variability, etc. This book is a compilation of chapters that discuss the diverse concepts like land economics, financial reforms and subsidies for agricultural sector, sustainable rural development, organic farming techniques, etc. Different approaches, evaluations, methodologies and advanced studies on this subject have been included herein. The book attempts to offer a multifaceted perspective for students and researchers engaged in this field.

I hope that this book, with its visionary approach, will be a valuable addition and will promote interest among readers. Each of the authors has provided their extraordinary competence in their specific fields by providing different perspectives as they come from diverse nations and regions. I thank them for their contributions.

Editor

1

Technical efficiency in the sheep dairy industry: an application on the Sardinian (Italy) sector

Roberto Furesi, Fabio A Madau* and Pietro Pulina

* Correspondence: famadau@uniss.it
Department of Science for Nature and Environmental Resources, University of Sassari, Via Enrico de Nicola 1, Sassari 07100, Italy

Abstract

Sardinia (Italy) is one of the most important European regions for sheep dairy and sheep milk cheese production. However the Sardinian sheep dairy industry is currently going through a dramatic crisis, and verifying whether it can recover part of its profitability is now a priority. Attention is now focused on estimating whether the sheep dairy firms can improve their productivity by more efficient use of their available technical resources. This paper aims to estimate technical efficiency in the Sardinian sheep dairy industry. A stochastic frontier analysis approach was used on panel data from 36 sheep dairy firms over the period 2004–2009 in order to assess whether there are some margins for technical improvements in productivity, given the existing level of technology. A comparative analysis of private firms and cooperatives was also carried out, in order to establish if there were differences in the technology they used and/or their efficiency in using technical inputs. Our findings suggest that there is technological homogeneity among the firms and between private firms and cooperatives. Technical efficiency is equal to 0.905 and it is significantly different between private firms (0.933) and cooperatives (0.877). Our findings have certain implications for what policies should be implemented in order to improve efficiency in the sector and on the orientation of decision makers strategies.

Keywords: Sheep dairy industry, Technical efficiency, Stochastic Frontier Analysis (SFA), Private firms and cooperatives, Sardinia

Background

Sardinia (Italy) is one of the most important EU regions for sheep dairy production. It has more than 3.5 million sheep (3.7% of the EU total in 2009) [a] that totally produce more than 300,000 MT of milk. This quantity corresponds to about 4% of total world production (Istat, 2011; FAO, 2011; Eurostat 2011) and the sheep milk is processed into different types of cheese. Sardinia produces approximately 50–60 thousand MT of cheese a year, manufactured by more than 50 dairy factories, about half of which are cooperatives.

More than half of the cheese produced is Pecorino Romano (PR), one of the most important sheep cheeses in the world. An average of 29,100 MT of PR a year was produced between 2000 and 2009 (Consorzio per la tutela del formaggio Pecorino Romano, 2011). Approximately 70% of PR production is exported (18,500 MT on average a year between 2000 and 2009), with 80% to 85% of the export going to the USA. As Sardinia is the largest supplier of PR to the USA, Sardinian PR represents about 70%

of the volume of the US sheep cheese market (FAS-USDA, 2011). Some processing firms are also exporters. They export their own cheese and what they buy from other dairies.

The US PR market is currently experiencing its worst crisis in years. This is due to a decline in domestic demand and fierce price competition from US cow milk cheese (Idda et al., 2010a). US imports of sheep milk cheese decreased dramatically between 2005 and 2009, from 26,000 MT to less than 20,000 MT (a fall of 24.3%). Imports from Sardinia fell from 18,200 MT to 13,100 MT (a fall of 28.3%). The average cheese prices peaked at $10.70 a kilo in November 2008, started deceasing in 2009, and fell to $6.62 a kilo in June 2010, a drop of 26.9% when compared to the average price in 2008.

Effects of the crisis in the US market were immediately felt in Sardinia. The dairy industry paid Sardinian breeders 20% less for their milk in 2010 than they did in 2008. Sardinian sheep farms operate with low profit margins (Idda et al., 2010b). As a result, if prices were to fall for a prolonged period, many of them would have to go out of business. This would have a serious impact on productivity and the Sardinian economy in general, as well as on externalities and public goods produced by sheep farming in each farm, such as protecting the environment and the landscape, and preserving the vitality of rural communities and the lifestyle of their inhabitants.

Hence, it is of value to see whether the profitability of the Sardinian milk cheese industry can be improved. Given the central role played by the cheese producers – both private firms and cooperatives produce and export cheese – it may be useful to focus on them.

This paper aims to evaluate technical efficiency in the Sardinian sheep dairy processing sector. A stochastic frontier analysis approach was used on a panel data from 36 sheep dairy processors over the period 2004–2009, in order to assess whether there are some margins for technical improvements in productivity, given the present state of technology. In other words, our analysis was focused on estimating whether sheep dairy processors could improve their productivity by using their available resources more efficiently, without considering the prices of outputs and inputs. Particular attention was paid to determine the input–output elasticities, the role of inefficiency in explaining different levels of productivity in different firms, and the effects of a number of particular variables on inefficiency. A comparative analysis of private firms and cooperatives was also carried out, in order to establish if there were differences in the technology they used and/or their efficiency in using inputs.

Methods

Technical efficiency and the Stochastic Frontier Analysis (SFA)

According to Farrell's well-known model (Farrell 1957), technical efficiency is defined as the measure of the ability of a firm to obtain the best production from a given set of inputs (*output-increasing oriented)*, or as a measure of the ability to use the minimum feasible amount of inputs, given a determined level of output (*input-saving oriented*) (Greene, 1980; Atkinson and Cornwell, 1994) [b] .

This sub-section briefly illustrates how technical efficiency output-oriented measures can be obtained from Stochastic Frontier Analysis (SFA) models. SFA was originally and independently proposed by Aigner *et al.* (1977) and Meeusen and van der Broeck (1977). A stochastic frontier production function for panel data can be written as:

$$y_{it} = f(x_{it}, t; \beta) \cdot exp\,(e); \tag{1a}$$

$$e = (v_{it} - u_{it}) i = 1, 2, \ldots .N t = 1, 2, \ldots .T \tag{1b}$$

where y_{it} denotes the level of output for the i-th observation at year t; x_{it} is the row vector of inputs; t is the time index, β is the vector of parameters to be estimated; $f(\bullet)$ is a suitable functional form for the frontier (generally Translog or Cobb-Douglas); v_{it} is a symmetric random error assumed to account for measurement error and other factors not under the control of the firm; and u_{it} is an asymmetric non-negative error term assumed to account for technical inefficiency (Kumbhakar and Lovell, 2000).

The v_i's are usually assumed to be independent and identically distributed $N\left(0,\ \sigma_v^2\right)$ random errors, independent of the u_{it}'s, which are assumed to be independent and identically distributed and with truncation (at zero) of the normal distribution $\left|N\left(0,\ \sigma_u^2\right)\right|$. The Maximum Likelihood Estimation (MLE) of (1) allows us to estimate the vector β and the variance parameters $\sigma^2 = \sigma_u^2 + \sigma_v^2$ and $\gamma = \sigma_u^2 / \sigma_u^2 + \sigma_v^2$; where $0 \le \gamma \le 1$ (Coelli, 1996) [c]. The technical efficiency (TE_i) measure is obtained from the ratio of y_{it} to the maximum achievable level of output (y*) that lies on the frontier ($u_{it} = 0$) and it is given by:

$$TE = \frac{yit}{y*} = \frac{f\,(xit, t, \beta) \bullet \exp(\varepsilon)}{f\,(xit, t, \beta) \bullet \exp(vi)} = exp\,(-u_i) \tag{2}$$

It means that technical efficiency depends only on the inefficiency component under the control of the firm (u_{it}). Assuming a semi-normal distribution for u_{it} and according to Jondrow et al. (1982), we are able to estimate the degree of technical efficiency of each firm.

Some authors have proposed SFA models in which the inefficiency effects (u_i) are expressed as a function of a vector of observable explanatory variables and all parameters – frontier production and inefficiency effects – are estimated simultaneously (Kumbhakar *et al.*, 1991; Reisfschneider and Stevenson, 1991; Huang and Liu, 1994). These approaches were adapted by Battese and Coelli (1995) to take into account panel data. They proposed a model where the functional relationship between inefficiency effects and the firm-specific factors was directly incorporated into the MLE. The inefficiency term u_{it} has a truncated (at zero) normal distribution with mean m_{it}:

$$u_{it} = m_{it} + W_{it} \tag{3a}$$

where W_{it} is a random error term which is assumed to be independently distributed, with a truncated (at -m_{it}) normal distribution with mean zero and variance σ^2 (*i.e.* $W_{it} \ge -z_{it}$ such that u_{it} is non-negative). The mean m_{it} is defined as:

$$m_{it} = Z\,(z_{it}, d) \qquad i = 1, 2, \ldots .N \qquad t = 1, 2, \ldots .T \tag{3b}$$

where Z is the vector (Mx1) of the z_{it} firm-specific inefficiency variables of inefficiency; and δ is the (1xM) vector of unknown coefficients associated with z_{it}. This allows us to estimate the inefficiency effects arising from the z_{it} explanatory variables.

Data and empirical model

Data were collected on a balanced panel data from 36 sheep dairy processors in Sardinia. The sample consisted of 18 private firms and 18 cooperatives [d].

The panel data covered a six-year period from 2004 to 2009 for a total of 216 observations. Information came from the Official Register of Accounts, specifically from the income statement, that firms have to submit to the *"Register of the Companies"* of the Italian Chambers of Commerce, Industry, Handicrafts and Agriculture (C.C.I.A.A.) [e]. In other words, we used economic and financial data reported in the available official balance sheets to describe the sheep dairy production process in Sardinia. We assumed a Translog functional form as the frontier technology specification for the sheep dairy firms. This model is similar to the Battese and Coelli (1995) model, with a non-neutral specification for the production frontier function. Basically, following Huang and Liu (1994), the model assumes that technical efficiency depends on both the method of application of inputs and the intensity of input use (Karagiannis and Tzouvelekas, 2005).

Thus the inefficiency term u_{it} explained by (3) is equal to:

$$u_{it} = d_0 + d_{it}z_i + d_m ln\, x_{mit} + W_{it} \qquad i = 1, 2, \ldots .N \quad t = 1, 2, \ldots .T \tag{4}$$

which allows us to evaluate the role of inputs in conditioning inefficiency.

The Translog stochastic function production model is specified as follows:

$$lnY_{it} = \beta_0 + \sum_{j=1}^{5} \beta_j \ln x_{jit} + \frac{1}{2}\sum_{j\leq}^{4}\sum_{k=1}^{4} \beta_{jk} \ln x_{jit} \cdot \ln x_{kit} + (vit - uit) \tag{5a}$$

where four explanatory variables are used to describe the production frontier (a fifth dummy variable is added in order to identify possible technological differences between private firms and cooperatives).

The dairy output was aggregated into one category (Y), which represents the *value of sheep cheeses* (and by-products) produced by each firm in a certain year (measured in Euros).

The aggregate inputs that were included as variables of the production function are as follows:

- X_1 *Intermediate inputs* used in the production process: the cost of intermediate inputs spent by firms (measured in Euros);
- X_2 *Labour* used in each firm in terms of total wages paid to workers (measured in Euros);
- X_3 *Capital:* measured in terms of annual depreciation rate so as to have a measure of the average annual use of the capital stock (measured in Euros);
- X_t *Time :* the year of observation, so that the technological progress component can be captured (2004 = 1; 2005 = 2; . . ..2009 = 6).

As mentioned above, a further variable was inserted in the model, a *dummy* variable X_p that describes the organisational form of the firms, *i.e.* private firms (X_p = 1) or cooperatives (X_p = 0). This variable was inserted in order to evaluate if there were technological differences between the two types of firm and so to analyse whether they are part of a single or two different production frontiers.

Taking into account formula (4), the inefficiency model (5a) has the following form:

$$u_{it} = \delta_0 + \delta_1 Z_{1it} + + W_{it} \tag{5b}$$

Explanatory firm-specific variables of the inefficiency effects were represented by Z_1. This reflects the *Age* of the firms and - according to the non-neutral model proposed by Huang and Liu (1994) – the same pool of variables (included time) used to describe the production frontier function (x_{it}). Thus the (5b) function is expressed by:

$$u_{it} = \delta_0 + \delta_1 Age_{it} + \delta_2 lnIntermediate\ inputs_{it} + \delta_3 lnLabour_{it} + \delta_4 lnCapital_{it} + \delta_t Time_{it} + W_{it} \tag{6}$$

The variable *Age* was selected in order to evaluate the role of experience in technical efficiency. We assume that older firms are more efficient than newer ones because they have accumulated experience and knowledge (*learning by doing*) that allow them to improve their technical and economic performance (Nelson and Winter, 1982).

Results and Discussion

The Frontier Model

The parameters for the function and inefficiency model were estimated simultaneously, using the computer programme FRONTIER 4.1, created by Coelli (1996). The results of the ML estimates on the Translog function model (5) are reported in Table 1. The original model was, however, repeatedly tested, in order to evaluate whether the adopted model was statistically significant and suitable for the data used.

We used the *Generalised likelihood-ratio test* as a testing procedure. This allowed us to compare a restricted model with the adopted model (Bohrnstedt and Knoke, 1994). The statistic associated with this test is defined as:

$$l = -2lnL = -2\left[\ln \frac{L(H0)}{L(H1)}\right] = -2[ln\ L(H_0) - ln\ L(H_1)] \tag{7}$$

where L(H_1) and L(H_0) are the log-likelihood value of the adopted model and of the restricted model respectively. The statistic test parameter λ has, approximately, a chi-square (or a mixed-square) distribution with a number of degrees of freedom equal to the number of parameters (restrictions). These are assumed to be zero in the null-hypothesis H_0 [f].

A first set of tests was applied to the frontier model. We tested the null-hypotheses of adopting a Hicks-neutral technical change function ($H_0 : \beta_{1t;}\ \beta_{2t;}\ \beta_{3t} = 0$), of adopting a non-technological change in the observed panel ($H_0 : \beta_{t,}\ \beta_{1t,}\ \beta_{2t,}\ \beta_{3t,}\ \beta_{tt} = 0$), and of adopting a Cobb-Douglas specification form to describe the frontier production ($H_0 : \beta_{ij} = 0$), respectively.

We also tested if there were technological differences between private firms and cooperatives. If the null hypothesis $H_0 : \beta_p = 0$ is rejected, this means that cooperatives and private firms use different technologies and lie on different frontier functions.

The results in Table 2 suggest that the former two hypotheses can be rejected, but that the hypothesis of the functional form of the frontier cannot be rejected. This implies that the Cobb-Douglas frontier is a more adequate representation of the data than the Translog frontier.

Table 1 ML estimates for SFA parameters and for TE

Variable	Parameter	Translog [(1)]	Cobb-Douglas [(2)]
FRONTIER MODEL			
Constant	β_0	2.888 *(0.121)*	1.161 *(0.168)*
Intermediate inputs	β_1	0.101 *(0.148)*	0.662 *(0.036)*
Labour	β_2	0.557 *(0.162)*	0.249 *(0.037)*
Capital	β_3	0.012 *(0.044)*	0.098 *(0.023)*
Private/ coop.	β_p	−0.013 *(0.063)*	-
Year	β_t	−0.114 *(0.074)*	0.011 *(0.008)*
(Inter. Inputs × Inter. inputs)	β_{11}	0.098 *(0.022)*	-
(Intermediate inputs × Labour)	β_{12}	−0.278 *(0.160)*	-
(Intermediate inputs × Capital)	β_{13}	0.049 *(0.034)*	-
(Intermediate inputs × Year)	β_{1t}	0.089 *(0.035)*	-
(Labour × Labour)	β_{22}	0.124 *(0.058)*	-
(Labour × Capital)	β_{23}	0.010 *(0.015)*	-
(Labour × Year)	β_{2t}	−0.018 *(0.015)*	-
(Capital × Capital)	β_{33}	−0.013 *(0.011)*	-
(Capital × Year)	β_{3t}	0.013 *(0.009)*	-
(Year × Year)	β_{tt}	0.008 *(0.003)*	-
EFFICIENCY EFFECTS			
Constant	δ_0	0.781 *(0.328)*	-
Age	δ_1	−0.344 *(0.185)*	−0.188 *(0.121)*
Intermediate inputs	δ_2	−0.299 *(0.183)*	−0.172 *(0.152)*
Labour	δ_3	−0.415 *(0.251)*	−0.363 *(0.237)*
Capital	δ_4	0.668 *(0.454)*	0.614 *(0.365)*
Year	δ_t	−0.026 *(0.019)*	−0.028 *(0.020)*
VARIANCE PARAMETERS			
$\sigma^2 = \sigma_u^2 + \sigma_v^2$	σ^2	0.081 *(0.054)*	0.062 *(0.041)*
$\gamma = \sigma_u^2/(\sigma_u^2 + \sigma_v^2)$	γ	0.885 *(0.085)*	0.864 *(0.104)*
$\gamma* = \gamma/\left[\gamma + \frac{1-\gamma}{\pi/(\pi-2)}\right]$	γ^*	0.954	0.946
Log-likelihood function		32.305	23.560
Mean TE		**0.898** *(0.090)*	**0.905** *(0.091)*
Returns to scale		1.022	1.009

[(1)] Adopted Model (Translog) [(2)] Preferred Model (Cobb-Douglas with no δ_0 intercept).
γ^* is described in the text below.

The null hypothesis on technological homogeneity between the two organisational forms is also not rejected, so private firms and cooperatives lie on the same frontier function.

A second set of tests associated with the inefficiency model was conducted using the Cobb-Douglas function model [f]. These tests were designed to verify the following: whether the neutral model is more suitable for the data than the non-neutral one (H_0 : δ_2; δ_3; δ_4 = 0); whether the inefficiency effects are absent from the model ($H_0 : \gamma = \delta_0$; δ_1;. . .δ_n = 0); whether the nature of the inefficiency effects are stochastic or not (H_0 : $\gamma = \delta_0$; δ_2; δ_3; δ_4 = 0): whether the inefficiency effects have an intercept ($H_0 : \delta_0$= 0); whether no firm-specific factors exist ($H_0 : \delta_1$;. . .δ_n = 0); and whether *Age* is a significant explanatory variable of the model ($H_0 : \delta_1 = 0$) [h].

Table 2 Hypothesis testing for the adopted model

Restrictions	Model	L(H_o)	λ	*d.f.*	$\chi^2_{0.95}$	Decision
None	Translog, non neutral	32.31				
$H_0 : \beta_{1t;} \beta_{2t;} \beta_{3t} = 0$	Hicks-neutral technical change	24.77	15.06	3	7.82	Rejected
$H_0 : \beta_{t,} \beta_{1t,} \beta_{2t,} \beta_{3t,} \beta_{tt} = 0$	No technical change	20.86	22.89	5	11.67	Rejected
$H_0 : \beta_{p,} = 0$	Technological homogeneity	30.85	2.90	1	3.84	Not reject
$H_0 : \beta_{ij} = 0$	Cobb-Douglas	23.71	17.19	10	18.31	Not reject
None	Cobb Douglas	23.71				
$H_0 : \delta_2; \delta_3; \delta_4 = 0$	Neutral	17.73	11.96	3	7.82	Rejected
$H_0 : \gamma = \delta_1;\ldots\delta_n = 0$	No inefficiency effects	20.82	5.77	6	11.91*	Rejected
$H_0 : \gamma = \delta_0; \delta_2; \delta_3; \delta_4 = 0$	No stochastic effects	21.06	5.30	5	10.37*	Rejected
$H_0 : \delta_0 = 0$	No intercept	23.09	1.24	1	3.84	Not reject
$H_0 : \delta_1;\ldots\delta_n = 0$	No firm-specific factors	16.86	13.68	4	9.49	Rejected
$H_0 : \delta_1 = 0$	No *Age* effects	21.59	4.04	1	3.84	Rejected

* Critical values with an asterisk are taken from Kodde and Palm (1986). For these variables the statistic λ follows a mixed χ^2 distribution.

Our results suggest that all the above hypotheses can be rejected, with the exception of the hypothesis on the absence of the intercept in the inefficiency model. The preferred model is thus a Cobb-Douglas frontier function in which the parameter δ_0 is equal to zero (see last column of Table 1).

Structure of production and technical efficiency

The estimated function satisfies both properties of *monotonicity* (all parameters are positive) and *diminishing marginal productivity* (the magnitude is lower than unity for each parameter) (Table 1).

The estimated production elasticities suggest that *intermediate inputs* are by large the most important input in conditioning productivity. The production elasticity associated with this variable is 0.662 and this suggests that a 1% increase (or reduction) in intermediate inputs would produce, all other things being equal, a corresponding increase (or reduction) in production of 0.66%. The estimated elasticities of the other two inputs are lower (0.249 for labour and 0.098 for capital). Returns to scale are estimated to be substantially constant (1.009)

The time variable is positive, but the low magnitude (0.011) means that time does not greatly affect production.

Analysis of the technical efficiency of sheep dairy firms shows that, on average, they use their technology with 90.5% efficiency. Since it is an output-oriented measure, this result suggests that, if technology remains at the present levels and sheep dairy firms use their disposable resources more effectively, they would only be able to increase output by about 10%.

Estimation of the ratio-parameter γ gives information on the weight of technical efficiency on production. The estimated γ is significant (for $\alpha = 0.01$) and this implies that differences in technical efficiency among firms are relevant for explaining variations in output in the sample (more than 80% of the variability on the whole). However, this parameter value could not be taken as a measure of the relative contribution of the inefficiency term to total output variance, because this measure can be obtained by estimating the ratio-parameter γ^*.

Parameter γ^* is also significant (for $\alpha = 0.01$), which indicates that about 95% of the general differential between observed and best-practice output is due to the existing difference in efficiency among firms ($\gamma^* = 0.946$), i.e., it suggests that technical efficiency might play a crucial role among the factors affecting productivity in the sheep dairy firms.

Finally, on the basis of the specific estimated scores reported by each firm, there was on average a significant difference (for $\alpha = 0.01$) in technical efficiency between private firms (TE = 0.933) and cooperatives (TE = 0.877) (Table 3).

Source of inefficiency

Our estimated findings suggest that, as expected, the age of the firms (−0.188) is positively (negatively) related to technical efficiency (inefficiency). They suggest that older firms tend to be more efficient, even if only slightly, than newer firms. This is probably because they have accumulated the experience and knowledge (*learning by doing*) that permits them to improve their technical performance over time. In addition, ML estimates of the relationship between technical efficiency and inputs show that all inputs play a significant role in determining efficiency. Using intermediate inputs (−0.172) and labour (−0.028) affects efficiency positively, while we found that a firm's inefficiency was strongly related to the capital endowment (0.614). In other words, technical efficiency tends to decrease when capital increases in sheep dairy firms.

In conclusion, the empirical findings suggest that sheep dairy firms tend to become more efficient over time, although the magnitude of the increase is extremely low (0.028), which indicates that the time factor has only a weak effect on efficiency levels.

Conclusions

The model's results allow us to present some arguments which need to be explored in more depth in further studies.

i. *Technological homogeneity*. 95% of the variance of performances is explained by differences in technical efficiency. In other words, results suggest that any significant differences exist in processing technologies in the sample. This technological homogeneity applies to all types of Sardinian firms and can be partly explained by the overwhelming dominance of PR cheese in the sheep dairy industry, where other products have to share the same factories with PR. In our opinion this is only part of the story, because the same performance levels were found in the few firms where PR was not produced or made up only a small part of the production. This technological homogeneity was found in both private firms and cooperatives, and sheds light on the well-defined strategic choices made by the

Table 3 Estimated technical efficiency in private firms and cooperatives

TE	Private	Cooperatives	TOTAL
Mean	0.933	0.877	0.905
S.d.	*(0.057)*	*(0.110)*	*(0.091)*
Max	1.000	0.980	1.000
Min	0.717	0.544	0.544
No. of firms	18	18	36
No. of observations (6 years)	108	108	216

Sardinian sheep industry as a whole when the question of adopting new technology arises. Constant returns to scale strengthen the idea of widespread homogeneity in technological levels in the region among both large and small firms, almost as though there was a modular processing technique.

ii. *Limited scope for recovering efficiency.* We cannot place great reliance on recovering efficiency. Despite the relative and not absolute meaning of the term "efficiency" used here, we have to recognize that the mean TE of the Sardinian sheep dairy industry is 90%, with private firms having a TE of 93% and cooperatives a TE of 87%. This means that full exploitation of the factors of production would allow each firm to increase its output by a mean value of 606,000 Euros. If one bears in mind that our survey covered 36 of the 50 Sardinian firms, then it is reasonable to assume that the total loss to the industry is 30 million Euros. Thus, this can be considered to have a significant but not decisive impact on the agri-food integrated system.

iii. *The role of raw material, capital and other inputs.* The raw material (milk) to be processed and labour inputs showed high production elasticities and TE. Capital input is, however, associated with a low production elasticity, and has a decreasing effect on TE. This is an unequivocal sign of excessive investment and under-exploitation of the plants' production capacities. As is well known, the firms stop processing activities in the summer due to the lack of milk, and the plants only work for a limited number of hours during the day. Recovering efficiency, where it is worth it, means increasing the volume of milk to be processed and more intensive employment of workers. There is a conflict between arguing for improving efficiency and the present market conditions, where producers are facing difficulties in selling accumulated surpluses of PR. The technical solution to the problem is therefore to change the conditions of the market. Another important factor in recovering efficiency is experience. We found a significant relationship between TE elasticity and the age of the firm. Experience is a strategic tool that can be used to develop the Sardinian sheep dairy industry.

Summing up the above arguments, we can state that the processing part of the Sardinian sheep milk agribusiness carries out its role effectively. In other words, farmers cannot blame industrial entrepreneurs for the differences in prices of the finished products and the prices paid to primary producers for the raw materials. On the other hand, there is some limited space for improving milk processing performances, and these can be exploited at present. Undoubtedly, oversized capital investment, rather than underemployment of raw materials and labour, seems to be the most important limiting factor in the performance of the industry.

Efficient use of processing plants needs to be supported in two ways: first, the raw material needs to be supplied for a longer period; second, the plants need to be used for longer periods of the day. The former objective encounters some problems related to the close link between the availability of natural forage resources and the reproductive activity of the animals. Refrigerating the milk could help to overcome this problem, as well as also having the additional advantage of making milk available for producing fresh dairy products, such as cottage cheese, which are in high demand during the summer. However, it must be recognised that this solution is in contradiction with the problem of the under-exploitation of the plants in the other seasons. This leads us to suggest an alternative solution. Given the accumulation of unsold PR surpluses, suggesting that the more plants should be used more intensively by increasing the volume of milk to be processed seems inappropriate. It is more realistic to envisage a solution where the number of processing plants and their production capacity is appropriate for the market demand for the final product. In other words, analysis of technical efficiency cannot be treated separately from marketing strategies.

The three issues addressed here – technical homogeneity, recovery of efficiency and overcapitalization – seem to suggest that decision makers should converge on the same strategic option, i.e. diversification. This does not necessarily or exclusively mean broadening the Sardinian dairy industry supply basket. It also has to be seen as a strong recommendation for re-organising the entire processing and marketing system by breaking the circle of technological homogeneity and finding new outlets for new as well as existing final products.

Endnotes

[a] After Wales, Scotland and Castile and Leon (Spain), Sardinia has the largest number of dairy sheep in the EU.

[b] When firm operates in a constant return to scale area the input and output-oriented measures coincide (Fare and Lovell, 1978).

[c] Following the suggestion made by Battese and Corra (1977), Battese and Coelli (1995) suggest replacing the parameter $\lambda = \sigma_u / \sigma_v$ originally proposed in the SFA models with γ because it can be searched for between zero and one. This allows us to obtain a suitable starting value for an iterative maximisation process.

[d] Ethical concerns: no experimental research on humans or animals have been carried out.

[e] The authors wish to thank the C.C.I.A.A. of Sassari for allowing us to collect the data. For more information about the Register of the Companies and data, see http://www.registroimprese.it.

[f] When λ is lower than the correspondent critical value (for a given significance level), we cannot reject the null-hypothesis.

[g] Tests on Translog specification produced similar results to these applied on the Cobb-Douglas specification. In Table 2 only the test estimations on the preferred specification of the frontier (Cobb-Douglas) are reported.

[h] If the null hypothesis $H_0 : \delta_m = 0$ is not rejected, the model can be substantially described by the neutral function model explained by (3) and originally proposed by Battese and Coelli (1995).

Competing interests
The authors declare that they have no competing interests.

Authors' contributions
The study was jointly carried out by the authors. However, RF wrote the paragraph 1, FA Madau wrote the paragraph 2 and 3, and PP wrote the paragraph 4. All authors read and approve the final manuscript.

References

Aigner DJ, Lovell CAK, Schmidt PJ (1977) Formulation and Estimation of Stochastic Frontier Production Function Models, J. Econom 6(1):21–37

Atkinson SE, Cornwell C (1994) Estimation of Output and Input Technical Efficiency Using a Flexible Functional Form and Panel Data. Int Econ Rev 35(1):245–255

Battese GE, Coelli TJ (1995) A Model for Technical Inefficiency Effects in a Stochastic Frontier Production Function for Panel Data, Empir. Econ 20(2):325–332

Battese GE, Corra GS (1977) Estimation of a Production Frontier Model: With Application to the Pastoral Zone of Eastern Australia. Austr J Agr Econ 21(3):169–179

Bohrnstedt GW, Knoke D (1994) Statistics for Social Data Analysis. F.E. Peacock Publishers Inc., Itasca

Coelli TJ (1996) A Guide to Frontier Version 4.1: A Computer Program for Stochastic Frontier Production and Cost Function Estimation. CEPA Working Papers 7/96. Department of Econometrics, University of New England, Armidale, of New England, Armidale

Consorzio per la tutela del formaggio Pecorino Romano (2011) Dati sulla produzione., http://www.pecorinoromano.com. Accessed February 8, 2011

Eurostat (2011) Database Statistics., http://epp.eurostat.ec.europa.eu/portal/page/portal/agriculture/data/database, Accessed February 8, 2011

FAO (2011) FAOSTAT, Production, Livestock Primary., http://faostat.fao.org/site/569/default.aspx#ancor. Accessed February 7, 2011

Färe R, Lovell CAK (1978) Measuring the Technical Efficiency of Production. J Econ Theory 19:150–162

Farrell MJ (1957) The Measurement of Productive Efficiency. J R Stat Soc 3:253–290

FAS-USDA (2011) FAS-USDA (2010) U.S., http://www.fas.usda.gov/ustrade, Accessed February 8, 2011

Greene WH (1980) On the Estimation of a Flexible Frontier Production Model, J. Econ 13(1):101–115

Huang CJ, Liu JT (1994) Estimation of a Non-Neutral Stochastic Frontier Production Function. J Prod Anal 5(1):171–180

Idda L, Furesi R, Pulina P (2010a) L'allevamento ovino in Sardegna tra crisi di mercato e politiche per il rilancio. Agriregionieuropa 23:65–68

Idda L, Furesi R, Pulina P (2010b) Economia dell'allevamento ovino da latte – Produzione, trasformazione, mercato. Franco Angeli, Milan

Istat (2011) Sistema Informativo su Agricoltura e Zootecnia., http://agri.istat.it/jsp/dawinci.jsp?q=plB040000030000103200&an=2009&ig=1&ct=204&id=8A|9A, Accessed February 8, 2011

Jondrow J, Lovell CAK, Materov IS, Schmidt P (1982) On the Estimation of Technical Inefficiency in the Stochastic Frontier Production Functions Model, J. Econometrics 19:233–238

Karagiannis G, Tzouvelekas V (2005) Explaining Output Growth with a Heteroscedastic Non-neutral Production Frontier: the Case of Sheep Farms in Greece. Eur Rev Agric Econ 32(1):51–74

Kodde DA, Palm FC (1986) Wald Criteria for Jointly Testing Equality and Inequality. Econometrica 54(5):1243–1248

Kumbhakar SC, Lovell CAK (2000) Stochastic Frontier Analysis. Cambridge University Press, New York

Kumbhakar SC, Ghosh S, McGuckin T (1991) A Generalized Production Frontier Approach for Estimating Determinants of Inefficiency in U.S. Diary Farms. J Bus Econ Stat 9:279–286

Meeusen W, van den Broeck J (1977) Efficiency Estimation from Cobb-Douglas Production Function with Composed Error. Int Econ Rev 18(2):435–444

Nelson RR, Winter SG (1982) The Schumpterian Tradeoff Revisited. Amer Econ Rev 72(1):114–132

Reifschneider D, Stevenson R (1991) Systematic Departures from the Frontier: A Framework for the Analysis of Firm Inefficiency. Int Econ Rev 32(3):715–723

2

Economic performance of small ruminant production in a protected area: a case study from Tell Ar-Rumman, a Mediterranean ecosystem in Jordan

Khalid M Al-Khalidi[1*], Amani A Alassaf[2], Mustafa F Al-Shudiefat[1] and Raed J Al-Tabini[1]

* Correspondence: kalkhalidi@royalbotanicgarden.org
[1]Community Based Rangeland Rehabilitation Programme, Royal Botanic Garden, P.O. Box 99, 11910-Amman, Jordan
Full list of author information is available at the end of the article

Abstract

This paper discusses the qualitative impact of a grazing/herding management program on the profitability of Small Ruminant (SR) herds near the protected area of the Royal Botanic Garden (RBG) in Tell Ar-Rumman, Jordan. As the ultimate goal of establishing a protected area is to rehabilitate and enhance vegetation cover in a sustainable manner, a unique program was set up to meet the needs of local herders and enable the RBG to pursue its biodiversity conservation goals. Financial data pertaining to five local herders were recorded using the accrual accounting system and analyzed for the years 2009 and 2010. Significant changes occurred in the productivity and profitability of the SR herds, with the herders' net incomes increasing by 6% to 159% as a result of direct and indirect support provided by the RBG in 2010. The grazing/herding management program is not only improving the income of participating pastoralists, but is also contributing to the growth and conservation of vegetation and wildlife in Tell Ar-Rumman.

Keywords: Economic performance; Small ruminant; Grazing-herding management

Report

Background

Small Ruminant livestock (SR) refers to sheep and goats. In 2010, the Al-Balqa Governorate in Jordan hosted 7.8% of Jordan's total population of SR (DOS, 2011), primarily produced under traditional systems characterized by inadequate feed and water supply, poor veterinarian services and weak management practices (Alassaf, 2012).

The SR population in Jordan dropped from 7.1 million in 2008 to 5.9 million head in 2009 (DOS, 2011), due to limited grazing options and low rainfall. Livestock owners generally have few resources to invest in technologies and other advanced equipment that would maximize herd productivity. So there was a need to record and analyze herd management decisions in order to find ways to raise profitability. A rise in profitability would be a clear indicator of success. The tangible benefits are regular cash income, meat, milk and manure, and the intangible benefit is the role of SR as insurance in case of emergency (Kosgey et al., 2008).

Economic performance has an important impact on the decision to continue farming in marginal areas (Alassaf et al., 2012; Hadjigeorgiou, and Zervas, 2009). In general, livestock owners in Tell Ar-Rumman (Jordan) are small herders and have adopted a low level of production technology. They have suffered from successive weak production seasons, causing a fragile economic status. But there is always room for improving herd economic efficiency and productivity. An economic and financial analysis can assist herders in making profitable decisions, especially in a context of agricultural sustainability in a protected area (Thomson and Nardone, 1999).

Grazing and herding practices have a great impact on the vegetation and biodiversity of protected areas. It has been found that plant cover in ungrazed plots was 20% greater than in grazed plots (Louhaich et al., 2009). However, the absence of grazing may not necessarily improve the variety of plant species available for livestock utilization, as rangeland genetic erosion can be an attribute of traditional production systems (Thornton et al., 1999).

A protected area, called the Royal Botanic Garden (RBG), was set up in Tell Ar-Rumman with the specific goal of enhancing the quantity and diversity of native vegetation cover. Local pastoralists initially voiced their opposition to the RBG, because a fence was erected around an area they had traditionally used for grazing, and they expected that their livelihood would be endangered.

However, the RBG initiated a collaborative grazing and land management program based on a participatory approach with members of the local community. This study tracks the economic results of the first group of farmers who cooperated with the RBG in this program, with particular attention to qualitative changes in herd management attributed to RBG recommendations, and reflected in the herders' revenue and production costs. An efficiently managed grazing system has been established on land in the protected area, with no negative impact on the herders' income.

Since pastoralism constitutes a deep cultural tradition in Jordan, simply removing animals from the range was never considered to be the right solution. It is well-documented in other parts of the world that removing grazing from a landscape which has traditionally been grazed often has a *negative* impact on species diversity (Al-Tabini et al., 2012; Louhaichi et al., 2009; Thornton et al., 1999). Therefore, after initial biomass studies and grazing trials, the RBG introduced a managed grazing system that allows local herds to graze on RBG land at certain times of the year and under specific conditions.

In 2008, the RBG conducted a rapid assessment of how local herders were managing their herds and the current herd status. It was found that the herds suffered from poor feeding management and weak veterinarian services, which resulted in low productivity. Low productivity was forcing herders to intensify grazing, in order to reduce the high costs of feed, and this was detrimental to the RBG's efforts to rehabilitate vegetation cover and restore biodiversity.

Through close cooperation and consulting between the RBG and the pastoralists, the latter have not only seen improvements in herd health, herd management, and increased profits, but have also adopted a conservation mindset and are now actively protecting the RBG site, having recognized the importance and benefits of conserving biodiversity. At the same time, plant surveys on RBG land, where managed grazing is being allowed, have revealed a significant change in plant species composition, with an increase from 436 species in 2006 to 580 species in 2010 (RBG, 2011).

Methods

Tell Ar-Rumman, like most areas in Jordan, suffers from excessive grazing pressure which, combined with drought, leads to the loss of desirable returning grasses and shrubs. The loss of plants and grasses causes increased soil erosion and reduced availability of grazing resources.

The present study was conducted at the RBG located 30 km northwest of Amman, Jordan, near the village of Tell Ar-Rumman (32°10'57" N, 35°49'38"E), on 2 km^2 of land donated by the Ministry of Agriculture for the establishment of the RBG, a new botanical garden and protected area. The RBG is intended to be a center for public education, scientific research, and biodiversity conservation.

Data were gathered in 2009 and 2010 from five herders who used to illegally graze their herds on the site. The RBG conducted training courses in 2009 aimed at introducing skills and practices in SR management, including herd management, veterinarian training on common disease diagnosis and treatment, and herd record keeping. The RBG provided additional veterinarian services: a vaccination program, training for a veterinarian extension worker, and a veterinarian clinic. The RBG also supplied supplementary feed, to compensate the loss of rangeland, to cover the daily feed requirements of the herds, especially in mating and late pregnancy periods.

The RBG staff trained the herders on how to record herd production data using structured records. The herders recorded their production data and related costs on a daily basis under the supervision of the RBG staff. Herd production data were collected and analyzed for each herd separately, and showed the productivity and net revenue per head. The collected data included the following:

- Production features and revenue: herd size and types (sheep, goat), lambs, milk production and prices, sales and prices of livestock production (lambs, sheep, and milk).
- Production costs: the expenses included feed cost, transportation, medical services, water, and labor.

The accounting system was designed to obtain reasonable accuracy for the net income (revenue) over a given period of time. It was assumed that 2009 was the second productive year for each herd. The net revenue calculated in this study included the opportunity costs of herders, in the case of family-based herding. Later the data were recorded and analyzed using MS Excel.

Results and discussion

Economic performance of SR production

Small Ruminants are raised for different reasons. Some people in Jordan are motivated to raise SR for additional income, while others depend on SR as a main source of income, for domestic food, and as simple investments they are familiar with.

Herd sizes depend on the purpose for raising the livestock. Small herds are mainly meat producing, as well as fiber and milk producing. More importantly they are considered as an in-hand cash source.

Table 1 represents the herd size, total net revenues and per head revenues for 2009 and 2010. The percentage of change was calculated to determine changes in herd size and revenues, and thus to represent improvements, if any, in performance due to RBG recommendations on herd management. The results showed a fluctuation in herd size between −3% and 3%, except in the case of Herder 4 who owns a goat herd. In the latter case, a 14% increase was seen in 2010, as goats are known for high twinning rates.

Changes in herd sizes were inconsistent with changes in revenue from 2009 to 2010. The herders who had either sheep or goats (but not both) generated higher revenues in comparison to those who owned both. This indicates that specialization in one type of SR might improve herd management and make it easier to meet expected revenue goals.

In detail, Table 1 shows that Herder 2's goat herd had the highest change in revenue (70%), in spite of having a negative change (−2%) in herd size for the two years 2009 and 2010. This change in revenue may be attributed to improved herd and grazing management practices and a consequent increase in lambs and milk per unit of production. It is important to emphasize that the changes in revenue are not attributed to changes in prices or market availability. The product prices and market outlets for these herders was the same in 2009 and 2010.

On other hand, the change in revenue for Herder 3 was the least (−3%), while Herder 5 had an exceptional reduction in revenue (−55%). Both of their herds were mixed, consisting of both sheep and goats.

The herders who paid most attention to the RBG's recommendations derived greater benefits than those who did not fully implement the recommendations.

Factors affecting economic performance of SR production

Production costs In general, there are two types of costs associated with producing an agricultural product: variable and fixed costs. Variable costs vary according to the size of the herd, whereas fixed costs occur regardless of the level of output. Examples of

Table 1 Herd size and revenue in 2009 and 2010

Herder name	Animal type	Herd size			Revenue				
		2009 head	2010 head	% of change	2009 JD®	2010 JD®	Per head (2009) JD	Per head (2010) JD	% of change
Herder 1	Sheep	250	243	−3	20,290.80	33,547	81.16	138.05	60
Herder 2	Goat	58	57	−2	2,541.63	3,626.80	43.82	63.63	70
Herder 3	Sheep	129	133	3	12,423.28**	12,093.40**	58.87	56.25	−3
Herder 3	Goat	82	82	0					
Herder 4	Goat	125	146	14	3,445.79**	6,213.47**	27.56	42.55	55
Herder 5	Sheep	35	34	−3	886.56	489.26	10.19	6.52	−55
Herder 5	Goat	52	41	−3					

® JD: Local currency, Jordanian Dinar.
**Revenue for sheep and goats.
Source: Income statements of 2009 and 2010.

variable costs include feed, medicine, labor, and supplies. Fixed costs include depreciation on the herd and buildings, land, equipment, and rents. In this group of herders, fixed costs represented a very low portion out of the total costs. The herding style is traditional, where concrete buildings, vehicles and equipment are in limited use, and the herders do not have to pay rent for the land they graze on. The traditional style of herding involves mainly variable costs.

Table 2 presents the percentages of variable cost items out of the total production costs in 2010. The main purpose of this calculation was to compare the relative costs, to reflect different management decisions and their impact on reducing the production cost and thus increasing the net revenue.

The records show that feed and labor accounted for the highest share of the total costs. The feed cost was 51% to 86%, while labor was 11% to 44%. Understanding this contribution to the cost shares gives insights on how herders could better manage costs to improve their profitability. Herders can control feed costs by maximizing the use of pasture, producing their own harvested feeds, buying and storing feed in bulk, and minimizing feed wastage. Even though the cost of veterinarian medications is very low, it accounted for 0% to 3% of the total variable cost, and had a tangible impact on improving herd productivity.

Lamb sales, breeding and mortality rates A short lambing season is a result of successful herd management. On the other hand, the number of lambs born per ewe is certainly an economically important attribute in an SR enterprise. Profitability is largely determined by the number of lambs sold per ewe, as selling lambs is the main return from SR farming. Therefore, a great deal of effort should be devoted to taking care of animal health before, during, and after birth. Knowledge of when and how lamb mortality occurs can be helpful in minimizing the mortality rate. The percentages of sales, breeding, and death losses from the total number of lambs give quantitative measures for herd management. For this group of herders, lambs were either sold, died or were replaced into the herd, as the herders decided.

The high animal mortality rate for some herders indicated poor herd management, specifically with respect to veterinarian services. In 2009, the RBG staff conducted trainings on different veterinarian topics, which helped control the mortality rate in almost all the herds. Herder 3, however, had the highest mortality rates due to a mismanagement decision to purchase sick lambs and kids. It is important to know when and

Table 2 Percentage of variable cost items out of total costs recorded in 2010

Item	Herder 1 sheep	Herder 2 goat	Herder 3 sheep	Herder 3 goat	Herder 4 goat	Herder 5 sheep	Herder 5 goat
Feed	86%	62%	79%	61%	62%	51%	53%
Labor	11%	34%	18%	36%	33%	42%	44%
Transportation	0%	0%	0%	0%	0%	0%	0%
Water	1%	0%	1%	0%	2%	3%	1%
Maintenance	0%	1%	1%	0%	0%	1%	1%
Medications	1%	3%	1%	1%	2%	2%	0%

Source: Calculated from 2010 records.

how lamb and kid mortality occurs, in order to avoid such losses in the future. This information should be included in the herd's records to handle future losses.

Table 3 shows the lambing rate in 2009 and 2010, and indicates an improvement for most of the herds. In general, the lambing rate was higher in 2010, except for Herders 3 and 5. Herder 3 had a slight decrease from 67% to 62%, while Herder 5 reduced the size of his goat herd from 52 head to 41 head in 2010, which negatively affected the lambing rate for the year.

Lambs and milk sales are the primary source of income for small herds in Jordan, with minor contributions from the sale of manure and wool. Milk productivity depends on a number of factors such as the nutrition and health of the milking flock. Herders need to give more attention to these factors to ensure a productive and constant milking season with high quality milk, and to gain increased revenues.

As demonstrated in Table 3, lamb sales counted for 26% to 73% of the total sales. These figures explore how the revenue from selling lambs could cover the variable costs of production, especially feed costs, leaving the revenue from milk as net revenue for the herders.

Variable costs as a percentage of revenue A high level of SR sales does not necessarily lead to high revenue. There is a need to examine the costs associated with these sales. This examination is helpful in determining how the sales could cover the variable costs used in production.

Table 4 shows that sales accounted for a 32% recovery of the variable costs out of total sales for Herder 1, indicating that Herder 1 had the most efficient herd management of all the herders. At the opposite end of the scale, Herder 5 used 93% of his total sales to cover variable costs of sheep production, and 86% of goat sales to cover variable costs. In other words, a major share of his sales revenue was used to recover variable costs, which decreased the herd's profitability.

Impact of the RBG's direct and indirect support on herders' net profits

The RBG introduced a grazing/feeding plan for the herders, to measure the impact of managed grazing and herding on both SR profitability and the biodiversity of the RBG site. The plan was based on direct and indirect feed supplements. RBG support meant a reduction in feed cost that would improve the net revenue. The direct supplement was the provision of barley based on the daily feed requirements for mating and late

Table 3 Lambing rate in 2009 and 2010, and lamb distribution and sales in 2010

Herder	Type	Total 2009	Total 2010	Lambing rate 2009	Lambing rate 2010	Lamb distribution in 2010			Milk and lamb sales as a % of total sales	
						Sales	Breeding	Mortality	Lamb %	Milk %
Herder 1	Sheep	250	243	79%	99%	75%	23%	2%	44%	21%
Herder 2	Goat	58	57	66%	79%	96%	0%	4%	73%	13%
Herder 3	Sheep	129	133	67%	62%	57%	27%	16%	44%	24%
Herder 3	Goat	82	82	47%	74%	58%	21%	21%	48%	24%
Herder 4	Goat	125	146	30%	98%	72%	23%	6%	58%	5%
Herder 5	Sheep	35	34	96%	112%	74%	11%	8%	58%	10%
Herder 5	Goat	52	41	82%	70%	38%	58%	5%	26%	21%

Table 4 Variable costs as a percentage of revenue in 2010

Name	*Type*	Variable costs as a percentage of revenue (%)
Herder 1	Sheep	32
Herder 2	Goat	53
Herder 3	Sheep	48
Herder 3	Goat	76
Herder 4	Goat	47
Herder 5	Sheep	93
Herder 5	Goat	86

pregnancy periods, while the indirect supplement referred to periods of managed grazing inside the RBG site. These grazing periods were defined by a specific estimation of the biomass and stocking rate in the RBG site on a yearly basis. The herders were allowed to graze for a specific number of hours per day on pre-determined days, for the equivalent of 50% of the daily requirements per SR head.

The RBG grazing/herding management assisted the herders to decrease their variable costs and improve profits in 2010. Table 5 shows the impact of RBG in-kind support, in the form of direct and indirect supplements, on the revenue of the herders. The results range between 6% and 159%. Since Herder 5 did not follow RBG recommendations and ended up with very low revenue, the RBG support appears extremely high in this table. Supplementary grazing (indirect support) improved the revenue by 4% to 34%, while grazing plus supplemental feed (direct and indirect support) resulted in a 6% to 159% improvement in revenue. Some herders would have made zero net revenue, or suffered a loss, if the RBG grazing/herding management had not been offered. Moreover, without the RBG's program, such a reduction in herd revenue would have been detrimental to the RBG's vegetation recovery rate, because the herders would have cut fences to enter RBG land and graze everywhere possible in an uncontrolled manner.

Conclusion

To understand economic performance, this study analyzed the profitability of SR production for five herders at close coordination with the RBG. Profits were driven by cost control and gross production. The study suggests that traditionally measured parameters such as lambing rate, cost per head and death loss, although they do have an effect on net profit, are not enough to determine best practice management in SR flocks.

Table 5 Impact of RBG support on net income of herders

RBG support	Herder 1	Herder 2	Herder 3	Herder 4	Herder 5
Indirect: Pasture (JD®)	1,346	269	673	875	404
Direct: Barley (JD®)	805	74	833	722	610
Total JD	**2,151**	**343**	**1,506**	**1,597**	**1014**
Indirect revenue (%)	4	7	6	14	34
Direct revenue (%)	2	2	9	12	125
Overall net revenue %	**6**	**9**	**15**	**26**	**159**

®JD: Local currency, Jordanian Dinar.

Herders are strongly advised to use a profit analysis method to determine and monitor the profitability of their flocks, as well as obtain clear insights for future business and management decisions.

In addition, the study suggests that SR herders should maintain historical records allowing them to measure progress over time. Herders have to understand how to increase the productivity of each head rather than just increase the number of head per herd. The role of specialists in animal health and nutrition is very important in helping the livestock owners reach this goal.

This multi-faceted support and managed grazing program for livestock herders, using a participatory approach, has been shown to raise the income of herding families while helping to conserve and augment the vegetation cover and wildlife in Tell Ar-Rumman. Future biodiversity conservation polices and initiatives should therefore consider introducing a similar program to promote better livelihoods for herders and enhance the sustainable growth and conservation of biodiversity.

Competing interests

The authors declare that they do not have any financial competing interests (political, personal, religious, ideological, academic, intellectual, commercial or any other) with other parties for this specific research topic.

Authors' contributions

KA had made a significant contribution to the concept and design of the research theme, he participated in data collection, analysis and interpretation the results. AA had performed data analysis and participated significantly in drafting manuscript. MA: had have been involved in drafting the manuscript and in interpretation the results. RA had participated in manuscript design, coordination and revision of this manuscript. All authors read and approved the final manuscript.

Acknowledgements

We would like to express our gratitude to Her Royal Highness Princess Basma bint Ali, founder of the Royal Botanic Garden (RBG) of Jordan, for her encouragement and continued support, to RBG Executive Director Tariq Abu Taleb, and the Association française de développement and Fonds français pour l'environnement mondial for funding the project. Thanks is also extended to Tell Ar-Rumman's community for their help and Habiba Dingwall for her kind support in editing the paper.

Author details

[1]Community Based Rangeland Rehabilitation Programme, Royal Botanic Garden, P.O. Box 99 11910- Amman, Jordan.
[2]Department of Agricultural Economics and Agribusiness, Faculty of Agriculture, University of Jordan, P.O. Box 791 11910- Amman, Jordan.

References

Al-Assaf A (2012) Economic implications of small ruminant diseases in the Northern Area of Jordan. Journal of Food, Agriculture & Environment 10(1):323–326

Al-Assaf A, Majadalawi M, Nawash O (2012) Factors affecting farmer's decision to continue farm activity in marginal areas of Jordan. African Journal of Agricultural Research 6(12):2755–2760

Al-Tabini R, Al Khalidi K, Al-Shudiefat M (2012) Livestock, medicinal plants, and rangeland viability in Jordan's Badia: through the lens of traditional and local knowledge. Pastoralism: Research, Policy and Practice 2:4

DOS (Department of Statistics) (2011) Annual Statistical Report. Amman Jordan 1:17–18

Hadjigeorgiou I, Zervas G (2009) Evaluation of production systems in protected areas: Case studies on the Greek Natura 2000 network. In: Pacheco F. (ed.), Moran d-Fehr P. (ed.). Changes in sheep and goat farming systems at the beginning of the 21st century: research, tools, methods and initiatives in favour of a sustainable development. Zaragoza: CIHEAM / DRAP-Norte / FAO, 2009. pp 101–111

Kosgey I, Rowlands G, Van Arendonk J, Baker R (2008) Small ruminant production in smallholder and pastoral/extensive farming systems in Kenya. Small Ruminant Res 77:11–24

Louhaichi M, Salkini K, Petersen S (2009) International Journal of Agricultural & Biology. Effect of Small Ruminant Grazing on the Plant Community Characteristics of Semiarid Mediterranean Ecosystems. 6:681–689

Garden RB (2011) Yearly plant survey. Unpublished Report, Amman Jordan

Thomson P, Nardone A (1999) Sustainable livestock production: Methodological and ethical challenges. In: Livestock Production Science, 61st edition, pp 111–119

3

Understanding the process of adaptation to climate change by small-holder farmers: the case of east Hararghe Zone, Ethiopia

Yibekal A Tessema[1*], Chanyalew S Aweke[2] and Getachew S Endris[2]

* Correspondence: yibekalab@gmail.com
[1]School of Agricultural Economics and Agribusiness, Haramaya University, Dire Dawa, Ethiopia
Full list of author information is available at the end of the article

Abstract

This study examined smallholder farmers' level of perception about climate change, source of information on climate change, types of adaptation strategies, factors influencing adaptation choices and barriers to adaptation in Eastern Hararghe Zone, Ethiopia. The surveyed farm households in the study area perceived at least one aspect of climate change primarily through their life experience. Planting trees is the major adaptation measure and 89.1 percent of the farmers took this adaptation strategy. Most farmers (96 percent) believe that deforestation is the main cause of climate change and the choice of farmers to plant trees as an adaptation strategy may be partly a mitigation strategy. However, the majority (49.6 percent) of the households employed at least one adaptation response on top of tree planting. The other adaptation strategies include: early planting, terracing, irrigation and water harvesting. The main source of information for these adaptation strategies for 58.4 percent of the respondents is from extension advice. Results of a multinomial logit model showed that non-farm income, farmer- to-farmer extension, access to credit, distance to selling markets, distance to purchasing markets, and income affect the choice of adaptation strategies. Finally, the study identified lack of information as the most important barrier to climate change adaptation. The other barriers include: lack of farm inputs, shortage of land, lack of money, lack of water and shortage of labor.

Keywords: Climate change; Ethiopia; Perception; Adaptation; Multinomial logit model

Background

Climate change has adversely affected the livelihoods of people in developing countries where a large proportion of the population is heavily dependent on agriculture. Similarly, climate change has exacerbated poverty, food insecurity and vulnerability of agrarian communities in Sub-Saharan Africa (Akponikpè et al. 2010; Bryan et al. 2009). Ethiopia, a country located in the Horn of Africa, where the agricultural sector accounts for about 52 percent of the GDP and 85 percent of the foreign exchange earnings, and employs about 80 percent of the population (CSA Central Statistics Authority 2004) could be a typical example of the impact of climate change on the vulnerable rural communities in the developing world.

A recent mapping on vulnerability and poverty in Africa has positioned Ethiopia as one of the most vulnerable countries in the continent with the least adaptive capacity to climate change. In fact, the country has suffered from at least five major national droughts since 1980, not mentioning numerous local droughts all over the country. In the years between 1999 and 2004 alone, more than half of all households in the country faced at least one major drought. These cycles of drought have effectively kept the population in vicious poverty nullifying the endeavors of the population in creating assets and increasing its food security status (ILRI 2006; Stige et al. 2006; ACCCA 2010).

However, the impact of climate change is not limited to the occurrence of droughts (Lautze et al. 2003). Other important climatic variables are also changing implying the multiple repercussions of the change in climate. Over the past decade, average minimum and maximum temperatures of the country have increased by around 0.25°C and 0.1°C respectively. Further, it is expected that in the year 2050, mean temperature will increase by 1.7-2.1°C (Tadege 2007). Most climate models support this increase in temperature. However, there is a contradiction as to the change in precipitation, where both increase and decrease in precipitation are forecasted depending on the climate models employed (Strzepek and Mccluskey 2006). Nevertheless, high variability in precipitation is observed in the country over the past decade (Deressa et al. 2011).

The change in climate is inevitable, at least in the near future, and Ethiopian farmers are now faced with adapting to this inevitable change in climate. Following IPCC (2001), adaptation to climate change refers to the "adjustment in natural or human systems in response to actual or expected climatic stimuli or their effects, which moderates harm or exploits beneficial opportunities". Adaptation could be effected at different scales: individual or farm-level, and national or international level. Even though there will be some autonomous adaptation at farm-level, this adaptation is inadequate and needs the intervention of institutions like governmenet (Maddison 2007; Smit and Pilifosova 2001). Adaptation actions at national or international level, however, entail an understanding of the process of autonomous adaptations at farm-level (Bryan et al. 2009).

Adaptation at farm-level involves two stages: perceiving the change in climate, and deciding whether to adapt or not, or which adaptation strategy to choose (Maddison 2007). There are still important questions on perception that need to be addressed, such as: Are farmers able to perceive the change in climate in the long run? Which changes are they able to perceive? What economic, social and institutional factors influence their level and speed of perception (Vedwan and Rhoades 2001; Falco et al. 2011; Deressa et al. 2009)? Perception is not, however, an adequate condition for adaptation. Farmers who have perceived the change in climate may not adapt or the nature of their adaptaion respones may vary as a result of a complex interplay between social, econmic and institutional factors (Maddison 2007; Maharjan et al. 2011; Bryan et al. 2009).

There is a large deficit of information on the process of adaptation in the developing world including Ethiopia (Smit and Pilifosova 2001; McSweeney et al. 2010). There are currently few research outputs in Ethiopia, almost all focused only to the Nile Basin of the country. Deressa et al. (2009), Deressa et al. (2011) and Falco et al. (2011) studied the perception and adaptation process in the Nile Basin of Ethiopia. They were able to identify factors that affect perception, adaptation decisions, and also highlighted the main barriers to adaptation in the study areas. On the other hand, Maddison (2007) and Bryan et al. (2009) compared the adaptation process in Ethiopia with other African countries, again based on the survey in the Nile basin of Ethiopia. All the studies carried out in the Nile Basin of Ethiopia were dependent on a single survey made during 2004/2005 on 1,000 farmers. However, ACCCA (2010) studied the nature of adaptation in the drought prone areas of Tigray, northern Ethiopia, based on a survey of 160 farmers.

The studies carried out so far are only limited to the Nile Basin of Ethiopia. There is a need to study the other parts of the country: the southern and eastern parts. There are regions in the country which could provide new insights on the adaptation process at farm-level. East Hararghe Zone of Oromia regional state could be a very good example in this respect. The Zone is characterized by farmers with very small average landholdings, even compared to the national and regional average, and follows a mixed crop-livestock production system consisting of cash crops (such as Khat and coffee) and cereals, unlike most of the areas in the Nile Bain. The Zone is also one of the food insecure areas of the country having a large population with frequent need for emergency food aid (Niemistö 2011). Hence, a study on adaptation to climate change in this area could supplement the current knowledge on the adaptation process in the country and could subtantially contribute to plan development interventions in the Zone. Moreover, it seems that most studies (Deressa et al. 2011; Deressa et al. 2009; Falco et al. 2011; Bryan et al. 2009) assumed that farmers have perceived climate change gradually by themselves and did not explicitly ask the source of their perception on climate change, which could be training given by extension agents. Unlike many studies, this study also seprately considers distance to selling markets and distane to purchasing markets as determinants of adaptation decisions.

This study, therefore, investigated the whole adaptation process starting from perception to adaptation responses in Eastern Hararghe Zone. It examined the process and degree of smallholder farmers' perception and adaptation to climate change in the study area. It also identified factors governing smallholder farmers adaptation responses and the most important barriers to adaptation.

Methods

Study area, sampling and data collection

The study area is East Hararghe Zone, Oromia National Regional State, Ethiopia. It is located in the eastern part of the regional state and the country. Following the national census in 2007, the Zone with an area of 17,930 km^2 has a total population of 2,723, 850 and 580,735 households (CSA 2007).

The study employed a multistage sampling technique. In the first stage, out of the 18 *Woredas* (districts) in the Zone, three *Woredas* were purposively selected to include different attributes of Eastern Hararghe Zone with respect to agro-ecological zones (traditional typology of agro-ecological zones in the country), and agricultural production systems. The selected *Woredas* are Haramaya, Kersa and Babile. Haramaya *Woreda* has an area of 561.6 km^2 and a rural population of 154,690 and is found in the northern central part of the Zone. Kersa is located in the northern part of East Hararghe Zone and has an area of 544.9 km^2 that is home to about 119,542 rural inhabitants. Both Haramaya and Kersa are characterized predominantly by *Woinadega* (middle land 1500–2500 m.a.s.l.) agro-ecology and mixed crop-livestock farming is the dominant production system. On the other hand, Babile *Woreda* is found in the eastern part of the study area with an area of 3,022 km^2 and with a rural population of 42,858. The largest part of this *Woreda* falls under *Kolla* (lowland, 500-1500 m.a.s.l) agro-ecological zone. In addition to the mixed crop-livestock farming, the agro-pastoral production system is an important characteristic of the *Woreda* (ONRSPO n.d.).

From the selected *Woredas*, 2 *Kebeles* (the smallest administrative unit) were randomly selected and from each *Kebele* 20 households for Haramaya and Kersa, and 15 households from Babile were randomly selected and included in the study. A smaller sample size was allocated to Babile as it represents a relatively smaller proportion of the population in East Hararghe Zone (ONRSPO n.d.). Hence, a total of 110 farm households were included in the study and the survey was carried out in 2012. However, during data processing, we dropped one of the questionnaires as it had missing values. The non-missing variables of the questionnaire were examined and no peculiar characteristics were found justifying the decision for list wise deletion, which is the best option under such situations (Allison 2002).

To collect the data, a semi-structured interview schedule was employed. Enumerators were trained for one day to familiarize them with the issues of climate change, farm-level adaptation to climate change, the significance of research on climate change adaptation, and the basic concepts in sampling, interviewing and processing data.

Data analysis

This study employed both qualitative and quantitative analysis techniques. The qualitative analyses used interpretations, comparisons and arguments. The quantitative analyses made use of both descriptive and inferential statistical techniques. The descriptive statistical techniques applied in the study include percentages and graphs. Inferences are made using a multinomial logit model (MNL). MNL is widely employed in climate change adaptation (e.g., Deressa et al. 2009; Sofoluwe et al. 2011; Gbetibouo 2009) and technology adoption (e.g., Uaiene et al. 2009; Paudel et al. 2011; Jariko et al. 2011) studies. It is employed when the dependent variable has more than two outcomes or, in our case, more than one adaptation response to climate change (Madalla 1983; Greene 2003).

To describe the MNL model, let y denote a random variable taking on the values {0;1;.....;J} where J is a positive integer and x denote a set of conditioning variables. In our study, y would be adaptation measures taken by households whereas x represents the explanatory variables hypothesized to influence the choice of the available adaptation options. The MNL model is employed to show how ceteris paribus changes in the elements of x influence the response probabilities, $P(y = j|x), j = 0, 1, \ldots, J$. $P(y = j|x)$ is known after determining the probabilities for $j = 0,1,2,\ldots,J$, which must sum to unity.

The MNL model has response probabilities given as Equation (1):

$$P(y = j1x) = \frac{\exp(x\beta j)}{1 + \sum_{i \neq j} exp(x\beta j)}, j = 1, \ldots, J \tag{1}$$

The parameter estimates of the MNL model only show the direction of the relationship between the dependent and independent variables. Therefore, to determine the actual magnitude of change or probabilities, the marginal effect of the explanatory variables, Equation 1 is differentiated over the explanatory variables to give Equation (2):

$$\frac{\partial P(y = jlx)}{\partial x_k} = P(y = jlx)\left[\beta_{ik} - \frac{\sum_{i \geq 1} \beta_{ik} exp(X\beta i)}{1 + \sum_{i \geq 1} exp(X\beta i)}\right] \tag{2}$$

The MNL, however, works under the assumption of the Independent Irrelevant Alternatives (IIA). Following this assumption, the odds of any two outcomes are independent of the remaining outcomes available. Hence, omitting or adding outcomes should not affect the odds of the remaining outcomes (Long and Freese 2001). The fitted MNL model was first checked to make sure that it does not violate this assumption.

Empirical specification of model variables

The dependent variable is farmers' choice of no adaptation, tree planting as a sole response, and tree planting and at least one additional measure. Farmers usually adopt more than one adaptation strategy at a time and the dependent variable is defined in a way that provides mutually exclusive outcomes. Farmers' choice of adaptation strategies is made in the context of households' socio-economic characteristics, institutions, and agro-ecological settings. Hypothesized factors are discussed below and the description of each explanatory variable is given in Table 1.

The agro-ecological settings of farmers such as climate and soil are expected to influence their adaptation to climate change. Evidence shows that farmers in drier and hotter climate are more likely to respond to climate change than farmers in cooler and wetter areas (Deressa et al. 2009). In Ethiopia, according to the country's traditional agro-ecological zones, areas categorized as *Kolla* (lowland, 500-1500 m.a.s.l) are characterized by relatively hotter and drier climate whereas *Woinadega* (middle land 1500–2500 m.a.s.l.) and *Dega* (highland, 2500–3500 m a.s.l.) are wetter and cooler (Deressa et al. 2009). In our study the sampled *Kebeles* fall under either *Kolla* or *Woinadega*. Hence, it is hypothesized that farmers living in *Kolla* are more likely to adapt to climate change than in *Woinadega*. Access to water for irrigation is hypothesized to negatively affect farmers' adaptation to climate change. This is based on the argument that farmers who have access to irrigation water are more resilient to climate change and are less likely to take adaptation measures (Gbetibouo 2009).

Female-headed households in Ethiopia in general and in East Hararghe in particular are expected to be less likely to adapt due to their limited access to land, information, inputs and institutions as a result of traditional social barriers (Wilson and Getnet 2011). However, Nhemachena and Hassan (2007) found that female-headed households are more likely to take up adaptation measures than male-headed households. The authors explain that females are engaged in more farm activities than males and have better farm experience and information. Although this finding has sound grounds, our study proposes that the argument in the contrary is more probable to prevail in the study area (Wilson and Getnet 2011) and hypothesized that male-headed households are more likely to adapt to climate change.

Farmers with more years of farming experience are more capable of assessing the available technologies and making adaptation decisions (Gbetibouo 2009). On the contrary, more experienced or older farmers tend to be risk-averse and lag behind in adoption decisions. There is no final consensus on how age affects adoption decisions (Adesina and Baidu-Forson 1995). The more a farmer is educated, the more likely he/she is to access information, perceive and adapt to climate change (Adesina and Baidu-Forson 1995; Maddison 2007). Hence, a positive relationship between level of education and adaptation decisions is expected.

Table 1 Description of the independent variables

Explanatory variables	Mean	S.D.	Description
Local agro-ecology (Mid-land)	0.69	0.47	Dummy takes the value of 1 if *Woinadega* and zero otherwise (Low land)
Non-farm income	0.71	0.46	Continuous
Educational level of the household head-able to read and write	0.20	0.40	Dummy takes the value of 1 if read and write and zero otherwise
Education level of the household head-primary school or higher	0.16	0.36	Dummy takes the value of 1 if primary and above and zero otherwise
Gender of the household head	0.92	0.28	Dummy takes the value of 1 if male and zero otherwise
Access to irrigation water	0.13	0.34	Dummy takes the value of 1 if there is access and zero otherwise
Farming experience of the household head in years	31.35	17.32	Continuous
Farm size in hectares	1.24	0.81	Continuous
Access to food aid of the household	0.39	0.49	Dummy takes the value of 1 if there is access and zero otherwise
Land ownership of the household	0.91	0.29	Dummy takes the value of 1 if owned and zero otherwise
Farmer-to-farmer extension	0.67	0.47	Dummy takes the value of 1 if there is access and zero otherwise
Access to credit services	0.30	0.46	Dummy takes the value 1 if there is access and zero otherwise
Access to extension services	0.83	0.38	Dummy takes the value 1 if there is access and zero otherwise
Distance from selling market in walking hours	2.93	4.75	Continuous
Distance from purchasing market in walking hours	1.67	1.68	Continuous
Family size of the household	7.28	3.05	Continuous
Annual income of the household	11548	12216	Continuous
Livestock ownership in tropical livestock units (TLU)	3.24	1.88	Continuous

Household wealth (land, livestock and income) highly influences adoption decisions of farmers (Deressa et al. 2009; Asfaw et al. 2011). Shortage of land is observed to be one of the major barriers in adaptation to climate change (Bryan et al. 2009; Maddison 2007) and land size (farm size) was hypothesized here to have a positive relation with adaptation to climate change. A related institutional issue with land is land tenure. Previous studies observed that farmers who own their farm land are likely to take adoption measures, particularly if the investment required for adaptation is tied to the land (Gbetibouo 2009; Nhemachena and Hassan 2007). It is, therefore, hypothesized that ownership of land encourages adaptation to climate change.

Apart from farm size, wealth of farmers in terms of higher income and livestock ownership facilitates adoption of technologies. Wealthier farmers are more likely to use their financial resources to acquire new technologies and are less risk-averse to experiment them. It is also argued that the more wealth a farmer has, the more likely he/she is to access information, credit and extension services (CIMMYT 1998). Hence, higher farm and non-farm income, and livestock ownership are expected to influence adaptation positively. According to Foster and Rosenzweig (2010), wealthier farmers are advantageous in adoption particularly due to lack of well-developed credit and insurance institutions. Thus, better access to credit hastens adoption (CIMMYT 1998) and it is hypothesized here to facilitate adaptation to climate change.

Household size can influence adaptation because of its association with labor endowment. It is argued that a larger household size enables the adoption of technologies by availing the necessary labour force in one hand (Croppenstedt et al. 2003) and enabling the generation of additional income from extra labor invested in off-farm activities (Yirga 2007). It is, therefore, expected to positively influence adaptation to climate change. Extension services foster adaptation through enhancing farmers' awareness of climate change and knowledge on adaptation measures (Falco et al. 2011; Doss 2003; Nhemachena and Hassan 2007). In this study, both formal (public) and informal (farmer-to-farmer) extension services are hypothesized to facilitate adaptation to climate change. Farmer-to-farmer extension is also used here to serve as a proxy for social capital. Enhanced social capital ensures better access to information, credit, and facilitates cooperation for actions that have positive externalities which are not internalized at the level of individual decision makers (Hogset 2005 cited in Katungi 2007). Isham (2000) and Abodi et al. (2007) showed the importance of this variable in encouraging adoption of technologies by farmers. Thus, the positive relation of farmer-to-farmer extension with adaptation is further supported in relation to social capital.

Markets serve as a means of sharing and exchanging information (Maddison 2007) and also enhance the returns from technology (Jack 2011). As a result, farmers close to input and output markets are in a better position to take adaptation measures, and distance to input and output markets is expected to be inversely related with adaptation. A final factor that is included in this study is food aid. Food aid has an insurance effect (Barrett 2006) and may encourage farmers to adopt new technologies. For example, Bryan et al. (2009) have shown that food aid increases the probability of adaptation among the poorest farmers. This study, therefore, expects that food aid enhances adaptive capacity to climate change.

Results and discussion

Perception of climate change

There are differences among farmers in how they perceive changes in temperature and precipitation (Figures 1 and 2). 91.2 percent of the respondents perceived an increase in temperature over the last 20 years while 5.3 and 3.5 perceived no change and decrease in temperature respectively. On the other hand, 90.3 percent of respondents perceived decrease in precipitation while 6.2 and 2.6 percent perceived no change and increase in precipitation respectively. Besides, 0.9 percent indicated precipitation is variable rather than agreeing either on an increase or decrease of rainfall. However, all the farmers have perceived at least one aspect of the change in climate indicating a high level of perception among farmers in the study area. A Heckman selection test was run to see whether farmers perceiving no temperature or precipitation change introduce selection bias in our analyses. The inverse Mills ratio term is statistically not different from zero at 10 percent level of significance ($p = 0.2985$ and $p = 0.2318$ for perceiving no precipitation and no temperature change respectively), indicating no statistically significant indication of selection bias.

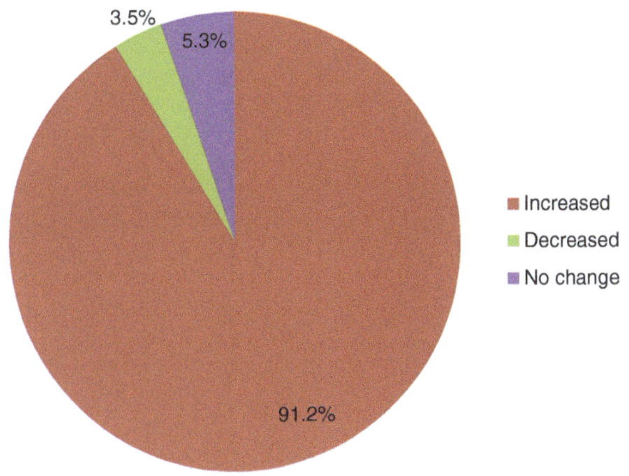

Figure 1 Farmers' perception of change in temperature due to climate change. The chart shows percent of respondents who perceived the different changes.

Sources of information on the existence of climate change

Farmers were explicitly asked how they first come to know about climate change (Figure 3). The majority of farmers (90 percent) perceived climate change through their life experience while 3.7 percent were informed by extension services. This result indicates that farmers are still heavily dependent on their gradual perception of climate change.

If Bayesian updating is assumed, this would mean that there is lack of information from extension services to update the perception of the farmers or the farmers attach less credibility (or probability of being true) to the information provided by extension staff.

Types of adaptation strategies

The majority of farmers (49.6 percent) adopted tree planting with at least one additional measure while 39.5 percent took tree planting as a sole response to climate change (Figure 4). A survey in the Nile Basin of Ethiopia (Deressa et al. 2011; Deressa et al. 2009) also found planting trees as the most common adaptation strategy. Planting trees has diverse ecological and economic benefits, which bring about adaptation to climate change. However, the responses of farmers to climate change can also be driven by their understanding of the causes of climate change. The majority (96 percent) of the households in this survey believe that the main cause for climate change is deforestation. Fosu-Mensah et al. (2010) also observed a similar understanding in Ghana. Therefore, planting trees may also be taken by the farmers partly as a mitigation strategy in addition to adaptation. Even though deforestation is one of the causes of climate change, it is known that carbon emission from industrialized countries is the main cause (IPCC 2007). Hence, the result shows the existing gap in the understanding of farmers on the main causes of climate change and a potential misconception on the performance of their major adaptation strategy- planting trees.

Although Figure 4 groups adaptation responses into three, there are actually 8 types of adaptation measures identified in the study area (Figure 5). The majority of farmers (89.1 percent) planted trees as an adaptation measure. Early planting, terracing, irrigation, and water harvesting follow tree planting as the next common adaptation strategies in the order they are listed. Praying was mentioned by 9.2 percent of the respondents as an adaptation strategy. Taking praying as a major adaptation measure may arise from associating

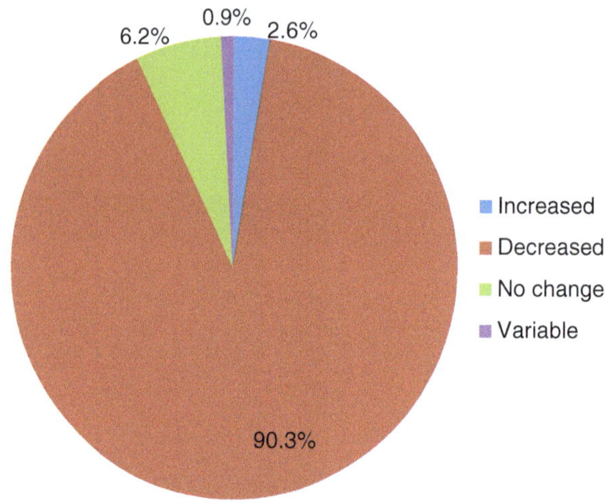

Figure 2 Farmers' perception of change in precipitation due to climate change. The chart shows percent of respondents who perceived the different changes.

climate issues to supernatural forces, which is usually the case in religious societies like Ethiopia (Nzeadibe et al. 2011). The use of early maturing plants and non-farm activity are rare responses to climate change in the study area. Generally, the adaptation measures taken by the farmers are similar to other findings in Ethiopia (Maddison 2007; Deressa et al. 2009; Deressa et al. 2011; Falco et al. 2011).

Source of information on adaptation strategies

Conley and Udry (2001) contend that farmers learn about new innovations from extension advice, from their own experimentation and from their neighbors' experimentation. In our survey, farmers were asked about their source of information on adaptation strategies. It was learned that extension advice took the lion's share (about 58.4%) followed by own experimentation, which accounts 23.9% (Figure 6). Furthermore, 17% of the framers interviewed used neighbors' experimentation and advice as a source of information for adaptation. The results indicate the importance of extension services in influencing farmers' adaptation decisions. The dominance of tree planting as

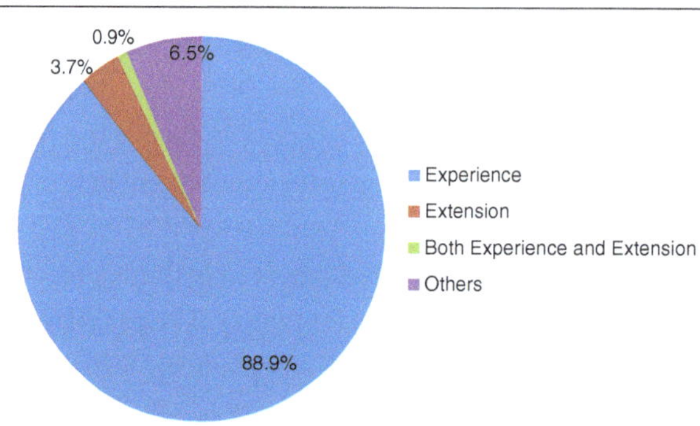

Figure 3 Source of information about climate change. The percent of respondents who mentioned the various sources of information is depicted in the figure.

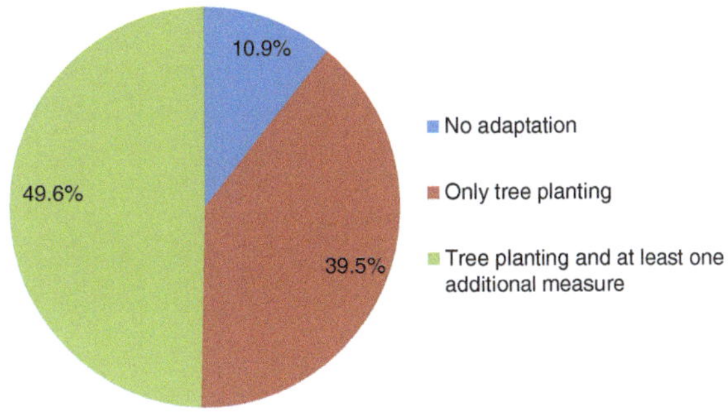

Figure 4 Adaptation strategies categorized into three. The chart shows the percent of respondents whose adaptation decisions fall in each of the three categories of adaptation.

an adaptation strategy may also be attributed to interventions by the public extension system, which is the main source of information on adaptation strategies.

Determinants of adaptation

The Multinomial Logit Model was run taking 'no adaptation' as the base category against which the remaining outcomes are compared with (Table 2). The model was checked for the endogeneity problem by fitting models with and without variables such as acess to extension services and acess to credit. Running Hausman specification tests showed that there is no significant systematic difference in coefficients as a result of exclusion of these variables.

An important assumption of the MNL is Independence of Irrelevant Alternatives (IIA) and the model was tested using the Hausman test to see if it fulfills the assumption. The Hausman test supported that IIA is not violated with χ^2 ranging from 0 up to –35.53 with probabilities almost equal to 1.0. LR tests and Wald tests also verified the power of the explanatory variables included in the model. To make sure that the explanatory variables do not pose a multicollinearity problem, OLS was fitted and VIF was calculated. All the VIF values are less than 10 (1.10 – 3.68) indicating that it is safe to assume the absence of multicollinearity.

For ease of interpretation, marginal effects were computed for each outcome in the dependent variable (see Table 3). The marginal effects show the change in probability of a particular choice of climate change adaptation for a unit change in the explanatory variables.

Non-farm income

Contrary to our hypothesis, non-farm income was showed to have a negative relationship with adaptation by employing tree planting with other measures. This could be explained by the enterprise diversification or risk reduction effect of non-farm income which may reduce the importance of urgent adaptation measures. In other words, the existence of non-farm income serves as an adaptation measure by itself and may delay other responses.

Farmer-to-farmer extension

As was expected, farmer-to-farmer extension enhances adaptation to climate change, whether by only planting trees or by taking measures on top of tree planting. This result underscores the importance of social capital and informal extension services in adaptation to climate change.

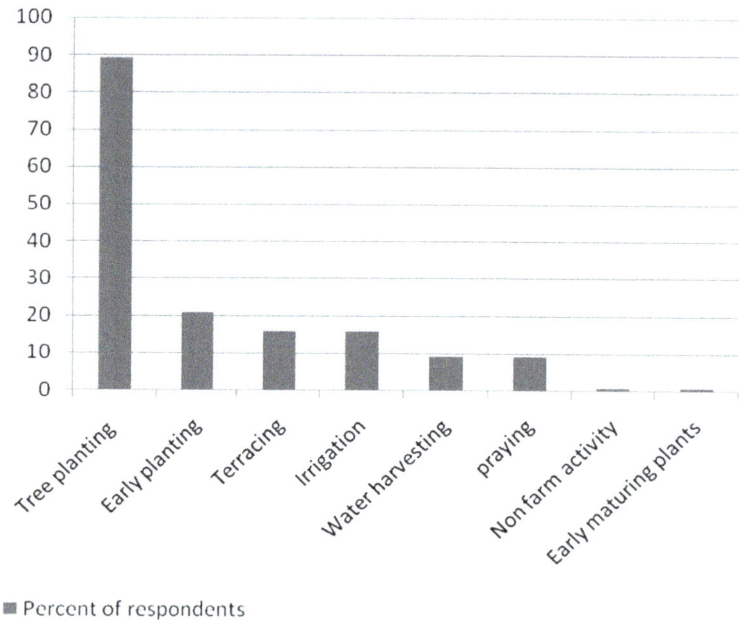

Figure 5 Adaptation strategies. The X-axis shows the different adaptation strategies taken by farmers while the heights of the bars correspond to the percent of respondents taking each measure.

Access to credit

Regardless of the expectations, households with better access to credit were found to be less probable to adapt to climate change by tree planting as the only response for adaptation. Although not significant, the coefficient for adaptation by employing tree planting plus other measures is also negative. This could be explained by the fact that farmers accessing credit are likely to be already engaged in different farm investment activities which improves their adaptive capacity and thereby delaying adaptation measures.

Distance to selling market

As was hypothesized, as distance from selling market increases the likelihood of adaptation by tree planting trees and taking at least one additional measure decreases by 13.5 percent. Even though not statistically significant, it has the same negative relationship with adaptation by only tree planting.

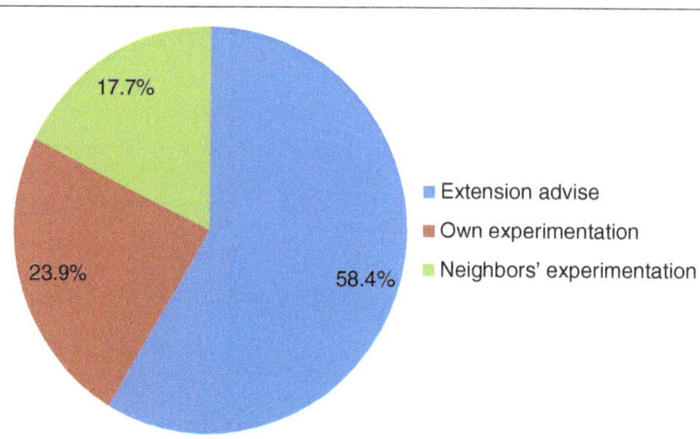

Figure 6 Source of information on adaptation strategy. The figure depicts the percent of respondents who use the different sources of information.

Table 2 Parameter estimates of the multinomial logit model

Explanatory variables	Only tree planting		Tree planting and other(s)	
	Coefficient	P > |z|	Coefficient	P > |z|
Local agro-ecology (Mid-land)	−1.636667	0.281	-.3587958	.807
Non-farm income	−2.708249	0.145	−3.288332	0.072*
Educational level of the household head-able to read and write	.0207302	0.989	-.5278419	0.723
Education level of the household head-primary school or higher	1.285579	0.479	1.445039	0.374
Gender of the household head	−1.903161	0.305	-.2514557	0.884
Access to irrigation water	2.221507	0.674	2.151656	0.683
Farming experience of the household head	-.0242545	0.539	-.0402274	0.287
Farm size in hectares	.9027755	0.410	.3203822	0.766
Access to food aid	1.694316	0.289	1.665551	0.288
Land ownership	−1.185355	0.806	−1.316571	0.783
Farmer to farmer extension	4.930387	0.013**	4.291278	0.027**
Access to credit services	−3.473911	0.070*	−2.471363	0.186
Access to extension services	.9831508	0.526	-.9210263	0.498
Distance from selling market	-.375467	0.135	-.9231534	0.003***
Distance from purchasing market	2.185086	0.058*	3.196014	0.007***
Family size	-.1358245	0.587	-.2319416	0.344
Annual income	.0002782	0.066*	.0002397	0.109
Livestock ownership (TLU)	-.2087889	0.549	-.2593202	0.444
constant	1.568256	0.811	4.27316	0.508
Number of observations = 109				
LR chi2(36) = 79.39				
Prob > chi2 = 0.0000***				
Log likelihood = −65.931038				
Pseudo R2 = 0.3758				
No adaptation is the base outcome				

***, **, * Significant at 1%, 5% and 10% significance level respectively.

Distance to purchasing market

Despite the expectations, the probability of higher level of adaptation increases with an increase in the distance of purchasing markets. With the increase in distance from purchasing markets, farmers are more probable (by 25.3 percent) to adapt by planting trees on top of at least one additional measure. It is also positively related with adaptation by only tree planting. The implication of this result is that farmers travelling further to purchasing markets may access larger markets where they are likely to be introduced with diverse inputs or technologies. The result further indicates that the markets from which distance is measured are not homogeneous, and underscores the need to verify the same while using the varibale as a measure for market access.

Income

In line with the hypothsis, income was found to be postively related with adaptation either by tree planting alone or by employing additional measures. Adaptation by tree planting plus at least one additional measure is, however, slightly insignificant (p = 0.109).

Table 3 Marginal effects from the multinomial logit model

Explanatory variables	Only tree planting		Tree planting and other(s)		No adaptation	
	Coefficient	P > |z|	Coefficient	P > |z|	Coefficient	P > |z|
Local agro-ecology (Mid-land)	-.3091525	0.053*	.304948	0.055*	.0042045	0.561
Non-farm income	.1317173	0.326	-.1437625	0.285	.0120452	0.455
Educational level of the household head-able to read and write	.1339774	0.454	-.1355246	0.447	.0015472	0.861
Education level of the household head-primary school or higher	-.0363438	0.870	.0411828	0.852	-.004839	0.505
Gender of the household head	-.3808178	0.041**	.3768494	0.042**	.0039684	0.532
Access to irrigation water	.0196342	0.941	-.0135092	0.959	-.006125	0.562
Farming experience of the household head	.0037983	0.398	-.0039727	0.376	.0001744	0.522
Farm size in hectares	.1424019	0.096*	-.139461	0.104	-.0029409	0.691
Access to food aid	.0104187	0.943	-.0022303	0.988	-.0081885	0.499
Land ownership	.0298371	0.899	-.034039	0.884	.0042019	0.711
Farmer to farmer extension	.1825314	0.228	-.0853452	0.598	-.0971861	0.354
Access to credit services	-.2390714	0.139	.2052233	0.218	.0338481	0.480
Access to extension services	.369668	0.005***	-.3712651	0.005***	.0015971	0.774
Distance from selling market	.1312373	0.013**	-.1348426	0.011**	.0036053	0.443
Distance from purchasing market	-.2389762	0.025**	.2533942	0.019**	-.014418	0.407
Family size	.0228787	0.354	-.0238753	0.333	.0009966	0.550
Annual income	9.89e-06	0.189	−8.56e-06	0.256	−1.33e-06	0.418
Livestock ownership (TLU)	.0117275	0.763	-.0129662	0.739	.0012387	0.536

***, **, * Significant at 1%, 5% and 10% significance level respectively.

Barriers to adaptation

Adaptation to climate change has many stumbling blocks. During the survey, farmers mentioned a number of barriers to adaptation emanating from different social, economic and institutional situations (Figure 7). The most important factor mentioned as barrier to adaptation by the surveyed farmers is lack of information (which accounts 22.5%) about climate change and adaptation strategies. This is attributed to various factors including lack of institutional support mechanisms and failure to mainstream the issue of climate change in the public extension system of the country. Lack of information was also identified as the main barrier to adaptation by a survey in the Nile Basin of Ethiopia (Deressa et al. 2009). Lack of farm inputs including seed (20.7 percent), chemical fertilizer (3.6 percent) and oxen (8.1%) are mentioned by the respondents as the second most important impediments. Out of the three farm inputs mentioned by the framers, the problem emerging from lack of seed (20.7 percent) outweighs the other two while the problem from lack of drought oxen rests as the second most important barrier. According to 17.1 percent of the sampled respondents, shortage of land plays a key role in diminishing their potential to adapt to climate change. An independent sample t-test showed that these respondents have significantly ($p < 0.05$) lower average land size compared to the rest of the respondents who did not mention shortage of land as a major barrier. The surveyed farmers also mentioned lack of money (14.4 percent) as another important climate change adaptation barrier. This result further

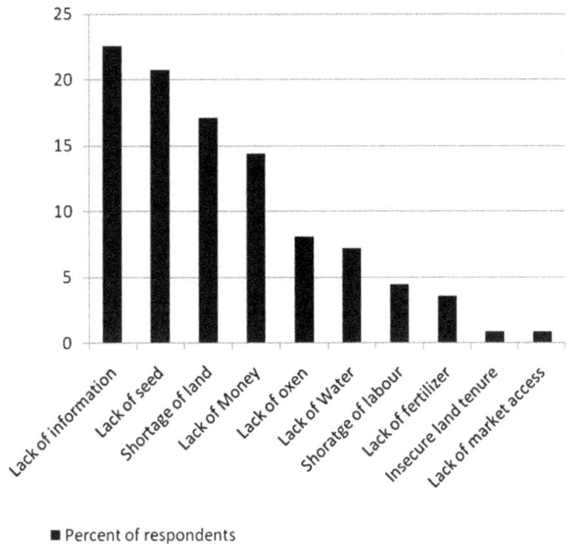

Figure 7 Barriers to adaptation. The X-axis shows the different barriers to adaptation faced by farmers while the heights of the bars correspond to the percent of respondents who mentioned each barrier.

signifies the importance of financial resources in adaptation to climate change. The other barriers mentioned by the farmers include: lack of water, lack of labor, insecure land tenure and lack of market access. In general, the findings are similar to other studies in the country (Deressa et al. 2009; Deressa et al. 2011; Bryan et al. 2009; Maddison 2007).

Conclusions

The research was carried out in three districts in East Hararghe Zone of Ethiopia where 110 farm households were randomly considered for the study. All of the surveyed farm households observed at least one type of climate change over the last 20 years primarily based on their life experience. Planting trees was found to be the single most common adaptation strategy in the study area. The majority of farmers (89.1 percent) adapted to climate change either by planting trees as a sole response or by taking at least one adaptation measure on top of it. Apart from tree planting, the main adaptation strategies taken by the households include: early planting, terracing, irrigation, water harvesting and praying. The majority of farmers learned about their adaptation strategies from extension services. Hence, the dominance of tree planting as an adaptation measure may be associated with the interventions of the extension system. Moreover, most farmers misconceived that the major cause of climate change is deforestation, which may indicate tree planting decisions could have been taken by the farmers partly as a mitigation strategy. There is a need to provide information to farmers and extension workers on the basics of climate change including its causes and what they should expect from tree planting as an adaptation strategy.

An MNL model was fitted on the data where the choices of adaptation to climate change make the dependent variable. Different socioeconomic, institutional and agro-ecological variables were hypothesized to influence adaptation decisions and were included in the model. Households with higher access to farmer-to-farmer extension, a proxy for social capital and access to informal extension services showed higher probability of taking adaptation responses. Farm households with higher income and better access to selling markets are more likely in taking adaptation measures. However, farmers tend to adapt more when they use purchasing markets located further. Farmers travelling to further input markets may be advantageous in being introduced with various technologies or inputs which increases their chance of adopting more adaptation strategies. This result also emphasizes the need to assess the homogeneity of markets before taking distance from markets as an indicator for market access.

With more non-farm income and credit, farmers will adapt less to climate change. Farmers with higher non-farm income may be slow in taking more adaptation measures as non-farm activities by themselves act as adaptation measures. Households who have access to credit may also be already engaged in different farm investments that improve their adaptive capacity to climate change, making less likely adaptive measures taken for climate change.

The surveyed farmers were asked about the most important barriers they face in adapting to climate change. The most frequently mentioned barrier is lack of information followed by lack of seed, shortage of land, lack of money, lack of oxen, lack of water, shortage of labor, lack of fertilizer, insecure land tenure and lack of market access, listed according to their importance. The result underscores the need for strengthening the extension system of the country in providing information on adaptation strategies and supplying agricultural inputs and particularly improved seeds. Enhancing the capacity of rural credit institutions is also an important element in supporting autonomous climate change adaptations at farm level.

In a nutshell, the study results indicate an information gap in the public extension system and among farmers about the causes of climate change, adaptation strategies and their performance. Planting trees as an adaptation measure should not be overemphasized and the extension system should introduce diverse adaptation options to climate change. Interventions that enhance income, access to finance, access to farm inputs, access to market and diversification of enterprises should be an integral part of climate change adaptation policies and strategies. It is also recommended that social capital should be utilized in national or regional interventions towards climate change adaptation of smallholder farmers.

Competing interests

The authors declare that they have no competing interests.

Authors' contribution

YAT came up with the initial idea for the research project and has been highly involved in data collection, analysis, interpretation and write up of the manuscript. CSA contributed ideas in the research proposal and has been also involved in data collection and write up of the manuscript. GSE contributed ideas in the research proposal and has been also involved in data collection and write up of the manuscript. All authors read and approved the final manuscript.

Acknowledgement

The authors are grateful for the valuable comments of three anonymous reviewers.

Author details

[1]School of Agricultural Economics and Agribusiness, Haramaya University, Dire Dawa, Ethiopia. [2]Department of Rural Development and Agricultural Extension, Haramaya University, Dire Dawa, Ethiopia.

References

Abodi PN, et al. (2007) An Economic Assessment of Banana Genetic Improvement and Innovation in the Lake Victoria Region of Uganda and Tanzania. In: Smale M, Tushemereirwe WK (ed). International Food Policy Research Institute, Washington, D.C

ACCCA (2010) Farm – Level Climate change Perception and Adaptation in Drought Prone Areas of Tigray, Northern Ethiopia. Improving decision-making capacity of smallholder farmers in response to climate risk adaptation in three drought-prone districts of Tigray, northern Ethiopia Vol 3. Advancing capacity to Support Climate Change Adaptation (ACCCA), Mekelle

Adesina AA, Baidu-Forson J (1995) Farmers' perceptions and adoption of new agricultural technology: evidence from analysis in Burkina Faso and Guinea, West Africa. Agric Econ 13:1–9

Akponikpè PBI, Johnston P, Agbossou EK (2010) Farmers' perception of climate change and adaptation strategies in Sub-Saharan West-Africa. In: 2nd International Conference: Climate. Sustainability and Development in Semi-arid Regions, Fortaleza - Ceará, Brazil. ICID + 18

Allison P (2002) Sage Monograph on Missing Data, Sage paper # 136. Sage, USA

Asfaw S, Shiferaw B, Simtowe F, Haile MG (2011) Agricultural technology adoption, seed access constraints and commercialization in Ethiopia. J Dev Agri Econ 3(9):436–447

Barrett CB (2006) Food Aid's Intended and Unintended Consequences. FAO, Rome

Bryan E, Deressa TT, Gbetibouo GA, Ringler C (2009) Adaptation to climate change in Ethiopia and South Africa: options and constraints. Environ Sci Policy 12:413–426

CIMMYT (1998) The Adoption of Agricultural Technology: A Guide for Survey Design. CIMMYT, Mexico, DF

Conley T, Udry C (2001) Social learning through networks: The adoption of new agricultural technologies in Ghana. Am J Agric Econ 83:668–673

Croppenstedt A, Demeke M, Meschi MM (2003) Technology adoption in the presence of constraints: the case of fertilizer demand in Ethiopia. Rev Dev Econ 7(1):58–70

CSA (2007) Statistical Tables for the 2007 Population and Housing Census of Ethiopia. Central Statistical Agency, Addis Ababa

CSA (Central Statistics Authority) (2004) The Federal Democratic Republic of Ethiopia Statistical abstract for 2003. CSA, Addis Ababa

Deressa TT, Hassan RM, Ringler C, Alemu T, d MY (2009) Determinants of farmers' choice of adaptation methods to climate change in the Nile Basin of Ethiopia. Global Environ Change 19:248–255

Deressa TT, Hassan RM, Ringler C (2011) Perception of an adaptation to climate change by farmers in the Nile basin of Ethiopia. J Agri Sci 149:23–31. 10.1017/S0021859610000687

Doss CR (2003) Understanding Farm-level Technology Adoption: Lessons Learned from CIMMYT's Micro Surveys in Eastern Africa. Economics Working Paper 03-07. CIMMYT, Mexico, DF

Falco SD, Yesuf M, Kohlin G (2011) What Adaptation to Climate Change? Evidence from the Nile Basin, Ethiopia. In: International Conference on Economics of Adaptation to Climate Change in Low-Income Countries. Ethiopian Development Research Institute and International Food Policy Research Institute, Washington, DC

Foster AD, Rosenzweig MR (2010) Microeconomics of Technology Adoption,Center Discussion Paper NO.984. Yale University, Newhaven, USA

Fosu-Mensah BY, et al. (2010) Farmers' Perception and Adaptation to Climate Change; A Case Study of Sekyedumase District in Ghana. World Food System — A Contribution from Europe, Zurich, Swizerland, Tropentag, International Research on Food Security, Natural Resource Management and Rural Development, Tropentag

Gbetibouo GA (2009) Understanding Farmers' Perceptions and Adaptations to Climate Change and Variability: The Case of the Limpopo Basin, South Africa (trans: Division EaPT). IFPRI-wide Discussion Paper. International Food Policy Research Institute, South Africa

Greene WH (2003) Econometric Analysis. Prentice-Hall, Upper Saddle RiverNJ

ILRI (2006) Mapping Climate Vulnerability and Poverty in Africa. International Livestock Research Institute, Nairobi, Kenya

IPCC (2001) Climate Change 2001: Synthesis Report. A Contribution of Working Groups I, II, III to the Third Assessment Report of the Intergovernmental Panel on Climate Change. R. T. Watson and the Core Team, Cambridge and New York, p 398

IPCC (2007) Climate Change 2007: The Physical Science Basis. Contribution of Working Group I to the Fourth Assessment Report of the Intergovernmental Panel on Climate Change, Cambridge, United Kingdom and New York, NY, USA

Isham J (2000) The Effect of Social Capital on Technology Adoption: Evidence from Rural Tanzania. Opportunities in Africa: Micro-evidence on Firms and Households. The Centre for the Study of African Economies, University of Oxford, Oxford

Jack (2011) Constraints on the adoption of agricultural technologies in developing countries. White paper, Agricultural Technology Adoption Initiative. J-PAL (MIT) and CEGA, UC Berkeley

Jariko GA, Junejo MA, Rahpoto MS, Shah MZ (2011) Socioeconomic factors affecting adoption of sunflower varieties in Sindh. Pak J Commer Soc Sci 5(1):192–201

Katungi EM (2007) Social Capital and Technology Adoption on Small Farms: The Case of Banana Production Technology in Uganda. PhD Dessertation. Department of Agricultural Econmics, Extension and Rural Development, University of Pretoria, Pretoria

Lautze S, Aklilu Y, Raven-Roberts A, Young H, Kebede G, Leaning J (2003) Risk and Vulnerability in Ethiopia: Learning from the Past, Responding to the Present. Preparing for the Future, Addis Ababa

Long JS, Freese J (2001) Regression Models for Categorical Dependent Variables Using Stata. A Stata Press Publication, STATA Corporation, College Station, Texas

Madalla G (1983) Limited Dependent and Qualitative Variables in Econometrics. Cambridge University Press, Cambridge

Maddison D (2007) The Perception of an Adaptation to Climate Change in Africa. Policy Research Working Paper. The World Bank, Development Research Group, Sustainable Rural and Urban Development Team, Pretoria, South Africa

Maharjan SK, Sigdel ER, Sthapit BR, Regmi4 BR (2011) Tharu community's perception on climate changes and their adaptive initiations to withstand its impacts in Western Terai of Nepal. Int NGO J 6(2):35–42

McSweeney C, Lizcano G, New M, Lu X (2010) The UNDP Climate Change Country Profiles. Available via http://journals.ametsoc.org/doi/abs/10.1175/2009BAMS2826.1. Accessed 26 Nov 2012)

Nhemachena C, Hassan RM (2007) Micro-Level Analysis of Farmers' Adaptation to Climate Change in Southern Africa. IFPRI Discussion Paper No. 714, vol 15–7. International Food Policy Research Institute, Washington, DC

Niemistö N (2011) The resilience of rural Ethiopian livelihoods, a case study from Hararghe zone. University of Jyväskylä, Eastern Ethiopia

Nzeadibe TC, Egbule CL, Chukwuone NA, Agu VC (2011) Climate Change Awareness and Adaptation in the Niger Delta Region of Nigeria. African Technology Policy Studies Network, Nairobi

ONRSPO (n.d) Socio-economic Profile of East Hararghe. Oromia National Reginal State President Office. Available via http://www.oromiyaa.com/index.php?option = com_content&view = category&layout = blog&id = 114&Itemid = 788. Accessed 10 Dec 2012

Paudel K, Pandit M, Mishra A, Segarra E (2011) Why Don't Farmers Adopt Precision Farming Technologies in Cotton Production? In: Agricultural & Applied Economics Association's 2011 AAEA & NAREA Joint Annual Meeting. Agricultural and Applied Economics Association, Pittsburgh, Pennsylvania

Smit B, Pilifosova O (2001) Adaptation to Climate Change in the Context of Sustainable Development and Equity. Contribution of the Working Group to the Third Assessment Report of the Intergovernmental Panel on Climate Change. Cambridge University Press, Cambridge, UK, pp 879–912

Sofoluwe NA, Tijani AA, Baruwa OI (2011) Farmers' perception and adaptation to climate change in Osun State, Nigeria. Afr J Agri Res 6(20):4789–4794

Stige LC, Stave J, Chan K, Ciannelli L, Pattorelli N, Glantz M, Herren H, Stenseth N (2006) The effect of climate variation on agro-pastoral production in Africa. PNAS 103:3049–3053

Strzepek K, Mccluskey A (2006) District Level Hydro-climatic Time Series and Scenario Analysis to Assess the Impacts of Climate Change on Regional Water Resources and Agriculture in Africa. CEEPA Discussion Paper No. 13, Pretoria, South Africa

Tadege A (2007) Climate Change National Adaptation Program of Action (NAPA) of Ethiopia. NMS (National Meteorological Agency), Federal Democratic Republic of Ethiopia Addis Ababa

Uaiene RN, CA, Masters WA (2009) Determinants of Agricultural Technology Adoption in Mozambique. Ministry of Planning and Development, Republic of Mozambique

Vedwan N, Rhoades RE (2001) Climate change in the Western Himalayas of India: a study of local perception and response. Climate Res 19:109–117

Wilson K, Getnet M (2011) Investigating how development interventions increase community-level adaptive capacity in Ethiopia. In the New Voices, Different Perspectives: Proceedings of the AfricaAdapt Climate Change Symposium 2011. Addis Ababa, Ethiopia

Yirga CT (2007) The dynamics of soil degradation and incentives for optimal management in Central Highlands of Ethiopia. University of Pretoria, Pretoria, South Africa

4

Institutional barriers to organic farming in Central and Eastern European countries of the Baltic Sea region

Markus Larsson[1,2*†], Louise Morin[1†], Thomas Hahn[1] and Johanna Sandahl[3]

* Correspondence:
markus.larsson@mdh.se
†Equal contributors
[1]Stockholm Resilience Centre, Stockholm University, Kräftriket 2B, Stockholm SE-106 91, Sweden
[2]Mälardalens högskola, Box 883, Västerås 721 23, Sweden
Full list of author information is available at the end of the article

Abstract

A window of opportunity to promote organic farming is open for the Central and Eastern European Countries (CEEC) that joined the EU in 2004. The development of organic farming has the potential to decrease the amount of nutrient leaching to the Baltic Sea and could help to stop the environmental degradation of the Sea. However, this requires a diverse set of institutions. This paper explores the institutions that are lacking to promote the full development of organic farming in the CEEC, using Sweden as a baseline reference. A case study approach, formalised by introducing a set of indicators, has been used to identify the missing institutions. Data have been obtained from a desktop study, including a literature review, interviews and a questionnaire. The case studies partially support previous studies proposing that the development of organic farming proceeds along six steps: establishment of an organic farming community; establishment of political recognition; establishment of financial support; establishment of non-competitive relationships between the organic sector and general agricultural institutions; establishment of an organic food market, and development of a discussion and coordination arena. The results show that market development is the least developed step and that there is a correlation between higher governmental engagement and a more developed organic sector.

Keywords: Organic farming, Institutions, Baltic sea, Poland, Baltic states

Introduction

Background

In 1987, all the countries around the Baltic Sea agreed that by 1995, there should be a reduction of 50% of the nutrient load reaching the Sea. This goal is still unachieved and the environmental degradation of the Baltic Sea continues (HELCOM 2011). From all the input sources, agriculture is the biggest one and accounts for almost 50% of all the nutrients leaching to the sea. Almost half of those agricultural run-offs come from Poland, Estonia, Latvia and Lithuania and their share is likely to increase. The dissolution of the Soviet block in 1989 and the consequent entry into market economy had a significant effect on the economies of the Central and Eastern European countries (CEEC). The subsequent crash in farm-gate prices of most agricultural products and the reorganisation of the agricultural sector resulted in a more extensive land use where the utilisation of pesticides and chemical fertilizers was dramatically reduced

(Petersen and Hoogeveen 2004). However, as their economies recover, their agricultural practices are predicted to intensify again, a fact that is noticeable in the Baltic States (HELCOM 2011).

According to a review by Stolze et al. (2000), organic farming achieves higher biodiversity than conventional agriculture, due to the bans on pesticide, higher habitat heterogeneity and more extensive land use overall. Furthermore, organic farming generally decreases soil erosion, and conserves soil fertility and soil system stability to a higher degree than conventional farming. These properties can decrease the amount of nutrients leaching from agricultural land that result in eutrophication (Stolze et al. 2000; Larsson and Granstedt 2010). In turn, this has indirect, yet significant, effects on fisheries and tourism sectors (HELCOM 2004; Diaz and Rosenberg 2008). HELCOM[a] stresses that in the future "the EU agricultural policy will play a major role in the development of the condition of the Baltic Sea" (HELCOM 2011, p. 86). Researchers point out that the environmental degradation of the Baltic Sea cannot be resolved without addressing the agricultural run-offs, and that these can be managed efficiently by organic farming (HELCOM 2002; Granstedt et al. 2005). Thus, organic farming provides an array of public goods to society.

A study by HELCOM (2002) reveals that in order to attain the 50% reduction goal of nutrients, there would be a need for all agricultural land in the Baltic Sea drainage basin to turn organic or for half of the land to be set aside. The first solution implies a 25% reduction in yield, the second a 50% reduction. From this point of view, mass conversion to organic farming seems more attractive than to maintain conventional farming on a smaller area. In turn, this implies a high potential for the development of organic farming in all the CEEC, a fact that is recognised in the rural development plans of all countries under study. The accession of some CEEC to the European Union makes the realisation of this potential even more plausible. By opening a market, and by contributing new knowledge and providing new subsidies devoted to organic agriculture through the Common Agriculture Policy (CAP), the EU enlargement has opened a window of opportunity in the CEEC to promote the growth of the organic sector (Fischler 2003). The outcome of the currently on-going (Feb 2013) negotiations of reform of the CAP will determine whether additional resources will be targeted towards organic agriculture.

Objectives, research questions and limitations

The overall aim of this paper is to understand which institutional factors may hinder the growth of the organic sector in CEEC[b]. The study mostly focuses on formal institutions and the role of governments in developing them. In doing so, we describe and compare the institutions and actors of the organic sector in four CEEC (Estonia, Latvia, Lithuania, and Poland) and one Western European country (Sweden) by using a comprehensive set of indicators. Sweden is used as a baseline country which has a complete organic sector with all necessary institutions (Michelsen & Søgaard 2001; Dabbert et al. 2004). Sweden was chosen because it has production conditions (e.g. climate) similar to the Baltic States if not to Poland and because it lies within the drainage basin of the Baltic Sea to a larger extent compared to Denmark or Germany. The studied CEEC also lie completely within the drainage basin of the Baltic Sea, making them good candidates in reducing the eutrophication rate. Poland and the three Baltic States were also selected because of their high potential for organic farming and the window of opportunity that is open for them.

The following research questions were identified as central to the objective: What organic farming institutions already exist in Sweden and the four CEEC? How do these correspond or differ between each country? What institutions need to be developed to support the organic sector in the CEEC?

Case study description

Estonia, Latvia, Lithuania, Poland and Sweden are all members of the EU (since 2004 for the CEEC and since 1995 for Sweden). The CEEC cover 28.5% of the Baltic Sea Drainage Basin (BSDB) area (see Figure 1), while being inhabited by 55% of the population. In contrast, Sweden covers 25% but is only inhabited by 10% of the population of the BSDB (Hannerz and Destouni 2006). Furthermore, 58% of the agricultural land of the BSDB is in the CEEC, while 7% is within Sweden. The Baltic States are similar to Sweden in terms of population and agricultural land, but Poland is more populated and has more agricultural land.

CEEC have a 33-42% lower GDP per capita than the average of the EU27, and half the income per capita of Sweden (Eurostat 2012a). The agricultural sector has a larger share of the economy and more people live in rural areas and work in the agricultural sector in CEEC than in the EU27. During the communist regime, Polish land ownership was kept mostly private, while the farms in the Baltic States were mostly converted to collectivised farms (Lerman 2001). Post-independence land privatisation resulted in small individual farms in Poland and Latvia while Estonia and Lithuania have kept about 20% of large private corporate farms (FAO 2002). Organic farming started in the 1930's in Sweden, although the first organic standard was crafted in 1985 (Källander 2000/2010). In CEEC, organic farming started after they regained independence when the Soviet Union collapsed (Prazan et al. 2004).

Methods and theoretical framework

Case study approach

A comparative case study approach has been used. This includes multiple sources of evidence – a mail-out questionnaire (described below), personal communications and a review of research and official reports, regulation texts and grey literature (Yin 2003). The analysis was standardised through the use of indicators. The case studies were compared to a framework provided by Michelsen et al. (2001).

Questionnaire and personal communications

The data gathered during the desktop study were complemented by e-mail or telephone interviews with experts working in the organic sector. The point of making these contacts was to confirm and fill any gaps in the collected data. Data were also gathered using a questionnaire distributed to a limited number of research institutions, farmers' associations, certification bodies and governmental agencies. The questionnaire covered six different topics - legislative support, financial support, market, production, social dynamics and infrastructure availability - that required detailed knowledge to be answered correctly. The topics were selected following the literature review. One

Figure 1 The Baltic Sea drainage basin and the studied countries.

constraint encountered with personal communication was that some respondents did not have good enough knowledge of English to answer all the questions. The first respondents that were contacted were singled out from the literature review or from lists of participants in organic agriculture seminars and workshops. A subsequent 'snowball' effect led to further people being identified (Schultz et al. 2007; Sandström 2008). Since some respondents had a vested interest in the promotion of organic farming, a possible bias in their responses should be noted. In all, ten senior officials or researchers from Poland, the three Baltic countries and regional organisations answered the questionnaire. Even though the number of respondents was limited, the questionnaire provides valuable information about the conditions for organic farming in the region due to the central positions of the respondents.

Conceptual and theoretical framework - Institutions of the organic sector

Ultimately, the development of organic farming depends on the willingness of individual conventional farmers to convert to organic farming practices. In turn, this willingness depends to a large extent on the institutional system in which the farm is embedded (Michelsen et al. 2001). Some systems contain adequate institutions to promote the development of organic farming, others do not. Michelsen et al. (2001) and Prazan et al. (2004) have identified institutions from three societal domains - civil society, market and state – as key components for the development of the organic sector. In this paper, civil society includes institutions and organisations that are not government-controlled (Baylis and Smith 2001), e.g., producers' associations, informal farming practice guidelines, and non-state organisations that participate in lobbying, research and education. The market domain includes marketing initiatives, consumers, and food chain actors, such as processors and retailers. The state includes agricultural regulations, standards for organic certification and labelling and different kinds of support. Here, institutions are defined as the norms and rules that steer the behaviour of individuals. Norms and rules can be formalised by being carried out by an organisation, either from civil society or by the state. They can also be informal, that is perpetuated by culture and routine (North 1990). Prazan et al. (2004) also emphasise that the immaturity of the organic sector means that it lacks many formal institutions available to conventional agriculture: a detailed regulatory framework; a wide range of policy measures; a complete research program; adequate and widespread training and advisory services; and a complete market.

Six steps of institutional development

Michelsen et al. (2001) suggest that the institutional development of organic farming proceeds along six steps. The first three are seen as essential for the initial growth of the sector, while the last three are seen as essential for the continuous growth. The steps can be undergone multiple times, a process that leads to further development. All the steps do not need to have been completed before a step is repeated (Michelsen et al. 2001; Moschitz et al. 2004). Each step includes formal and informal institutions. Following a literature review, thirty four indicators were developed to assess the degree of completion of the six steps suggested by Michelsen et al. (2001) as well as the characteristics of the civil society, state and market domains, see Table 1 below. Throughout the text, indicator numbers are given in parentheses. These indicators refer to the property of the institution or actor that is being described and have been used to assess the degree of completion of the steps in the result analysis.

Step 1: The establishment of an organic community starts by the self-organisation of a group of organic farmers into a producers' association. Michelsen et al. (2001) identify the acceptance of a formalised common standard that specify the requirements to be considered organic as the first step in developing the organic community identity (indicator 1). Moschitz et al. (2004) argue that maintaining the validity of private standards helps the organic community to keep a strong identity after the involvement of the

Table 1 The six steps of institutional development and indicators with which they can be assessed

Characteristics of the completed step	Indicators used to assess the characteristic
Step 1: Establishment of an organic community	
a) The organic community has a strong identity	1: Introduction of first standard (date)
	2: Number of national producers' associations
	4: Private standard can still be used
b) Producers' associations have the power to influence other parts of society	3: Power of producers' association
	5: Proportion of organic farmers that are members of an association (%)
Step 2: Establishment of political recognition	
a) Regulatory support exists	6: Introduction of first national regulation (date)
	7: Introduction of current regulation (date)
b) Financial support is high	10: Introduction of area payment (date)
	11: Area payment (euro/ha)
	12: % of agri-environmental scheme devoted to organic farming
	13: Total amount devoted to OF (millions of euro)
	15: Loans available for improvement of organic farms
	16: Lower taxes for organic farmers
	17: Certification/inspection costs reimbursement
	18: Research support (million euros/year)
c) Strategic support is high	20: Action plan for organic farming
	21: OF incorporation in strategic plan of government
	22: Quantitative target for proportion cultivated organically (%)
Step 3: Establishment of financial support	
a) Financial support has been introduced	10: Introduction of area payment (date)
b) Area payments are high enough to off-set the cost incurred by conversion and lower productivity	11: Area payment (euro/ha)
	13: Total amount devoted to organic farming (millions of euro/year)
	14: % uptake of area support
c) Research program is well-funded and complete	18: Research support (million euro/year)
	19: Estimate % of agricultural research funding to organic farming
	25: Range of research program
d) There are financial schemes to help organic farmers	15: Loans available for improvement of organic farms
	16: Lower taxes for organic farmers
	17: Certification/inspection costs reimbursement
	18: Research support (million euro/year)
Step 4: Development of non-competitive relationships	
a) Partnership	8: Partnerships strength
b) Integration of organic farming in state institutions	23: Specific office in agricultural ministry
	24: Degree of integration of OF in training establishments
	26: Degree of integration of OF in advisory services
c) Discussion arena existence	9: Discussion arena existence

Table 1 The six steps of institutional development and indicators with which they can be assessed *(Continued)*

d) Strength of the organic community identity	1: Introduction of first standard (date)
	2: Number of national producers' associations
	4: Private standard can still be used
e) The power a producers' associations has	3: Power of producers' association
	5: Proportion of organic farmers that are members of an association (%)
Step 5: Establishment of organic food market	
a) A complete supply chain exist	27: Number of certified processing facilities (2004)
	28: Promotion by large retailers
	29: Distribution of sales channels
b) Other structural conditions are favourable	34: Perception of the quality of certification and control system
c) The behaviour of the general consumers is favourable	31: Proportion of consumer that can correctly define organic farming (%)
	32: Acceptable price premium by 50% of the population
	33: % of the population that buy OP > once a week/>once a month/< once a month
d) The market share of organic farming is large	30: Market share of organic products (%)
Step 6: Development of a committed institutional setting	
a) An inclusive discussion arena exists	9: Discussion arena existence
b) An inclusive coordination arena exist	20: Action plan for organic farming

Note that some indicators are relevant for more than one step.

state in certification and standard setting (indicator 4). The power of the producers' association increases with its level of inclusiveness (indicator 5) (Boström 2006).

Step 2: The introduction of private standards is strengthened by **political recognition**. This gives credibility and ensures that a minimum set of requirements is uniformly applied nationally (Michelsen et al. 2001; Dabbert et al. 2004). At this stage, organic farming is recognised under law as an alternative way of practicing agriculture (indicators 6–7). However, political recognition is usually further anchored by implementing other types of support, such as financial (indicators 10–19) and strategic support (indicators 20–22) (DFAF 2001; Moschitz et al. 2004).

Step 3: Financial support usually takes the form of area payments that can be differentiated by crops and regions, depending on the policy objectives (Michelsen et al. 2001). These subsidies are paid for each hectare that is cultivated organically (indicators 11 and 13). The introduction of area payment triggers the initial growth of the organic sector (indicator 10). Area payments for organic farming are co-financed by the EU via the funding of CAP and the higher the payment support, the more profitable organic farming becomes, and the larger the rate of growth (Dabbert et al. 2004). Other forms of financial support, such as reimbursement of certification fees or guaranteed loans, are particularly relevant in the CEEC where most farmers survive on very low profit margins (indicators 15–17) (Prazan et al. 2004). Furthermore, financial support should involve research and development because this is what trigger innovations in any market domain and makes it competitive (indicators 18 and 25) (EU 2004).

Step 4: Establishment of **non-competitive relationships** between the organic sector and general agricultural institutions (from both the civic and state domains). This interaction can take three forms: cooperation, competition and creative conflict (Michelsen et al. 2001). Under purely cooperative relationships, the differences between each sector are not stressed and there is a risk that the organic sector loses its identity by being incorporated into the conventional sector (indicators 1, 4, 23, 24 and 26). Under purely competitive relationships, the organic sector is hindered by a lack of contact between the organic and the conventional sector. Political, financial and strategic support (e.g. action plan) is missing. The middle point on this continuum is what Michelsen et al. (2001) call "creative conflict", where discussion takes place between each sector without one being subordinate to the other. The organic sector in this case must have a strong identity and potential to influence society (indicators 1–5).

Step 5: To maintain the growth of the organic sector there is a need for a **mature and independent organic food market** (Michelsen et al. 2001). This necessitates the establishment of a complete supply chain, which includes producers, processors, distributors and retailers (indicators 27–29). A complete supply chain will favour supermarkets as sale channel and will increase the availability of organic products (indicator 29), which in turn will increase demand (indicator 33) and supply (indicator 30) (Terra Nord 2005). Another important characteristic of the organic market is the price premium of organic products (indicators 32). Consumers that are more knowledgeable about organic farming will be willing to pay a higher price premium (indicator 31) (DFAF 2001). The demand for organic products increases when certification authority is considered reliable (indicator 34) (DFAF 2001).

Step 6: Establishment of an institutional setting including a discussion arena, which eases the coordination among all the actors. This could be the administrative committee and council of the certification organisation (Boström 2006) or an advisory council in the ministry of agriculture (Zerger et al. 2005) (indicator 9). Coordination is eased by the creation and implementation of national action plans (indicator 20) (DFAF 2001).

Definition of organic farming

In this paper, organic farming is defined according to the EU regulation (EC) 834/2007[c] that specifies the basic requirements needed in terms of production, certification, labelling and processing that must be implemented in each member state. The regulation defines organic farming as a way of producing agricultural goods that restrict the use of off-farm inputs in favour of other farming practices (cultural, biological and mechanical) that can be established on any farm after a period of conversion.

Results

Factors influencing organic farming in CEEC

The results from the questionnaire are summarised in Table 2. Market development and financial support are deemed more important than the social dynamic of the rural region (e.g., ageing population or out-migration), and infrastructure problems (e.g., rural services, state of road network), which did not qualify among the top 15 most

Table 2 Factors with the largest influence on the development of organic farming (OF) in CEEC

Rank	Type	Factors	Scale	
1	F	Public investment in processing of organic products	Domestic	++
2	M	Consumer demand for organic products	Domestic	++
3	F/M	Public investment for marketing of organic products	Domestic	++
4	S	Integration of OF in national rural development plan	Domestic	+
5	F	Area payment for conversion to OF	Domestic	+
6	F	Public expenditure in extension services and training	Domestic	++
7	F	Public investment in research and development in OF	Domestic	++
8	M	Consumer concerns about food quality, food safety, environmental protection and animal welfare	Domestic	++
9	F	Area payment for maintenance of OF	Domestic	+
10	M	Link between producers and retailers	Domestic	+
11	M	Consumer demand for organic products	EU	+
12	P	Availability of organic producers' association	Domestic	++
13	P	Political power of organic producers' association	Domestic	+
14	M	Consumer concerns about food quality, food safety, environmental protection and animal welfare	EU	+
15	F	Other financial help (e.g. guarantee loan, compensation for lost crop, etc.)	Domestic	+

Based on questionnaire results. The type (F = financial support, M = market development, S = strategic support, P = production characteristic), the scale at which they function (domestic or EU) and the estimate of how much they need to change ("–"decrease significantly, "-"decrease, "=" do not need to change, "+" increase, "++" increase significantly) are also shown.

important factors. Production factors that concern the producers' associations and integration of organic farming within strategic documents are also considered important. 13 out of 15 factors concern domestic obstacles.

Characteristics of the civil society domain

Producers' associations first appeared in the 1990's in CEEC and in the 1980's in Sweden (indicator 1, Table 3). The questionnaire reveals that although they appeared long ago, their power needs to increase significantly (++) in order to promote the organic sector development (Table 2; indicator 3 Table 3). In Latvia, Sweden and Lithuania there is one main producers' association for organic farmers, making them largely representative (indicator 2 and 5). In Poland, there are numerous small and local producers' associations (indicator 2), and no official record keeping of the membership is done, preventing the measurement of indicator 5. In Estonia, there are two national producers' associations and many smaller, local ones (indicator 2). Membership has been decreasing (Moschitz et al. 2004) and an estimate suggests that as few as 10-15% of organic farmers are part of any organic association, making the Estonian producers' associations hardly representative (indicator 5).

Characteristics of the state domain

In CEEC, new regulations were adopted in 2000 or 2001 in anticipation of their accession to the EU. Today, organic farming in all the studied countries is regulated based on regulation (EC) No 834/2007[c] of the EU (indicator 7). In Sweden and Lithuania, partnerships are frequent and thorough, while in Estonia and Latvia they are infrequent

Table 3 Indicator 1–5 describing the institutions of the civil society domain

		Estonia	Latvia	Lithuania	Poland	Sweden
1	Introduction of 1st private standard	1989	1994	1992	1994	1985
2	Number of national producers' associations	2	1	1	7	3
3	Power of producers' associations	Need to be increased	Need to be increased	Need to be increased	Need to be increased	Good
4	Private standard can still be used	No	No	Yes	Yes	Yes
5	Proportion of organic farmers in producers' associations	10-15%	≈ 60%	Most	nd	90-100%

1) EE: Milkk 2005, LV: Zarina 2009, LT: Kaspersen et al. 2004, PL: Metera 2005, personal communication, Josef Tyburski, Mazury University, Olsztyn, Poland, SE: Källander 2000/2010.
2) EE: Vetemaa & Milkk 2012, LV: Zarina 2009, Drozdovska 2005, LT: Kaspersen et al. 2004, PL: Kilcher et al. 2011, SE: Schusseleder 2009.
3) CEEC: based on questionnaire, SE: Kilcher et al. 2011.
4) CEEC: Tyburski & Zakowska-Biemans 2003, SE: Boström & Klintman 2003.
5) EE: personal communication Eve Ader, Estonian Plant Production Inspectorate, LV: Drozdovska Drozdovska 2005, LT: Kaspersen et al. 2004, PL: personal communication, Josef Tyburski, Mazury University, Olsztyn, Poland, SE: estimate by comparing number of organic farms in Figure 2 and the number of organic farms member of different associations, Schusseleder 2009.

and irregular. In Poland, they are increasingly frequent (indicators 8–9). The main discussion arena of Sweden is the administrative board and committees of KRAV, a certification body which includes members of most interest groups (e.g., food industry, environmental and animal protection organisations, and producers' associations) involved in the development of the organic sector (Boström 2006). In Poland, the discussion arena is a council, hosted by the Ministry of Agriculture and Rural Development. In Latvia, actors within the government sometimes participate in the board meeting of the organic producers' association. However, no formal discussion arena has been established that would group all the actors (Table 4).

Financial support appeared early in Lithuania and Sweden, long before their accession to the EU and before funds were made available for member states through regulation (EEC) 2078-92[d] (indicator 10). The other countries implemented payment schemes later, through the pre-accession structural funds, such as SAPARD and PHARE. Area payments for organic farming vary between 40 and 855 euro per hectare (indicator 11) and represent between 8% and 80% of the money invested in protecting the

Table 4 Indicator 6–9: institutions of the state domain related to general political recognition

		Estonia	Latvia	Lithuania	Poland	Sweden
6	First national regulation	1997	1996	1994	2001	1995*
7	Current regulation	2009	2009	2009	2009	2009
8	Strength of partnership	Weak	Weak	Good	Medium, increasing	Good
9	Discussion arena existence	No longer	No	Nd	Yes	Yes

6) CEEC: Tyburski & Zakowska-Biemans 2003, EE: Vetemaa & Milkk 2012, LV: Zarina 2009, LT: Kaspersen et al. 2004, PL: Metera 2005, SE: Lampkin et al. 1999. * In Sweden, the regulation for organic farming is not detailed and is limited to the requirement of EU regulations, Kilcher et al. 2011.
7 All: The basic rules of organic farming are uniform in all European Union member states. From January 1st, 2009 the new EU Organic Farming Regulations (EC) No 834/2007 and (EC) No 889/2008 came into effect (Vetemaa & Milkk 2012).
8) EE: personal communication Eve Ader, Estonian Plant Production Inspectorate; Moschitz et al. 2004, LT: personal communication, Leviana Sturite, State Stende Cereals Breeding Institute, Latvia, PL: personal communication, Josef Tyburski, Mazury University, Olsztyn, Poland; Moschitz et al. 2004, SE: Kilcher et al. 2011.
9) EE: personal communication Eve Ader, Estonian Plant Production Inspectorate, LV: personal communication, Leviana Sturite, State Stende Cereals Breeding Institute, Latvia; unpublished observation, Latvian Organic Agriculture Association, PL: personal communication, Josef Tyburski, Mazury University, Olsztyn, Poland; Moschitz et al. 2004, SE: Källander 2000/2010.

environment in agricultural land (indicator 12). The uptake rates for area payments vary between 50% and 240% (indicator 14). The 240% uptake rate of Sweden is due to the possibility of obtaining area payment without being certified.

Area payments are not the only types of possible financial support (indicators 15 to 19). In Lithuania and Poland, there is a larger diversity of financial measures than in Estonia and Latvia. Lack of financial support was the number one factor hindering the development of the organic sector in CEEC in the 1990's (Zobena 1998), and although it has increased a lot since the accession to the EU, it is still listed as a serious hindering factor in the questionnaire and other case studies (see for example Buciene and Eidukeviciene 2005; Hajduk and Staniszewska 2005). The questionnaire identified the lack of public investment in processing, marketing, area payments, advisory services and research as being among the top 10 factors hindering the development of the organic sector, see Table 2. In CEEC, it seems that there is a lack of information for farmers about agri-environmental schemes (including organic area payments) and that measures for intensive agriculture have priority over agri-environment schemes (IUCN 2004) (Table 5).

The rural development plan of each country includes provisions for organic farming (indicator 21). However, the questionnaire reveals that in CEEC integration of organic farming in rural development is the fourth most important factor hindering organic farming and needs to be increased (see Table 2). All the countries except Poland have an action plan for coordinating organic farming development (indicator 20). In Sweden and Lithuania the quantitative target set up in these action plans is relatively high: at

Table 5 Indicator 10–19: Institutions of the state domain related to financial support

		Estonia	Latvia	Lithuania	Poland	Sweden
10	Introduction of area payment	2000	2001	1993	1999	1989
11	Area payment (Euro/ha)	80	82-139	118-734	55-382	40-855
12	% of agri-environmental scheme devoted to organic farming	16	80	72	8	25
13	Total amount devoted to organic farming (millions of euros/year)	3.1	4.9	15	7	57.8
14	% uptake of area support	80%	93%	50-60%	88%	240% *
15	Loans available for improvement of organic farms	No	No	Yes	Yes	nd
16	Lower taxes for organic farmers	No	No	Yes	Yes	nd
17	Certification/inspection costs reimbursement	No	No	Yes	Yes	No
18	Research support (million euros/year)	0.1	0.1	0.046	0.27	5.9**
19	Estimate % of agricultural research funding devoted to organic farming	1%	19%	6.3%	0.2%***	10%

10) EE: Vetemaa & Milkk 2012, LV: Zarina 2009, LT: Kaspersen et al. 2004, PL: Metera 2005, SE: Källander 2000/2010.
11) EE: Prazan et al. 2004, LV: Latvian Ministry of Agriculture 2006a, LT: Rush 2006, Lithuanian Ministry of Agriculture 2006, PL: Porter 2006, Polish Ministry of Agriculture and Rural Development 2006b, SE: Swedish Board of Agriculture 2012.
12) EE, LV, LT, SE: Tuson & Lampkin 2006, Hrabalova et al. 2005, PL: Polish Ministry of Agriculture and Rural Development 2006b.
13) EE, SE: Tuson & Lampkin 2006, Hrabalova et al. 2005, LV: Drozdovska 2005, PL: Polish Ministry of Agriculture and Rural Development 2006a, LT: Lithuanian Ministry of Agriculture 2005.
14) EE, LV, PL, SE: Tuson & Lampkin 2006, Hrabalova et al. 2005, LT: Lithuanian Ministry of Agriculture 2006. * In Sweden, a farm does not need to be certified to get payment for organic farming.
15) All: Tuson & Lampkin 2006, Hrabalova et al. 2005.
16) All: Prazan et al. 2004.
17) All: Tuson & Lampkin 2006, Hrabalova et al. 2005.
18) All: Slabe 2004. ** The total research funding includes forestry and fisheries. Thus, the number is likely to be smaller than this.
19) EE: Statistics Estonia 2007, LV: Latvian council of science 2006, LT: Slabe 2004, PL: Polish Central Statistical office 2007, SE: Slabe 2004. ***Approximation based on two different years.

Table 6 Indicator 20–22: Institutions of the state domain related to strategic support

		Estonia	Latvia	Lithuania	Poland	Sweden
20	Action plan for organic farming	Yes	Yes	Yes	No	Yes
21	OF incorporation in strategic plan of government	Yes	Yes	Yes	Yes	Yes
22	Quantitative target for the proportion of agricultural land that should be organic	Yes	No	Yes	No	Yes

20) All: Slabe Slabe 2004, EE: Estonia ministry of agriculture 2002, LV: Zarina 2009, Prazan et al. 2004, LT: Jansen & Simon 2005, Rush 2006, Prazan et al. 2004, PL: Prazan et al. 2004, SE: Kilcher et al. 2011.
21) Estonian Ministry of Agriculture 2006, Latvian Ministry of Agriculture (2006b), Lithuanian Ministry of Agriculture (2006), Polish Ministry of Agriculture and Rural Development (2006b), Swedish Ministry of Agriculture 2008
22) EE: Kalm & Laansalu 2002, LV: personal communication, Leviana Sturite, State Stende Cereals Breeding Institute, Latvia, LT: Jansen & Simon 2005, Rush 2006, PL: Prazan et al. 2004, SE: Statistics Sweden 2012.

least 15-20% of the agricultural land should be cultivated organically by some given year (indicator 22) (Table 6).

The integration of organic farming within the state domain varies greatly (Table 7). It is most integrated in Latvia, Poland and Sweden, least integrated in Estonia and Lithuania. Research is almost exclusively carried out by institutes subsidised by the state. The implementation of a full research programme occurs only in Sweden and Lithuania, although Poland has a full programme for crop research (indicator 25). In Estonia, integration of organic farming is weak. In all other countries, advising services are provided by the same advisory and training centre as conventional agriculture. In Poland and Sweden, the main agricultural universities offer organic farming courses in their curricula.

Characteristics of the market domain

Indicators 27 to 29 show that all the actors of the supply chain are present in Sweden, while both processors and distributors are very limited in CEEC. The lack of processing facilities is one of the most important factors that hinder the development of the organic sector in CEEC according to the questionnaire and to previous studies (Milkk Milkk 2005; Terra Nord 2005; Buciene and Eidukeviciene 2005; Hajduk and Staniszewska 2005). Indicator 29 shows that supermarkets account for more than 75% of all the sales in Sweden. Favoured sales channels in CEEC are typically direct sales and market places. The market share of organic products is low but still substantially higher in Sweden (4.1%) than in other countries (indicator 30). However, this somehow contradicts the results from the literature review which suggests that people in

Table 7 Indicator 23–26: Integration of the organic sector within agricultural institutions of the state domain

		Estonia	Latvia	Lithuania	Poland	Sweden
23	Specific office in agricultural ministry	Yes	No	Yes	Yes	No
24	Integration of OF in training establishments	Weak, almost fully segregated	Largely integrated	Nd	Largely integrated	Largely integrated
25	Range of research program	Production only	Production only	Almost full range*	Almost full range*for crop	Full range* for crop and livestock
26	Integration of OF in advisory services	Almost none	Largely integrated	Little integration	Largely integrated	Largely integrated

23) CEEC: Tyburski 2003, Ministry of agriculture websites, SE: Swedish Ministry for Rural Affairs 2012.
24) EE: Milkk 2005, PL: Moschitz et al. 2004, Brent 1999, LV: Zarina 2009, SE: Källander 2000/2010.
25) All: Slabe 2004. *Full range: research programme covers topics of production, food quality, and market development.
26) EE: Milkk 2005, personal communication Eve Ader, Estonian Plant Production Inspectorate, PL: Moschitz et al. 2004, Brent 1999, LV: Zarina 2009, Drozdovska 2005 LT: Tatulos programa 2005, SE: Källander 2000/2010.

Table 8 Indicator 27–30: Maturity of the organic market based on physical structures

		Estonia	Latvia	Lithuania	Poland	Sweden
27	Number of certified processing facilities	35	8	19	99*	±500
28	Promotion by large retailers	No	No	Beginning	Beginning	Yes
29	Distribution of sales channels ⊞ Direct sales ■ Supermarket ▣ Organic/health food shop ▤ Market place □ Other				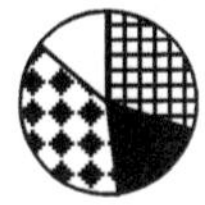	
30	Market share of organic products	< 1%	< 1%	1.5%	< 0.5%	4.1%

27) EE: Vetemaa & Milkk 2012, LV: Latvian Ministry of Agriculture 2006b, Zarina 2009, LT: Lithuanian Ministry of Agriculture 2006, PL: Polish Ministry of agriculture and rural development 2006a, SE: KRAV 2006. *In 2005.
28) EE: Moschitz et al. 2004, Milkk 2005, LV: Gulbe & Hazners 2005, Terra Nord 2005, LT: Rutkoviene & Abraityte 2006, PL: Moschitz et al. 2004, SE: FAS 2004.
29) EE: Pehme et al. 2007, LV: Gulbe & Hazners 2005, Terra Nord 2005, LT: Kaspersen et al. 2004, Rutkoviene & Abraityte 2006, PL: Bakula & Smoluk 2005, Kilcher et al. 2011, SE: Dabbert et al. 2004, FAS 2004, Kilcher et al. 2011.
30) EE, LV: based on numbers from countries with similar market characteristic from Dabbert et al. 2004 and Ritcher et al. 2006. LT: Terra Nord 2005. PL: Kilcher et al. 2011. SE: KRAV 2012.

Lithuania and Poland more frequently buy organic products compared to people in Sweden (indicator 33) (Table 8).

At first sight, indicator 30 (market share of organic products) does not correspond with indicator 32 (acceptable price premium for organic products). Half the population is willing to pay a substantial price premium but very few choose to do so. However, the price difference between conventional and organic products that consumers face might be higher than what is stated in indicator 32. This is not examined in the present study. Sweden shows the highest acceptable price premium which corresponds with Sweden having the highest market share of organic products. The questionnaire identified consumer demand for organic products in CEEC as the second most important factor that hinders organic farming (see Table 2). In Sweden and Poland, the reliability of the certification body is good (indicator 34), while it is poor in Estonia and Latvia, and medium in Lithuania (Table 9).

Discussion

Present and lacking institutions

According to the results the most important institutions are private and governmental standards for organic production and certification, strategic and financial support measures from the government, and market demand and supply. Missing institutions are identified using the six-step process for the institutional development of the organic sector proposed by Michelsen et al. (2001). The degree of completion, used to identify missing institutions, has been measured by using the indicators describing the characteristics of the steps (see Tables 1, 3, 4, 5, 6, 7, 8, 9). Sweden has been used as a baseline country and the indicators should resemble those of Sweden for a step to be considered completed. A country performing worse than Sweden in one-third of the indicators scores half-way. A country performing worse than Sweden in two-thirds of the

Table 9 Indicator 31–34: Maturity of the organic market based on consumer behaviour

		Estonia	Latvia	Lithuania	Poland	Sweden
31	Proportion of consumer that can correctly define organic farming	Medium	Medium	Medium	Low 30%	High
		65%	70%	70%		93% **
32	Acceptable price premium (in% higher) by half the population	10%	Very low*	25%	10-20%	30%
33	% of the population that buy OP > once a week/>once a month/< once a month	nd	nd	35/32/33	17/48/33	6/48/45
34	Perception of the quality of certification and control system*	Poor	Poor	Medium	Good	Good

31) EE: Pehme et al. 2007, LV: Gulbe & Hazners 2005, LT: Rutkoviene & Abraityte 2006, PL: Kucińska et al. 2006, SE: KRAV 2006. ** 93% recognise KRAV logo and associate it with organic farming.
32) EE: Milkk 2005, LV: Gulbe & Hazners 2005, LT: Rutkoviene & Abraityte 2006, PL: Kucińska et al. 2006, Bakula & Smoluk 2005, SE: Dabbert et al. 2004. * 4% are ready to pay 30% price premium, 21% are ready to pay a price premium (Gulbe & Hazners 2005).
33) LT: Rutkoviene & Abraityte 2006, PL: Bakula & Smoluk 2005, SE: Terra Nord 2005.
34) CEEC: Prazan et al. 2004 * This study does not depict the point of view of the consumer directly, but the point of view from ministry of agriculture, organic producers' association, organic farmer advisors, and organic farmers. However, the deficiencies that are perceived by major actors are likely to be transmitted to the general public. SE: Michelsen & Søgaard 2001.

indicators, or more, scores beginning. This is a rough estimate and it should not be interpreted as the situation in Sweden being perfect from the perspective of organic farming. Michelsen et al. (2001) acknowledge that a step can begin without the previous one being completed. This was also observed in the present study (Table 10).

The data also suggest that step 5 is the most difficult step to complete and might be dependent on the completion of the other steps to be completed. For all CEEC, step 6 is half-way while step 5 is still beginning. Furthermore, even Sweden which has completed all the steps still has a small market (step 5). The result that market development is the main hindering factor for the development of the organic sector, is supported by Terra Nord (2005) and DFAF (2001). The development of the organic market in CEEC reflects the economic situation compared with the wealthier EU15. Furthermore, although infrastructures are not considered the most important factors hindering organic farming, they need to improve to allow its full development according to the questionnaire. For example, the quality of the road network and the availability of water and electricity services would benefit organic as well as conventional production.

Table 10 Degree of completion of the six-step process of the institutional development of the organic sector

	Step 1 Organic community	Step 2 Political recognition	Step 3 Financial support	Step 4 Type of relationship	Step 5 Organic market	Step 6 Discussion and coordination arena
EE	***Half-way (eroded)***	*Beginning (eroded)*	*Beginning*	*Competition (eroded)*	*Beginning**	***Half-way***
LV	***Half-way (near completed)***	***Half-way***	*Beginning*	***Cooperation/ Competition***	*Beginning**	***Half-way***
LT	Completed	Completed	***Half-way***	***Cooperation/ Competition****	*Beginning/* ***Half-way***	***Half-way****
PL	***Half-way****	***Half-way (increasing)***	***Half-way***	***Cooperation/ Competition***	*Beginning/* ***Half-way***	***Half-way***
SE	Completed	Completed	Completed*	Cooperation/ Creative conflict	Completed (although still small)	Completed

Italic letters indicates that the step is only beginning, ***italic bold letters*** that the step is half way, and normal type letters that the step is completed. The evaluation of the completion of each step is based on the indicators described in the result section. The * symbol means that data for one indicator is missing.

The political recognition and willingness to financially promote organic farming is reflected in the degree of completion of step 2 to 4. Table 10 suggests that the government of Lithuania has a high political willingness, followed by Poland and Latvia. Latvia invests proportionally more in organic farming and has higher strategic support, while Poland focuses more on partnership and discussion. In Estonia the political willingness is lower. When considering the amount of money invested in organic farming, it should be kept in mind that the CEEC have a lower GDP per capita than EU15 or Sweden, which limits what they can invest. This is also relevant in relation to the CAP support, since the support devoted to organic farming (pillar II of CAP) requires co-finance from the member states.

Figure 2 illustrates the number of certified organic farms in the studied countries. Over time the number increases in all countries. The introduction of area payment (indicator 10) seems to have had a positive effect in Poland (area payments introduced in 1999) and to some extent in Estonia and Latvia (area payments introduced in 2000 and 2001 respectively) whereas for Lithuania there is a ten year lag (area payments introduced in 1993). For both Poland and the Baltic states the accession to the EU in 2004 seems to be have been more important and the number of certified farms accelerated in the period 2003–2005. The best explanation for the rapid increase in reported Swedish organic farms, followed by a decrease, is that different sources have been used. In practice, the number of certified organic farms have increased slowly, and rather steadily, over time.

Data over organically cultivated area follow a similar pattern with two differences. Whereas the number of organically certified farms increased fast in Poland, the amount of organically cultivated land increased at a lower rate indicating that smaller farms to a larger degree converted to organic standards (EkoConnect 2006). In contrast, Estonia, showing moderate growth rate of organically *certified* farms experienced a faster than average growth of organically *cultivated* land. In Sweden 12.6% of the agricultural land was organically cultivated in 2011, an increase from 10.7% in 2010. Another 3.5% of agricultural land was under conversion (Statistics Sweden 2012) which is not enough to reach the politically set target of 20% organically cultivated land by 2013. This can be compared with a rather modest market share of organic production (indicator 30) of 4.1%. Why the two figures differ is not investigated here.

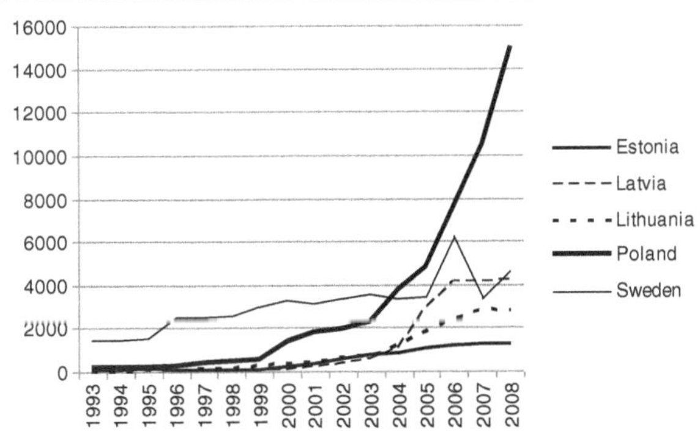

Figure 2 Number of certified organic farms over time in the five case-study countries. This might differ from the number of farms that are cultivated organically. Compiled from different sources. CEEC: EkoConnect 2006; Eurostat 2012b. Sweden: Källander 2000/2010; Statistics Sweden 2002, 2005; Eurostat 2012b.

Ways to promote institutional development through the six steps

Step 1 is the only step that pertains to the establishment of institutions within the civil society domain. The step begins by the self-organisation of a group of organic farmers into a producers' association (Michelsen et al. 2001). Step 1 can be promoted by governments with the money of the European Agricultural Fund for Rural Development (EAFRD) which has a provision for the establishment and operation of producers' associations (EEC 1698-2005[e]). However, Moschitz et al. (2004) argue that the identity of the community should develop by itself, otherwise its acceptance by most actors is difficult. In a similar vein Milestad (2003), in a study of Austrian farmers, argues that subsidies on the one hand makes a transition to organic farming possible but on the other hand influence farmers in a direction they might not have chosen if it was not for the subsidy. They experience a loss of independence.

Steps 2 and 3 are where governments have the most power to act. To increase the support stemming from political recognition, Dabbert et al. (2004) suggest that governments should acknowledge the possibility of organic farming to fulfill multiple policy objectives. This can be fostered by research, by discussion, and by partnership, which all increase the understanding of organic farming.

Step 5 is influenced by the actors in all three societal domains. The example of Sweden has shown that producers' associations can have significant impact, especially in the beginning, because they are the first to lobby large retailers and to provide marketing initiatives. In addition, market actors, such as large retailers, have proven that they can have a clear leadership role in promoting organic farming in Sweden (Källander 2000/2010; FAS 2004). Finally, governments can favour the establishment of processors by providing guaranteed loans or lower income tax to this type of enterprise (Terra Nord 2005). Governments can also stimulate the market by, for example, increasing public procurement of organic food (Larsson 2012; Larsson et al. 2012) or even making the use of organic products in public catering mandatory (Edman 2004).

The development of steps 4 and 6 are closely related because the development of cooperative relationships often involves discussion and partnerships, which may in turn involve the development of official discussion and coordination arenas to facilitate this. A stakeholders' survey at the European level identified the establishment of a national organic network or a national advisory committee grouping farmers, market actors and the state as a good way to facilitate coordination (Zerger et al. 2005). The government is not the only actor who can promote this step. Boström (2006) proposes that the discussion arena can be the administrative board and committees of a certification body. Effectively, certification and labelling play a central role for all the actors of the organic sector, creating incentives for discussion and coordination (Boström 2006).

Conclusion

Organic farming is proposed as a way to improve the environmental situation of the Baltic Sea, including reduced eutrophication. This paper furthers the understanding of the institutional factors that hinder the development of organic farming in the CEEC. The results support previous work of Michelsen et al. (2001) that proposes that the development of organic farming proceeds along six steps. According to the results, the market (step 5) is the least developed step. A well functioning market would further promote the development of organic farming, but this step is identified as being difficult to complete. The results also suggest a correlation between higher political willingness to promote the organic sector and more advanced development of the sector.

In all studied CEEC, the window of opportunity offered by the entry to the EU has been used to some extent when establishing a regulatory framework and financial help. However, there is still a long way to go before the organic sector in these countries is fully developed. Lithuania is the CEEC with the most steps of institutional development completed. Latvia and Poland are half-way through most of the steps. Estonia is the country with the least completed steps.

The results also provide support for decision makers that want to promote organic farming effectively. We chose not to give country specific recommendations but there are some general recommendations to be drawn from the results. The first recommendation, drawn from the questionnaire, for all CEEC is to ensure support, including financial support, to organic farming that develops the independence and self-organisation of the organic sector. The second recommendation, following the poor performance of step 5 in combination with its importance, is that support should be directed towards a market-oriented approach. The third recommendation is to foster discussion, collaboration and coordination between all the actors. This is central to many steps in the development of organic farming and it can further the understanding of each actor and increase the level of support to organic farming.

Endnotes

[a]The Helsinki Commission, or HELCOM, works to protect the marine environment of the Baltic Sea from all sources of pollution through intergovernmental co-operation between Denmark, Estonia, the European Community, Finland, Germany, Latvia, Lithuania, Poland, Russia and Sweden. See www.helcom.fi.

[b]From this point, the term CEEC refers only to Estonia, Latvia, Lithuania and Poland.

[c]On January 1 2009 the new EU regulation (EC) No 834/2007 came into effect in all EU member states, replacing the previous regulation EEC 2092–91: Council Regulation (EEC) No 2092/91.

[d]EEC 2078–92: Council Regulation (EEC) No 2078/92. Official journal of the European Union. L 215, 30/07/1992, p 85–90.

[e]EEC 1698/2005: Council Regulation (EEC) No 1698/2005. Official journal of the European Union. L 277, 21/10/2005, p 1–40.

Competing interests
The authors declare no conflict of interests.

Authors' contributions
ML and LM both took part in drafting the text, analysing and interpreting data. LM collected the bulk of the data and ML, TH and JS finalised the text.

Acknowledgements
We are most thankful to two anonymous reviewers for valuable comments and the research programme Baltic Ecological Recycle Agriculture and Society (www.beras.eu) for supporting this work.

Author details
[1]Stockholm Resilience Centre, Stockholm University, Kräftriket 2B, Stockholm SE-106 91, Sweden. [2]Mälardalens högskola, Box 883, Västerås 721 23, Sweden. [3]Swedish Society for Nature Conservation, Box 4625, Stockholm 116 91, Sweden.

References

Bakula WL, Smoluk J (2005) The perception of Polish organic food consumers. Paper presented at the 8th Wissenschaftstagung Ökologischer Landbau - Ende der Nische. Hohenheim University, Kassel, http://orgprints.org/3943/01/3943.pdf. Accessed June 26 2012

Baylis J, Smith S (2001) The globalization of world politics. An introduction to international relations. Oxford University Press, New York

Boström M (2006) Regulatory credibility and authority through inclusiveness: standardization organization in cases of eco-labelling. Organization 13(3):345–367

Boström M, Klintman M (2003) Framing, debating, and standardising "natural food" in two different political contexts: Sweden and the U.S. Score Rapportserie. Stockholm University, Stockholm

Brent JK (1999) Extension, research and farm competitiveness in Central and Eastern Europe. Paper presented at the 2nd EU Accession Workshop on the Rural Sector, Structural change in the farming sectors of Central and Eastern Europe: lessons and implications for EU accession", June 27–29 1999. Warsaw, http://www.fao.org/sd/exdirect/EXan0037.htm. Accessed June 26 2012

Buciene A, Eidukeviciene M (2005) How viable is the local organic food market in Klaipeda town and district from the consumer's point of view? P 9–20. In: Kakriainen S, von Essen H (eds) Obstacles and solutions in use of local and organic food. Baltic Ecological Recycling Agriculture and Society (BERAS) 4. Ekologiskt lanbruk 44. Centre for sustainable agriculture, Swedish University of agricultural sciences, Uppsala

Dabbert S, Häring AM, Zanoli R (2004) Organic farming: policies and prospects. Zed books, London and New York

DFAF (Danish ministry of food, agriculture and fisheries) (2001) *Organic food and farming: toward partnership and action in Europe*. Conference proceedings, Copenhagen

Diaz RJ, Rosenberg R (2008) Spreading dead zones and consequences for marine ecosystems. Science 321:926–929

Drozdovska L (2005) Organic agriculture in Latvia, Environmental friendly food production system: requirements for plant breeding and seed production. Conference proceedings, Talsi, pp 42–45

Edman S (2004) Hållbara laster, Konsumtion för en ljusare framtid. Government Commission Report/SOU 2004:119. Swedish Ministry of Agriculture, Stockholm

EkoConnect (2006) Development of the total organic area in countries of Central and Eastern Europe (in ha). http://www.ekoconnect.org/pdf/06-02-27_development_cee.pdf. Accessed June 15 2012

Estonian Ministry of Agriculture (2002) Estonia agriculture, rural economy and food industry. Tallinn, http://www.agri.ee

Estonian Ministry of Agriculture (2006) Estonia Rural Development Plan: 2007-2013. Tallinn

EU (2004) European Action Plan for Organic Food and Farming, Commission Staff Working Document. Annex to COM (2004)415 final, SEC(2004) 739. Brussels

Eurostat (2012a) Newsrelease 97/2012 – 20 June 2012: GDP per capita in the Member States ranged from 45% to 274% of the EU27 average in 2011. http://epp.eurostat.ec.europa.eu/cache/ITY_PUBLIC/2-20062012-AP/EN/2-20062012-AP-EN.PDF. Accessed June 26, 2012

Eurostat (2012b) Number of certified organic operators [food_act2]. http://appsso.eurostat.ec.europa.eu/nui/show.do?dataset=food_act2&lang=en. Accessed June 26, 2012

FAO (2002) Central and Eastern Europe and the commonwealth of independent states, The state of food and agriculture. Rome

FAS (USAD Foreign Agricultural Service) (2004) Organic Products: Swedish Market for Organic Products. GAIN Report: SW4012. http://www.fas.usda.gov/gainfiles/200409/146107434.pdf. Accessed June 27, 2012

Fischler F (2003) Future perspectives for organic farming in an enlarged EU. Speech given at the conference Perspectives of Organic Agriculture in an Enlarged EU, November 21s 2003, Plovdiv., SPEECH/03/562 http://europa.eu/rapid/pressReleasesAction.do?reference=SPEECH/03/562&format=PDF&aged=1&language=EN&guiLanguage=en. Accessed June 26, 2012

Granstedt A, Thomsson O, Schneider T (2005) Environmental impacts of Baltic ecological food systems: final report from BERAS work package 2. Ecological Recycling Agriculture and Society (BERAS) 5. Ekologiskt lantbruk 46. Centre for Sustainable Agriculture, Swedish University of Agricultural Sciences, Uppsala

Gulbe I, Hāzners J (2005) Development and implementation of collective trademarks and food labels in Latvia: label for organic food "ECOPRODUCT". Paper presented at NJF seminar 381, The Northern European food industry - challenges and transitions from an economic perspective, seminar of the Nordic association of agricultural scientists. Helsinki

Hajduk E, Staniszewska M (2005) Development of local organic markets and other environmentally oriented activities in Poland: Cases of Kluzborg and Bochnia and Zbiczno. In: Kakriainen S, von Essen H (eds) Obstacles and solutions in use of local and organic food. Baltic Ecological Recycling Agriculture and Society (BERAS) 4. Ekologiskt lantbruk 44. Centre for Sustainable Agriculture, Swedish University of Agricultural Sciences, Uppsala, pp 20–35

Hannerz F, Destouni G (2006) Spatial characterization of the Baltic Sea Drainage Basin and its unmonitored catchments. Ambio 35(5):214–219

HELCOM (2002) Sixth meeting of working group on agriculture, October 21–22 2002. Helsinki Commission, Kiel

HELCOM (2004) 30 years of protecting the Baltic Sea: HELCOM 1974-2004. Helsinki commission, Helsinki

HELCOM (2011) Baltic Sea Environment Proceedings No. 128. Fifth Baltic Sea Pollution Load Compilation (PLC-5), Helsinki

Hrabalova A, Handlova J, Koutna K, Zdrahal I (2005) Further development of organic farming policy in Europe with particular emphasis on EU enlargement. Report on the development of organic farming in CEE accession states with national report cards (final report). QLK5-2002-00917. Research Institute of Agricultural Economics, Brno

IUCN (2004) Gaps and Limitations of the Rural Development: Plans of the CEE New Member States. Foundation IUCN Poland (IUCN Programme Office for Central Europe). http://data.iucn.org/dbtw-wpd/edocs/EEP-041.pdf. Accessed June 26, 2012

Jansen B, Simon S (2005) Country Report: Organic agriculture in Lithuania. EkoConnect information letter organic agriculture of Central and Eastern Europe. http://www.ekoconnect.org/pdf/Infobrief_5/Infobrief-5-Englisch.pdf. Accessed June 26, 2012

Källander I (2000/2010) Organic Agriculture in Sweden. FiBL Country Reports (edited 2010). http://www.organic-world.net/fileadmin/documents/country_information/sweden/kaellander-2000-sweden-2000.pdf. Accessed June 27, 2012

Kalm U, Laansalu A (2002) Estonia agriculture, rural economy and food industry status report. Ministry of agriculture, Republic of Estonia, Tallin

Kaspersen BS, Mikkelsen KD, Jacobsen C (2004) Organic agriculture in the light of EU enlargement – A window of opportunity. Thesis (#2193), Institut for Miljø, Teknologi og Samfund, Roskilde University. http://dspace.ruc.dk/bitstream/1800/217/1/Organic_Agriculture_in.pdf. Accessed June 27, 2012

Kilcher L, Willer H, Huber B, Frieden C, Schmutz R, Schmid O (2011) The Organic Market in Europe, 3rd edn. FiBL and Sippo, Frick and Zürich

KRAV (2006) Annual Report 2005. KRAV Incorporated Association, Uppsala

KRAV (2012) Marketing Report. www.krav.se. Accessed June 15 2012

Kucińska K, Pelc I, Artyszak A (2006) Development of Organic Agriculture in Poland. Paper presented at Joint Organic Congress, Organic Farming and European Rural Development, Odense. May 30-31, 2006

Lampkin N, Foster C, Padel S, Midmore P (1999) The policy and regulatory environment for organic farming in Europe: synthesis of results. Organic farming in Europe, Economics and policy, vol 1. Hohenheim University, Stuttgart

Larsson M (2012) Environmental entrepreneurship in organic agriculture in Järna, Sweden. J Sustain Agric 36:153–179

Larsson M, Granstedt A (2010) Sustainable governance of the agriculture and the Baltic Sea – Agricultural reforms, food production and curbed eutrophication. Ecol Econ 69:1943–1951

Larsson M, Granstedt A, Thomsson O (2012) Sustainable Food System – Targeting Production Methods, Distribution or Food Basket Content? In: Reed M (ed) Organic Food and Agriculture - New Trends and Developments in the Social Sciences. InTech. http://www.intechopen.com/books/organic-food-and-agriculture-new-trends-and-developments-in-the-social-sciences/sustainable-food-system-targeting-production-methods-distribution-or-food-basket-content-. Accessed June 27, 2012

Latvian Council of Science (2006) Projects of fundamental and applied research for the year 2006: agricultural sciences. Latvian Council of Science, Riga

Latvian Ministry of Agriculture (2006a) Agriculture and Rural Area of Latvia. Riga, pp 115–117, http://www.zm.gov.lv

Latvian Ministry of Agriculture (2006b) Latvian Rural Development National Strategy Plan: 2007–2013. Riga

Lerman Z (2001) Agriculture in transition economies: from common heritage to divergence. Agric Econ 26:95–114

Lithuanian Ministry of Agriculture (2005) Annual report: agriculture in Lithuania 2005. http://www.zum.lt

Lithuanian Ministry of Agriculture (2006) Rural Development Programme 2007-2013. http://www.zum.lt

Metera D (2005) Organic Farming in Poland: Update July 2005. http://www.organic-europe.net/fileadmin/documents/country_information/ARCHIVE/poland-2005-organic-europe.pdf. Accessed June 27, 2012

Michelsen J, Søgaard V (2001) Policy instruments promoting conversion to organic farming and their in Europe 1985–1997. Department of Political Sciences and Public Management, University of Southern Denmark, Odense

Michelsen J, Lynggaard K, Padel S, Foster C (2001) Organic farming development and agricultural institutions in Europe: a study of six countries. Organic farming in Europe: economics and policy, vol 9. University of Hohenheim, Stuttgart

Milestad R (2003) Building farm resilience. Prospects and challenges for organic farming, Dissertation. The Swedish University of Agricultural Sciences, Uppsala

Milkk M (2005) Organic agriculture in Estonia. In: Belicka I, Viklante M, Arbidans M (eds) Proceedings of Environmental friendly food production system: requirements for plant breeding and seed production, Talsi. May 31–June 3

Moschitz H, Stolze M, Michelsen J (2004) Further development of organic farming policy in Europe with particular emphasis on EU enlargement, Report on the development of political institutions involved in policy elaboration in organic farming for selected European states (deliverable 7). Available via http://orgprints.org/4799/1/EUCEEOFP_D7_final_report.pdf. Accessed 16 April, 2013

North DC (1990) Institutions, Institutional Change and Economic Performance. Cambridge University Press, Cambridge

Pehme S, Luik A, Liivaauk P (2007) Estonian consumers' attitudes to organic food, Paper presented at the 3rd QLIF Congress, Improving Sustainability in Organic and Low Input Food Production Systems. University of Hohenheim, Germany

Petersen J-E, Hoogeveen Y (eds) (2004) Agriculture and the environment in the EU accession countries. Environmental issue report No 37. European Environmental Agency, Copenhagen

Polish Central Statistical Office (2007) Concise Statistical Yearbook of Poland 2006. Warsaw

Polish Ministry of Agriculture and Rural Development (2006a) Annual report: Agriculture and food economy in Poland. Warsaw, http://www.minrol.gov.pl

Polish Ministry of Agriculture and Rural Development (2006b) National Strategic Plan for 2007-2013 Rural Development. Warsaw

Porter E (2006) Agricultural Situation Organic farming in Poland – update 2006.Global Agriculture Information Network of the USDA Foreign Agricultural Service. GAIN Report. PL603. http://www.fas.usda.gov/gainfiles/200606/146197948.pdf. Accessed June 30, 2012

Prazan J, Koutna K, Skorpikova A (2004) Report on the development of organic farming and the policy environment in Central and Eastern European accession states, 1997–2002 (deliverable 6). EU-CEEOFP project report

Richter T, Padel S, Lowman S, Jansen B (2006) The European Market for Organic Food 2004/2005. In: Willer H, Yussefi M (eds) The world of organic agriculture: statistics and emerging trends 2006. International Federation of Organic Agriculture Movements (IFOAM), Bonn Germany & Research Institute of Organic Agriculture (FiBL), Frick, Switzerland

Rush C (2006) Lithuania organic products - organic farming in Lithuania 2006. Global Agriculture Information Network of the USDA Foreign Agricultural Service. GAIN Report. LH6002. http://www.fas.usda.gov/gainfiles/200606/146198110.pdf. Accessed June 30, 2012

Rutkoviene V, Abraityte G (2006) Organic markets and consumers in Lithuania. Paper presented at the Joint Organic Congress, Organic Farming and European Rural Development, Odense, Denmark. http://orgprints.org/8552/01/naujas_danija.doc. Accessed June 30, 2012

Sandström A (2008) Policy Networks: The relation between structure and performance, Dissertation. Luleå University of Technology, Luleå
Schultz M, Folke C, Olsson P (2007) Enhancing ecosystem management through social-ecological inventories: lessons from Kristianstads Vattenrike, Sweden. Environ Conserv 34(2):140–152
Schusseleder G (2009) Konsten att undvika GMO. Ordfront Magasin, 3:2009. http://www.ordfront.se/Ordfrontmagasin/Artiklar%202009/Goran%20Schusseleder%20om%20GMO%203_09.aspx. Accessed June 26, 2012
Slabe A (2004) How to improve trans-national cooperation? Consolidated report on 2nd seminar on organic food and farming research in Europe. European Commission, Research Directorate General, Brussels
Statistics Estonia (2007) Research and development expenditures in non-profit institutional sectors, Tallinn
Statistics Sweden (2002) Swedish agricultural year book. Chapter 11: Ekologisk produktion/ Organic production. http://www.scb.se/statistik/JO/JO1901/2003M00/JO01S%C3%850201_19.pdf. Accessed June 30, 2012
Statistics Sweden (2005) Swedish agricultural year book. Chapter 11: Ekologisk production/Organic production. http://www.scb.se/statistik/_publikationer/JO1901_2005A01_BR_JO01SA0501.pdf. Accessed June 30, 2012
Statistics Sweden (2012) Swedish agricultural year book. Chapter 11: Ekologisk produktion/Organic production. http://www.scb.se/statistik/_publikationer/JO1901_2012A01_BR_17_JO01BR1201.pdf. Accessed June 30 2012
Stolze M, Piorr A, Häring A, Dabbert S (2000) The environmental impacts of organic farming in Europe, Organic farming in Europe: economics and policy, vol 6. Hohenheim Univeristy, Stuttgart
Swedish Board of Agriculture (2012) http://www.jordbruksverket.se/jordbruksverketslattlastasidor/stodtilllandsbygden/miljoersattningar/certifieradekologiskproduktionellerkretsloppsinriktadproduktion.4.795c224d1274198ffc280003186.html. Accessed June 27, 2012
Swedish Ministry for Rural Affairs (former Ministry of Agriculture) (2012) http://www.sweden.gov.se/sb/d/2064, Accessed June 30, 2012
Swedish Ministry of Agriculture (2008) Swedish rural development programme 2007–2013. http://www.regeringen.se/sb/d/8723/a/82731. Accessed June 30, 2012
Tatulos programa (2005) Web portal about organic agriculture in Lithuania. www.organic.lt. Accessed March 17, 2007
Terra Nord (2005) Development of the Nordic-Baltic Market for Organic Food. Nordic Council of Ministers, Copenhagen, www.norden.org/pub/miljo/jordogskov/sk/TN2005548.pdf. Accessed June 30, 2012
Tuson J, Lampkin NH (2006) Further development of organic farming policy in Europe with particular emphasis on EU enlargement, Report detailing national and regional OF policy measures in EU states and Switzerland (deliverable 2). EU-CEE-OFP project report
Tyburski J (2003) Organic farming in Poland: Past, present and future perspectives, Organic agriculture: sustainability, market and policies. CABI publishing, Wallingford
Tyburski J, Zakowska-Biemans S (2003) Further development of organic farming policy in Europe with particular emphasis on EU enlargement, Report on Harmonisation of Regulations and Standards for Organic Agriculture in the Central and Eastern Europe Accession Countries and the European Union (deliverable 1). QLK5-2002-00917
Vetemaa A, Milkk A (2012) Organic Farming in Estonia. http://www.organic-europe.net/estonia.html?&L=0#c2927. Accessed June 27, 2012
Yin R (2003) Case study research: Design and methods, 3rd edn. Sage Publications, Thousand Oaks
Zarina L (2009) Organic Farming in Europe: country report for Latvia. http://www.organic-europe.net/latvia.html?&L=0. Accessed June 30, 2012
Zerger C, Vairo D, Häring AM, Zanoli R, Dabbert S (2005) Implementation of policy recommendations on organic farming: A cross-country synthesis of national policy workshops in 11 European countries. EU-CEE-OFP project report available via https://www.uni-hohenheim.de/i410a/EUCEEOFP/WP7_WS3_Synthesis_report_final.pdf
Zobena A (1998) Organic farming in the Baltic countries social aspects of development. In: Vlavianos-Arvanitis A, Morovic J (eds) Biopolitics, vol VI, The Bio-Environment. The Biopolitics International Organisation, Athens, Available at http://www.biopolitics.gr/BIOPOLITICS/HTML/PUBS/VOL6/HTML/fpage.htm

5

The demand of urban residents for the biodiversity conservation in U Minh Thuong National Park, Vietnam

Huynh Viet Khai[1,2] and Mitsuyasu Yabe[2*]

* Correspondence:
yabe@agr.kyushu-u.ac.jp
[2]Laboratory of Environmental Economics, Department of Agricultural and Resource Economics, Faculty of Agriculture, Kyushu University, Hakozaki 6-10-1, Higashi-ku, Fukuoka city 812-8581, Japan
Full list of author information is available at the end of the article

Abstract

The total economic value of biodiversity provides policy-makers reliable information to estimate welfare losses caused by biodiversity reductions and perform cost-benefit analysis of biodiversity conservation projects. Searching literature reviews on economic-valuation techniques, the contingent valuation (CV) method has been popularly applied to estimate the economic value of biodiversity. This approach is based on a hypothetical scenario in which respondents are requested through questionnaires to reveal their maximum willingness to pay (WTP) for biodiversity conservation. The article used CV to study the Mekong Delta urban households' preferences and their willingness to pay for the program of biodiversity conservation in U Minh Thuong National Park, one of the largest peat swamp forests in Vietnam. Logistic regression was used to predict the probability of respondents willing to pay for the conservation program. The mean WTP was calculated approximately VND16,510 ($0.78) per household per month for all respondents and about VND31,520 ($1.49) after excluding the protest zero and scenario rejecting respondents. Aggregately, Mekong Delta urban residents agreed to contribute about $10.97 million annually for the project of biodiversity conservation.

Keywords: Contingent valuation; Genetic resources; Wetland conservation; WTP

Background

Natural wetlands play an important role in terms of controlling water flow, preventing from the damage of flood and storm, supporting fisheries, absorbing waste, especially maintaining biodiversity, etc. In addition, wetland regions are places for water transport and recreation while the diverse resources in wetlands could be directly exploited for agriculture, fishing, water supply, wood and wildlife products. The aggregate economic benefit of a wetland's ecological functions, resources and services could surpass the economic value received from the wetland conversion for alternative uses (Wattage 2002).

There are a number of previous studies using the contingent valuation (CV) method to estimate the total benefit of wetlands. Hanemann et al. (1991) applied the dichotomous choice CV method to estimate willingness to pay (WTP) for wetlands maintenance program in San Joaquin Valley. The study showed that the single-bounded response estimated the mean WTP of $250 and $151 from the estimation of double-bounded question. Beran (1995) used CV questionnaires to estimate the passive use values associated with preservation of freshwater wetlands in South Carolina. The study revealed that mean WTP ranged from a one-time contribution of $6.03 to $45.4 according to the estimation technique and the type of wetland valued, and a total value of WTP for South Carolina residents was $3.7 million. The other study by Chen and Jim (2010) analyzed Guangzhou residents' motivations and WTP for an urban biodiversity conservation program in the National Baiyun Mountain Scenic area. The research estimated the median WTP of $19.5 per household per year and the total value of about $38.2 million annually to contribute the urban conservation project.

There is great biodiversity in the wetlands of the Mekong Delta, the largest wetlands in Vietnam, assisting a large number of herons, egrets, stocks, ibises and some rare species such as sarus cranes, black necked storks, lesser adjutants and great adjutants. Specially, the mature semi-natural Melaleuca forest and seasonally inundated grasslands of the Mekong Delta wetlands are the living environments of about 14 globally threatened bird species. Therefore, preserving these wetlands is important or beneficial not only for Vietnam but also for the world. Moreover, there are also a lot of unknown flora and fauna, microorganisms, and genetic resources that are expected to contribute to, for example, the future development of new medicines or coenzymes, which are necessary for biochemical reaction.

However, there are so many issues related to environmental pollution and an increase in wetland destruction due to an overheated economic development. Recently, the Mekong Delta wetlands are dealing with the serious biodiversity loss and degradation. For example, the numbers of endangered species (sarus cranes) in Tram Chim National Park - one of the largest national wetland parks - have rapidly decreased from 1,057 in 1987 to 93 in 2005 (Do and Bennett 2007; Khai and Yabe 2014). The degradation of wetland biodiversity is due to an increase in shrimp farming, the conversion of wetlands to agriculture and construction land, war destruction and excessive fuel wood collection. The development of dykes in the Mekong Delta has altered hydrologic conditions and also hence wetland health (Do and Bennett 2007). To prevent from the biodiversity loss and degradation, the local authorities have proposed plans to use public funding to improve the protection of biodiversity. However, up to now there is little information on the values of biodiversity as well as studies on nature and biodiversity conservation in the Vietnam's literature. Thuy (2007) applied the CV method with five bid-level questionnaires to conduct the study on willingness to pay for the conservation of Vietnamese Rhinoceros and estimate the mean WTP of $2.5 per household. Environmental choice modeling was applied by Do and Bennett (2007) to identify the biodiversity benefits of Tram Chim National Park. The study estimated total benefits of wetland conservation program were about $3.9 million.

Because of the limitation of information and studies related to the benefit of biodiversity conservation, policymakers cannot answer the question of whether the change in current management practices would generate net social benefits. It is relatively easy to calculate costs of biodiversity conservation program, but hard to estimate benefits. The benefits or design of biodiversity policy could be estimated by studying public preferences on conservation program. However, this is complicated because of the generally low level of awareness and understanding of what biodiversity means on the part of the general public (Christie et al. 2006). Moreover, although there are a lot of conservation activities especially in biosphere reserves of the Mekong Delta recognized by UNESCO, these are not strong or powerful enough to enlarge or improve the quantity and quality of biosphere reserves because of government budget constraint or the low level of support from local residents and authorities. The studies are needed to be done to answer the question of whether is worthy investing more financial resources in conserving biodiversity in these biosphere reserves. In this study, we try to estimate total economic values of proposed biodiversity conservation program in U Minh Thuong National Park using the approach of CV method. The study might partly seek to answer the above questions, and also provide policy makers and concerned people more information about residents' attitudes toward environment and natural resources as well as the benefits of biodiversity conservation.

The article is outlined as follows. The next part presents the empirical model, research area and data collection. The following part discusses about the results of willingness to pay for biodiversity conservation, including the discussions of refusals and zero responses, and the estimation of willingness to pay. The final part withdraws some conclusions of the study.

Empirical method

This study used CV method to identify the total value of wetland conservation. The CV method was first used at the beginning of the 1960s to estimate the benefits of outdoor recreation in Maine backwoods by Robert (1963). After that, Ridker (1971) applied the CV method for air pollution problems. From the 1970s up to now, this CV approach has been applied by many economists to measure the benefits of a wide variety of goods such as recreation, hunting, water quality, decreased mortality risk from a nuclear power plant accident and toxic waste dumps, etc. (Wattage, 2002).

The basic theory of the dichotomous CV approach was proposed by Hanemann (1984). The idea favored by current CV practice is to ask each respondent a closed from question, namely whether they would accept to pay a given amount to obtain a given change in their status quo. Hence the answers obtained are the type of 'yes' or 'no', necessitating a theory of how to translate these discrete responses into meaningful WTP estimates (Khai and Yabe 2011). Suppose that a respondent is asked to consider the change from the status quo Q^0 to Q^1, where Q^1 refers to the quality of environment and presumably the latter choice is preferable to the former. Denote the indirect utility of respondents as $V(P, Q, I, Z)$, where P is a vector of prices, I is the respondent's income and Z is a vector of respondent characteristics. Then if the respondent is asked whether he would be willing to pay an amount M to obtain Q^1, his answer would be 'yes' if the following condition holds (where Pr denotes the probability):

$$\begin{aligned} Pr(Yes) &= Pr\{V(P, Q^1, I-M, Z) +_1 \geq V(P, Q^0, I-0, Z) +_0\} \\ &= Pr\{V(P, Q^1, I-M, Z) - V(P, Q^0, I-0, Z) +_1 -_0 \geq 0\} \end{aligned} \tag{1}$$

where ε_0 and ε_1 are unobservable components of the utility and identically and independently distributed (*i.i.d*) random variable with zero mean. If we define $\Delta V = V(P, Q^1, I - M, Z) - V(P, Q^0, I - 0, Z)$ and $\gamma = \varepsilon_1 - \varepsilon_0$, and then write:

$$Pr(Yes) = Pr(\gamma \geq -\Delta V) = 1 - F_\gamma(-\Delta V) = F_\gamma(\Delta V) \tag{2}$$

where $F_\gamma(\Delta V)$ represents the cumulative density function (cdf) of the respondent's true maximum WTP.

The dichotomous CV technique estimates the mean and median WTPs based on the coefficients related the WTP responses against a constant and the bid (*BID*). Additional coefficients (*X*) of other variables like responses to attitude questions or the respondents' demographic information may also be factored into the model. The probit and logit model are commonly applied to analyze the dichotomous choice format of contingent valuation method. In this study, we used the logistic function, which is relatively easier to compute than the approach of probit function. The following form of logistic function could estimate these coefficients:

$$\Pr(Yes) = F_{\gamma}(\Delta V) = \frac{1}{1 + \exp(-\Delta V)} = \frac{1}{1 + \exp-(\alpha + \beta_1 BID + \beta_2 X)} \tag{3}$$

where α and β are coefficients to be estimated and *BID* is the amount of given money the respondents were asked to pay.

This logistic function was estimated using maximum likelihood. Let R_k be an indicator variable for observation *k*, with

$$\Pr(Yes) = \Pr(R_k = 1) = Pr(\gamma_k \leq \Delta V_k) = F_{\gamma}(\Delta V_k) \tag{4}$$

$$\Pr(No) = \Pr(R_k = 0) = Pr(\gamma_k \leq \Delta V_k) = 1 - F_{\gamma}(\Delta V_k)$$

So the log-likelihood function is formed as:

$$logL = \sum_{k=1}^{N} \{R_k F_{\gamma}(\Delta V_k) + (1 - R_k) \ln(1 - F_{\gamma}(\Delta V_k))\} \tag{5}$$

In this case, the mean and median WTP are equal to each other and calculated by the following estimator:

$$Mean/Median\ WTP = -\frac{(\hat{a} + \hat{\beta_2}\bar{X})}{\hat{\beta_1}} \tag{6}$$

Research area and data collection

U Minh Thuong National Park is one of two largest areas of peat swamp forest in Vietnam. Biodiversity conservation in this region has been assigned a national priority since it can buffer the negative effects of the Mekong River floods, recharge aquifers and provide a unique environment for many wetland species. With a total area of 8.038 ha and a buffer zone of 13.069 ha, the national park is a home of many diversified plants and animals including 243 plant, 32 mammal, 151 bird, 34 reptile, 7 amphibian, 34 fish and 181 insect species. Forty of these are listed as endangered species in the Vietnam Red Book (Dang 2009).

Although the government has declared the protected zone, U Minh Thuong National Park is still under serious threats to biodiversity such as an increase in human encroachment on and disturbance of wildlife habitats by converting the forest land into agriculture and construction land, environmental pollution caused by subsistence wastes, industrial wastes, use of insecticides, herbicides and toxic rat baits, illegal wildlife hunting and trade, etc.

Primary data were collected by randomly interviewing local citizens face-to-face in the urban area of Can Tho city, representing the largest city in the Mekong Delta and the urban region of Kien Giang province where U Minh Thuong National Park is located. The time of survey took about 3 mouths from January to March 2013 and was divided into two main periods. The first one was the pilot-survey, in the first week of January 2013. The pilot surveying and pretesting are essential elements in any contingent valuation study (Bateman et al. 1995). The aims of this interview are to refine the questionnaire, format bid starting point more clearly and concisely, and also help interviewers get used to and understand the content of questionnaire. After the interviewers are trained, there are about 50 households interviewed in this period. The revised questionnaire is used in the second period. The sample was composed of 366 respondents, 215 in Can Tho and 151 in Kien Giang. To make a good CV questionnaire without cheap talk bias, the content of questionnaire is formed based on the suggestion of Carson (1991), and Cummings and Taylor (1999). We first examine how the respondents concern environmental problems in the country and recognize respondents' attitude toward biodiversity conservation project. Secondly, the plan of proposed project is introduced with the payment vehicle and provision rule. Finally, socioeconomic information of respondents are collected.

The single-bound dichotomous choice contingent valuation questions were used in the study. Before the CV question was asked, the current conditions, biodiversity as well as the benefits of U Minh Thuong National Park were introduced. Then, we identified some threats of biodiversity loss such as encroachment, disturbance of vegetation, environmental pollution, wildlife hunting, and trade occurring in this area. A hypothetical conservation program was proposed to prevent from biodiversity loss in the national park. Biodiversity will continue to degrade more seriously without this conservation project. The study proposed funding a biodiversity conservation project to increase the number of plants and animals in U Minh Thuong National Park or at least keep them from declining every year. The conservation fund could then request international organizations to provide the same amount of money or more compared with the contributions of residents. The raised money would be only used for conservation activities in U Minh Thuong National Park (See the section of CV question for more details in the Appendix). The payment vehicle was used as voluntary continuous donation, contributed through a monthly water bill for three years, which could catch the present value of preferences for biodiversity conservation and also prevent potential protests due to compulsory payment like taxes (Rolfe et al., 2000). Each household was interviewed whether he/she would be willing to contribute the biodiversity conservation fund a certain amount of money every month as surcharge on his/her household water bills for the next three years. The admissible answers were 'yes' or 'no'. Five different bid values of VND10,000, VND35,000, VND60,000, VND85,000 and VND110,000 were chosen for the study. These values are equivalent to values in US dollars[a] of $0.47, $1.66, $2.84, $4.21 and $5.20, respectively. Each household was randomly interviewed whether he/she would be willing to contribute one of these bid values and answered whether he/she accepted only one bid value. For example, the respondent was asked 'Would you be willing to pay VND10,000 every month for biodiversity conservation program in U Minh Thuong?' If the answer was 'yes,' the list of 'yes' reasons was then presented. If the response was 'no', the list of 'no' reasons was then introduced and an open-end question was also asked whether a respondent would like to contribute another lower amount of money to identify the categories of protest zero, valid zero and WTP less bids.

Results and discussion

Refusals and zero responses

Table 1 shows the probability of answering 'no' increases as the amount of money the respondents are asked to pay increases. About 68.6% of respondents disagree to pay the given discrete choice amount, and 39.3% of those are not willing to pay any positive amount. Possible explanations for the high number of refusals and zero responses may be that respondents are not familiar with this kind of survey, the study might be too hypothetical or the scenarios could be too unspecific and simplistic. Interviewees who do not agree to pay the amount of money given in the questionnaire were asked why they had responded 'no'. Table 2 gives the overall frequencies of answers from a given list of reasons.

Table 2 reveals reasons for refusing to pay the given discrete amount of money. About 51 percent of respondents cannot afford to pay or have not space income to contribute towards the fund of biodiversity conservation (reason 1). Such high percentage of unwilling-to-pay reason due to the income constraint is consistent with other survey study in developing countries. The study on saving the Philippine Eagle by Harder et al. (2006) showed a 62 percent while the study on Vietnamese Rhinoceros conservation by Thuy (2007) revealed about 41 percent of respondents refusing to contribute the conservation program due to income constraint. The second highest-cited reason accounted for 16.3 percent of respondents who think that the biodiversity conservation program should be the government' responsibility since it has money from tax revenues while about 12 percent of unwilling-to-pay respondents with the reason of distrusting the institutions that will handle the money for this conservation work.

This study follows the other previous CV studies to discriminate between valid and 'protest' zero bids. The respondents who are not willing to pay any positive bids with the reason 4 (I do not believe paying will solve the problem), 6 (I do not trust the institutions that will handle the money for this conservation work) and 7 (It should be the

Table 1 Respondents refusing the discrete choice amount or refusing to pay any amount for biodiversity conservation

Bid value (VND)	Observations	Refusals to the discrete choice question		Refusals to make any positive payment$^{\Psi}$	
		Number	Percent (%)	Number	Percent (%)
10,000	76	33	43.4	28	36.8
35,000	77	48	62.3	28	36.4
60,000	74	55	74.3	29	39.2
85,000	72	59	81.9	32	44.4
110,000	67	56	83.6	27	40.3
Total	**366**	**251**	**68.6**	**144**	**39.3**

Note: $^{\Psi}$The respondents answered 'no' in the discrete choice question and also did not contribute any lower money in the open-end question.
Source: Own estimates; data appendix available from authors.

Table 2 Respondents giving reasons for refusing to pay the given discrete choice amount

Reason	Number	Percent (%)
1 = I cannot afford to pay/I have no spare income.	128	51.0
2 = I feel the environmental improvement of U Minh Thuong is unimportant.	3	1.2
3 = Being far from the place, I feel paying anything is irrelevant to me.	13	5.2
4 = I do not believe paying will solve the problem.	18	7.2
5 = I feel this improvement will take place without my contribution.	7	2.8
6 = I do not trust the institutions that will handle the money for this conservation work.	30	12.0
7 = It should be the government's responsibility since it has money from tax revenues.	41	16.3
8 = Other reasons.	11	4.4

Source: Own estimates; data appendix available from authors.

government's responsibility since it has money from tax revenues) are listed into the category of 'protest vote' or none-zero value. The percentages of respondents in the protest bid, valid no and positive bid categories are performed in Table 3.

Table 3 shows about 60.7 percent of respondents willing to pay some positive amount for conservation program while about 29.2 percent of respondents agree to pay less than the bid values (The respondents who disagreed to pay the amount of money given in the questionnaire but accepted to contribute a smaller amount of money) and around 24.6 percent of respondents are categorized as valid zero bids (the respondents were not willing to pay any amount of money for conservation program with any other reasons excepting the reason 4, 6 and 7). The study also shows that the proportion of protest bids averages 14.8 percent of all responses. Such figure is somewhat consistent with other results in the literature. Kirkland (1988) has 18 percent of protest responses in his study, about 24 percent are determined by Sappideen (1992) and 28 – 31 percent of protest bids in the study by Jakobsson and Dragun (1996). According to Moser and Dunning (1986), the high level of protest bids reveals that some questions could be misunderstood; the respondents have troubles to understand the study scenario or are not convinced that the proposed project becomes real.

Estimation of willingness to pay

Table 4 shows the socio-demographic description of the respondents. The age of surveyed respondents ranges from 20 to 87 years with the average age of 50 years and about 62 percent of respondents are female. The high percentage of female respondents

Table 3 The category of protest bid, valid no and positive bid respondents

	Number	Percent (%)
Protest zero bids (1)	54	14.8
Valid zero bids (2)	90	24.6
WTP less bids (3)	107	29.2
Willing to pay discrete choice amount (4)	115	31.4
Willing to pay some positive amount (5)	222	60.7

Note: (5) = Total observations – Protest zero bids (1) – Valid zero bids (2)
Source: Own estimates; data appendix available from authors.

may be due to the survey mainly done in the daylight of weekdays when it is difficult to interview working male households. Regarding the level of education, around 23 percent of respondents stated a university and higher degree. The average household income is over VND7.5 million per month with the most frequent category of below VND5 million (40%), followed by VND5 million – 10 million (about 39%), and over VND10 million (nearly 21%). The higher household income level of the sample in comparison with the average household income of Mekong Delta residents (about 5.2 million per month in 2010[b]) does not show a problem in terms of a sample selection bias because the difference could be explained by the inclusion of rural households who earn lower income than urban residents in the population average.

In the contingent valuation literature, there are some ways to solve the problem of zero bids. Imber et al. (1991) treat all 'no' responses as real 'no' answers. This may result in wrong policy implications (Carson 1991) or difficultly estimate the willingness to pay function correctly if the number of protest responses is high (Romer 1992). Other strategy is to eliminate all zero bids, but this may cause a sample selection bias, since the remaining bids from no longer originate from a random sample of the basic population (Romer 1992). The most common way is to identify and exclude protest bids from estimates of willingness to pay (Mitchell and Carson 1989). The estimates without protest responses give the higher value of WTP than those with all 'no' responses. In this study, we use the WTP estimate without protest bids as the mean WTP and the WTP calculation for all 'no' responses as the low bound of willingness to pay.

The one of the key questions in a contingent valuation survey is whether WTP is affected by important variables suggested by economic theory, for example income, education, age, etc. or whether the coefficients of these variables have signs that are consistent with expectations. These relationships analyzed by logistic function are performed in Table 5.

Table 5 shows the results of a logistic analysis of the dichotomous choice responses to the contingent valuation questions. The model 1 is the estimates for all respondents and the model 2 shows the results after excluding protest bids. The predictive powers of models are relatively high, with nearly 76 percent in model 1, and over 74 percent in model 2. The correlation matrix across explanatory variables supports the absence of

Table 4 Descriptive statistics of variables in the logistic function

Variables	Description	Mean	Std.Dev.	Min.	Max.
WTP	Willingness to pay for conservation program (1 = yes, 0 = No)	0.314	0.465	0	1
Bid	Bid value (thousand VND)	58.421	35.107	10	110
Age	Age of respondents (years)	49.713	14.281	20	87
Gender	Gender of respondents (1 = Male, 0 = Female)	0.385	0.487	0	1
Graduate	Educational level of respondents (1 = Graduate or higher, 0 = otherwise)	0.232	0.423	0	1
Status	Civil status of respondents (1 = married, 0 = otherwise)	0.872	0.335	0	1
Income	Monthly household income of respondents (thousand VND)	7,547	4,073	4,500	19,500
Location	Location of respondents (1 = Can Tho, 0 = Kien Giang)	0.587	0.493	0	1
Donation	Whether respondents have made any donations (1 = yes, 0 = no)	0.702	0.458	0	1

Source: Own estimates; data appendix available from authors.

Table 5 Logistic regression estimate of willingness to pay for conservation project

	Model 1 (All respondents included)		Model 2 (Protest bids excluded)	
	Coefficient	Standard error	Coefficient	Standard error
Bid	−0.0230***	0.004	−0.0263***	0.004
Age	−0.1133**	0.055	−0.0997*	0.058
*Age*Age*	0.0010*	0.001	0.0010*	0.001
Gender	0.3580	0.262	0.7380**	0.292
Graduate	−0.1298	0.327	−0.1577	0.359
Status	0.7818*	0.459	0.3833	0.473
Log(*Income*)	1.0777***	0.303	1.3601***	0.344
Location	0.1688	0.266	0.2740	0.288
Donation	0.7656***	0.296	0.8066**	0.317
Constant	−7.6845***	2.809	−10.0204***	3.142
Pseudo R^2	0.156		0.198	
Log likelihood	−192.230		−164.808	
Correct prediction (%)	75.956		74.359	
Observations (N)	366		312	

Notes: ***, **, * indicate statistical significance at the 0.01, 0.05 and 0.1 level respectively.
Source: Own estimates; data appendix available from authors.

multicollinearity because there are no correlation indices that are higher than 70 percent (Khai and Yabe 2013). The coefficients of the bid value in the two models are statistically significant and negative as expected, revealing that an increase in bid amount could reduce the 'yes' response and the existence of the WTP for conservation program. Consistently with other studies by Subade (2005) and Jianjun (2007), the coefficients of *Age* variable in the two models are statistically significant and have negative signs, implying the older respondents have more tendency to say 'no' to the WTP question. However, the residents in the sample from the age[c] of 57 are more likely to recognize the necessary of biodiversity conservation and agree to support the project since the parameters of *Age*Age* variable are significantly positive at 10 percent level. The coefficients of household *Income* are significant at the level of 1 percent. The positive sign of *Income* suggests higher income could increase the probability of 'yes' answer to the contingent valuation question. Regarding the motivations for conservation, the variable of *Donation* is defined as whether a respondent have ever donated for any charitable or environmental funds. The coefficients of *Donation* variable in two models are significantly positive, revealing respondents who have made any donation in the past are more willing to pay for the biodiversity conservation project. From the findings in Table 5 and applying the estimator (6), the median and mean WTP for the proposed biodiversity conservation project are calculated by utilizing the directly estimated values of the coefficients. The results of mean and median estimates of willingness to pay are presented in the Table 6.

Table 6 performs the mean WTP for all respondents is estimated to be VND16,510 with the 95% confident interval between VND-4,670 and VND29,340 while the mean WTP after excluding protest vote respondents increases up to VND31,520 with the 95% confident interval between VND17,860 and VND43,670. The two mean WTPs are significantly positive at the level of 10 percent for all respondents and 1 percent for the sample with protest bids excluded.

Table 6 Willingness to pay for biodiversity conservation program unit: thousand VND

	Mean/Median of WTP	95% Confidence Interval		
		Lower bound	Upper bound	ASL
All respondents included	16.51	−4.64	29.34	0.053
Protest bids excluded	31.52	17.86	43.67	0.0008

Note: ASL: Achieved Significance Level for testing H_0: WTP < =0 vs. H_1: WTP > 0.
Confidence intervals are estimated by the Krinsky and Robb (1986) method.
Source: Own estimates; data appendix available from authors.

Conclusions

This study tried to extend the understanding of Vietnamese households' preferences for biodiversity conservation to the context of their WTP. The mean WTP for the proposed conservation project was calculated to be approximately VND16,510 per household per month, which was about 0.32 percent of the average household income at VND5.2 million per month in the Mekong Delta region in 2010. Moreover, the study also estimated the mean WTP of VND31,520 per household per month after excluding the protest zero and scenario rejecting respondents. An aggregate welfare measure can be derived by multiplying the mean WTP by the total urban households in the Mekong Delta. According to the general statistics office (GSO 2013), there were 4,329,100 urban people in the Mekong Delta in 2012, which were equivalent to 1,170,027 urban households with the average of 3.7 persons per urban family (Binh 2011). The calculation discovered that Mekong Delta urban residents were willing to pay about VND19.32 billion (VND16,510 * 1,170,027 urban households) every month for biodiversity conservation. Therefore, the total value of annual urban resident's contribution is approximately VND231.81 billion ($10.97 million) which is relatively big enough for the government, policy makers and concerned people to pay more attention or give more financial invests in conserving and improving wildlife habitats and biodiversity in U Minh Thuong National Park. This total value could be the useful and trustworthy information for decision makers to allocate funds for the biodiversity conservation project while the results of this contingent valuation study could also be valuable for environmental evaluation or suggest applying this approach to the cost-benefit analysis of this project as well as other or future projects in Vietnam. In addition, the public evaluation of biodiversity conservation could help society more awareness on the important role of biodiversity, have an impact on rational behaviors or wide support of residents to improve the quantity or quality of biodiversity which benefits the present and future generation (Chen and Jim 2010).

Endnotes

[a]1 USD = 21,140 VND at the date of 05/12/2013.

[b]The average household income of Mekong Delta residents are calculated by multiplying per capital income at 1.25 million (GSO 2010) with the average member of 4.16 per family (Binh 2011) in the Mekong Delta region.

[c]At the minimum of *Pr*(Yes) function, $Age_{min} = -(-0.1133)/(2 * 0.001) = 56.65$.

Appendix: the format of CV question

Section 3: A conservation plan for the protection of U Minh Thuong National Park

We will now provide you some information about U Minh Thuong in Kien Giang:

U Minh Thuong National Park is one of two large areas of peat swamp forest in Vietnam. Biodiversity conservation in this region has been assigned a national priority since it could buffer the negative effects of the Mekong River floods, recharge aquifers and provide a unique environment for many wetland species. With a total area of 8.038 ha and a buffer zone of 13.069 ha, the Park is a home of many diversified plants and animals including 243 plant, 32 mammal, 151 bird, 34 reptile, 7 amphibian, 34 fish and 181 insect species. Forty of these are listed as endangered species in the Vietnam Red Book.

Threats to biodervisty in U Minh Thuong National Park

Although the government has declared the protected zone, U Minh Thuong is still under serious threats to biodiversity:

- Encroachment and disturbance of habitats: due to high density of local peoples, habitat area often undergoing encroachment for agriculture, aquaculture and other purposes. This activity is especially in the areas outside protected areas. Illegal encroachment causes lots of disturbance for habitat security.
- Environmental pollution caused by subsistence wastes, industrial wastes, use of insecticides, herbicides and toxic rat baits.
- Wildlife hunting and trade occurs commonly in the area. Wild animals in general and mammals in particular are hunted for household use and also for sale in urban centers.

A conservation plan for biodiversity in U Minh Thuong National Park

The goal of the program is to increase the number of plants and animals in U Minh Thuong or at least keep them from declining every year. To accomplish this goal, it should be carried out the following activities:

- Rationally planning ponds, shrimp ponds and rice farming land around the buffer zone to prevent water pollution and scarcity of food.
- Planting more trees suitable for nesting and reproducing, improving ponds, swamps and grassland within and outside the buffer zone to create a food source and the better living environment for wild animals. There is also promoting tree planting in rural areas, industrial zones and urban areas to create the good habitat conditions for the biodiversity conservation.
- Enhancing the coverage of forest to protect the soil from erosion, landslides and runoff.
- Conducting education and training activities to improve the awareness of the local people to conserve biodiversity and the professional skills of the management group staff.

While the program contains many good ideas, implementing it would require money. So far, the program has not received any funding or carried out any activities. A number of international organizations do provide financial support to protect this biodiversity area. However, they usually require that counterpart funds be made available. In other words, people from region must also contribute money to protection effort.

Therefore, the practical implementation of this program would require much more concerted efforts from all households.

Conservation fund

Suppose a conservation fund for biodiversity in U Minh Thuong would be set up, all citizens could contribute to the fund. The fund could then request international organizations to provide the same amount of money or more, according to the money raised locally. The money raised by the fund would be ONLY used for the conservation activities mentioned earlier to conserve biodiversity in U Minh Thuong.

The purpose of our survey is to find out if your household would be willing to contribute < ***bid level*** > every month as a surcharge on your household water bills for the next three years. The payment is a fixed amount and it doesn't change with the volume of water used. This money would go to the conservation fund for biodiversity in U Minh Thuong.

[The recent study shows that people generally accept to contribute value more than the ability of actual contribution. This survey tries to get information on the ability of your actual contribution. So, require you to think carefully with your decision].

11) Would you be willing to pay < bid level > every month for biodiversity conservation in U Minh Thuong?

(1) Yes (2) No

Competing interests

The authors declare that they have no competing interests.

Authors' contribution

HVK designed the study questionnaire, collected and analyzed the data, and drafted the manuscript. MY supervised the research and made critical revisions to the manuscript. All authors read and approved the final manuscript.

Acknowledgments

A previous version of the paper was presented at the 3th International Seminar of JSPS Core-to-Core Program: Collaborative Project for Soil and Water Conservation in Southeast Asian Watersheds, Water Resources University, Hanoi, Vietnam, 7–8 August 2013. We would like to express our gratitude to Japan Society for the Promotion of Science (JSPS) for sponsoring the study. Our warmest thanks go to the staff in the department of Agricultural, Resource and Environmental Economics, SEBA, Can Tho University for assisting in data collection.

Author details

[1]Department of Agricultural, Resource and Environmental Economics, School of Economics and Business Administration, Can Tho University, Can Tho City, Vietnam. [2]Laboratory of Environmental Economics, Department of Agricultural and Resource Economics, Faculty of Agriculture, Kyushu University, Hakozaki 6-10-1, Higashi-ku, Fukuoka city 812-8581, Japan.

References

Bateman IJ, Langford IH, Turner RK, Willis KG, Garrod GD (1995) Elicitation and Truncation Effects in Contingent Valuation Studies. Ecol Econ 12(2):161–179, doi:10.1016/0921-8009(94)00044-V

Beran LJ (1995) Measuring the Economic Benefits of Provision of Nonmarket Goods: Freshwater Wetlands in South Carolina. Dissertation, Clemson University, Clemson, South Carolina, US

Binh NT (2011) The trend of Vietnamese household size in recent years. In: 2011 International Conference on Humanities, Society and Culture IPEDR Singapore. IACSIT Press, Singapore, pp 47–52

Carson RT (1991) The Resource Assessment Commission's Kakadu Conservation Zone contingent valuation study. Remarks on the Brunton, Stone and Tasman Institude critiques. In: Commission ARA (ed) Commentaries on the Resource Assessment Commission's contingent valuation survey of the Kakadu Conservation Zone. San Diego, March, Canberra, Department of Economics, University of California

Chen W, Jim CY (2010) Resident Motivations and Willingness-to-Pay for Urban Biodiversity Conservation in Guangzhou (China). Environ Manage 45(5):1052–1064, doi:10.1007/s00267-010-9478-2

Christie M, Hanley N, Warren J, Murphy K, Wright R, Hyde T (2006) Valuing the diversity of biodiversity. Ecol Econ 58 (2):304–317
Cummings RG, Taylor LO (1999) Unbiased value estimates for environmental goods: A cheap talk design for the contingent valuation method. Am Econ Rev 89(3):649–665, doi:10.1257/Aer.89.3.649
Dang NX (2009) Biodiversity survey - Rapid assessment of flora and terrestrial aninimals in the key areas of Kien Giang Biosphere Reserve. Technical Report 09E0809DANG. Conservation and Development of the Biosphere Reserve of Kien Giang Provine Project, Rach Gia city, Kien Giang Province, Vietnam
Do TN, Bennett J (2007) Would wet biodiversity conservation improve social welfare? A case study in Vietnam's Mekong River Delta. In: The Proceedings of the GMSARN International Conference on Sustainable Development: Challenges and Opportunities for GMS, Pattaya, Thailand
GSO (2010) Result of Viet Nam: Household living standards survey 2010. Statistical Publishing House, General Statistics Office, Ha Noi, Vietnam
GSO The average of urban population by location. General Statistics Office. http://www.gso.gov.vn/default.aspx?tabid=387&idmid=3&ItemID=12869. Accessed 6/11 2013
Hanemann M, Loomis J, Kanninen B (1991) Statistical Efficiency of Double-Bounded Dichotomous Choice Contingent Valuation. Am J Agr Econ 73(4):1255–1263, doi:10.2307/1242453
Hanemann WM (1984) Welfare Evaluations in Contingent Valuation Experiments with Discrete Responses. Am J Agr Econ 66(3):332–341, doi:10.2307/1240800
Harder DS, Labao R, Santos FI (2006) Saving the Philippine Eagle: How Much Would It Cost and are Filipinos Willing to Pay for It? The Economy and Environment Program for Southeast Asia (EEPSEA), Singapore
Imber D, Stevenson G, Wilks L (1991) A contingent valuation survey of the Kakadu conservation zone. RAC Research Paper No. 3, AGPS, Canberra
Jakobsson KM, Dragun AK (1996) Contigent Valuation and Endangered Species. Edward Elgar Publishing Limited, Cheltenham, UK
Jianjun J (2007) Economic Valuation of Black-faced Spoonbill Conservation in Macao. The Economy and Environment Program for Southeast Asia (EEPSEA), Singapore
Khai HV, Yabe M (2011) An Analysis of Willingness to Pay for Commodity Insurance of Rice Production in Vietnam. Asian Afr J Econ Econometrics 11(2):485–493
Khai HV, Yabe M (2013) Impact of Industrial Water Pollution on Rice Production in Vietnam. In: Quinn NWT (ed) International Perspectives on Water Quality Management and Pollution Control. InTech Publishing, Rijeka, Croatia, pp 61–85
Khai HV, Yabe M (2014) Choice modeling: assessing the non-market environmental values of the biodiversity conservation of swamp forest in Vietnam. Int J Energy Environ Eng 5(1):1–8, doi:10.1007/s40095-014-0077-5
Kirkland WT (1988) Preserving the Whangamarino Wetland - An Application of Contingent Valuation Method. Master thesis, Massey Uiversity, New Zealand
Krinsky I, Robb AL (1986) On Approximating the Statistical Properties of Elasticities. Rev Econ Stat 68(4):715–719, doi:10.2307/1924536
Mitchell RC, Carson RT (1989) Using surveys to value public goods: The contingent valuation method. RFF Press, Washington DC
Moser DA, Dunning CM (1986) A guide for using the contingent value methodology in recreation studies. National Economic Development Procedures Manual Recreation, vol Vol II. US Army Corps of Engineers, Water Resources Support centre, Institute for Water Resources, Washington DC, US
Ridker RG (1971) Economic Costs of Air Pollution Studies. Praeger Publishers, USA
Robert KD (1963) The Value of Outdoor Recreation: An Economic Study of the Maine Woods. Harvard University, Cambridge, US
Rolfe J, Bennett J, Louviere J (2000) Choice modelling and its potential application to tropical rainforest preservation. Ecol Econ 35(2):289–302, doi:10.1016/S0921-8009(00)00201-9
Romer AU (1992) How to handle strategic and protest bids in contingent valuation studies. An application of the two-step Heckman procedure. Paper presented at the international conference on 'Econometrics of Europe 2000' of the Applied Econometric Association, Brussels
Sappideen B (1992) Valuing the recreational benefits of the Sale wetlands using the contigent valuation method. In: Lockwood MD,T (ed) Valuing natural areas: Applications and problems of the contingent valuation method. Johnstone Center, Charles Sturt University, Albury
Subade RF (2005) Valuing Biodiversity Conservation in a World Heritage Site: Citizens' Non-use Values for Tubbataha Reefs National Marine Park, Philippines. The Economy and Environment Program for Southeast Asia (EEPSEA), Singapore
Thuy DT (2007) Willingness To Pay for Conservation of the Vietnamese Rhino. The Economy and Environment Program for Southeast Asia (EEPSEA), Singapore
Wattage P (2002) Effective Management Biodiversity Conservation in Sri Lankan Coastal Wetlands: CVM1 - Literature Review. University of Portsmouth Cemare, United Kingdom

6

Off-farm labour supply and production efficiency of farm household in rural Southwest Nigeria

Adebayo M Shittu

Correspondence: shittuam@funaab.edu.ng
Department of Agricultural Economics and Farm Management
Federal University of Agriculture, Abeokuta, P.M.B. 2240, Abeokuta, Ogun State, Nigeria

Abstract

This study examines the influence of off-farm employment on production efficiency of farm household in Southwest Nigeria. The study was based on primary data collected from a cross-section of 489 rural farm households, drawn by multi-stage random sampling from Ogun and Oyo States, in Southwest Nigeria. The data was analysed by descriptive and econometric techniques, including specification and estimation of a conditional revenue frontier and a production inefficiency equation. The results showed that 47.1% of the rural farm households had some of their members involved in off-farm activities with a typical member devoting 34.3% of his/her work efforts to off farm activities, while off farm activities contributed 27.6% of a typical farm household's labour income (N282, 263.54 in 2005/2006). Increase in off-farm labour supply was found to be associated with significant ($p < 0.01$) reduction in production in-efficiency among the rural farm households. The study also found that farm household production efficiency is significantly enhanced by increasing the share of tree crops and livestock in farm household farming activities, having access to land by leasing and operating a much more consolidated landholdings. However, remittances from rural out-migrants do not significantly affect rural farm household production efficiency.

JEL code: J2, Q1, R2

Keywords: Off-farm labour supply; Production efficiency; Nigeria

Background

Nigeria, a nation once flaunted as the "Giant of Africa", is at present battling with a rising incidence of unemployment and poverty, with as much as 21.1% of the economically active people reported as unemployed in 2010 (NBS, 2011) while 64.4% of the populace lived below the US$1.25 income poverty line in 2008 (UNDP, 2010). The poor in Nigeria are reported to be predominantly rural dwellers and households that rely mainly on agricultural means of sustenance (World Bank, 2000; FOS 1999; NBS (2012); Babatunde, *et al.* 2008). Moreover, socio-economic conditions in most rural communities in Nigeria are generally poorer that what obtains in the cities: hence rural–urban migration has been a strategy adopted by many in a bid to escape poverty (Okali *et al.*, 2001).

Evidence in literature suggest that rural urban migration in Nigeria is massive, with as much as about 38% of the economically active members of the rural farm families in Southwest Nigeria reported to have migrated to the urban centres (Shittu, 2011). This massive rural out-migration has been a major cause of rapid urbanisation and congestion of Nigerian cities; leading to urban squalor and poverty, housing shortages, and pollution (DFID, 2004), while many that could not find job sometimes resort to crime making most Nigerian cities very dangerous, especially at nights (USAID, 2002). Moreover, rural out-migrants in Nigeria are predominantly the youths, male-folks and educated members of the rural farm households as well as artisans and other skilled workers in the rural sector (Okali *et al.*, 2001; DFID, 2004, Shittu, 2011). Thus, rural–urban migration in Nigeria has meant that the rural areas are often left with a demographically unbalanced population of women, younger children, and older people (Okali *et al.*, 2001; DFID, 2004). It also denies the rural sector the much-needed human capital, reduces availability of farm labour (Ogwumike and Aramolaran, 2000), and thereby tends to weaken productivity and income levels in the sector. The fact that unemployment level in Nigeria is now much higher in the urban sector than what obtains in the rural areas is also worrisome.

Against the above background, there is a rising believe among policy analysts, academia and government functionaries that provision of urban-type employment opportunities in the rural areas may be a veritable means of addressing the multifaceted problem of poverty, urbanisation and unemployment in Nigeria. This view is supported by FAO (1998), Matshe and Young (2004), and many others that have observed that given the limitations imposed by the fixed stock of land and increasing urbanisation, provision of opportunity for involvement of members of rural farm households in rural non-farm activities might turn out to be a means of creating favourable conditions to reduce poverty in the rural areas. Goldsmith, *et al.* (2004) also observed that growth in the rural non-farm activities might also be used to stem rapid rural–urban migration and the attendant urban poverty in most developing countries.

This paper presents empirical evidences on the types of off-farm employment opportunities available to rural farm households in Southwest Nigeria, returns to household labour supply use within the rural farm and non-farm sectors, and implications of the labour use patterns on farm household production efficiency in the study area. The remaining part of the paper is organised thus: A brief review of literature is presented in Section 2, followed by the theoretical framework Section 3. The study methodology is presented in section 4 followed by the results and their discussions in section 5, while the final section summaries the main evidences and conclusions from the study.

Literature review

A wide range of empirical studies have examined issues relating to off-farm labour supply and the implications on household welfare. Lanjouw and Lanjouw (2000), in a review of some of these studies, observed that while the rural non-farm sector was traditionally viewed as a low-productivity sector producing low quality goods that are expected to wither away as a country develops, recent years have seen a shift away from this position towards recognition of the fact that the rural non-farm sector can, and often does, contribute to economic growth, rural employment, poverty reduction, and a more spatially balanced population distribution.

The most common evidence from most studies of off-farm work among farm households has been that income from off-farm work accounts for significant and increasing proportion of total income of farm households in the developing countries (Anderson and Leiserson, 1984; Jacoby 1993; Newman and Gertler, 1994; Lanjouw, 2000; Escobal, 2001; Shittu, *et al.*, 2006; Haggblade, *et al.* 2010; Shittu, 2011; and many other). Recent estimates by Haggblade *et al.* (2010) put the non-farm share of the total income of rural households in the developing countries in the range of 35% and 50%, with the contributions among rural households in sub-Saharan Africa expected to rise significantly in the coming years given the increasing population growth and limited agricultural productivity growth in the region.

Evidences in literature suggest that a key motivation leading to off-farm labour supply among farm households in both the developed and the developing country has been the desire to have a diversified sources of income and manage risk (Chang and Mishra (2008). Hazell and Hojjati (1995) as well as Chavas, *et al.*, (2005), among others, have also reported that given the very weak capital market in most developing countries, many farm households in the often resort to off-farm work to raise cash with a view to relaxing their cash flow and liquidity constraints. This view is supported by evidences in Stampini and Davis (2009) as well as Pfeiffer, *et al.* (2009) that reported that households engaged in off-farm activities were able to spend significantly more on seeds, services, hired labour, and livestock inputs, which confirms that off-farm income relaxes credit constraints in agriculture.

Focusing on impacts of off-farm work, available evidence suggests that increased participation in off-farm work among members of farm households is associated with higher incomes as well as improved food consumption, nutrition and food security (Chang and Mishra, 2008; Babatunde and Qaim, 2010; Owusu *et al.*, 2011). It was also reported as linked to significant reduction in variance of total income (Schultz, 1990) as well as reduction in intensity of agricultural production (Phimister and Roberts, 2006; Huang *et al.*, 2009; Shi *et al.*, 2011; Owusu *et al.*, 2011) with positive environmental impacts due to reduction in the use of certain agrochemicals that impact negatively on the environment (Phimister and Roberts, 2006).

Despite the common evidences that income from non-farm sources helps in relaxing financial constraints on farm households and enhancing farm investment, evidences on the impacts on domestic food supply, production efficiency and household welfare, in general, remain quite conflicting. For example, while Lien *et al.*, (2010) reported that off-farm income had a positive effect on farm output but no systematic effect on farm technical efficiency, Pfeiffer, *et al.* (2009) reported that off-farm income has negative effect on agricultural output and the use of family labour on the farm, but positive impact on use of purchased inputs and confer a slight efficiency gain on farm households participating in off-farm activities. Shi *et al.* (2011), however, found that the negative lost-labour effect is much stronger than the (small) positive income effect while Holden *et al.* (2004) reported that access to non-farm income in less favoured Ethiopian highlands reduces farm households' incentives to invest in conservation and this leads to more overall soil erosion and more rapid land degradation even though intensity of production is reduced.

While Chang and Wen (2011) reported that off-farm work is not necessarily associated with lower (or higher) technical efficiency, they noted that farmers with off-farm work face higher production risk. They reported, however, that for farmers in the lower percentiles of the efficiency distribution, those with off-farm work are more efficient than their counterparts without off-farm work. Similarly, Chavas, *et al.* (2005), in a study of farm households in Gambia found that a significant part of a substantial allocative inefficiency that exists in the production systems of the farm households comes from inefficiency in labour allocation between farm and non-farm activities. They noted however, that in the presence of weak capital market in Gambia, off-farm activities acts to relax cash flow and liquidity constraints.

Wandschneider (2003) in a review of several studies of non-farm employment in developing countries of Africa and South Asia as well as the Transition Economies observed that a significant proportion of rural households and entrepreneurs in the studied regions do not only lack many of the required assets to successfully engage in non-farm employment, but also operate in a relatively adverse environment, characterised by limited opportunities both within and outside the farm economy. Consequently, he concluded that diversification into non- farm economic activities in all studied regions were largely out of necessity (distress-push) rather than as a response to remunerative wage employment and high return business opportunities (demand-pull). Similarly, Lanjouw (2001) in a study in rural El Salvador found that the poor were mainly engaged in "last resort" non-farm activities.

In summary, evidences in existing body of literature seem to suggest that while off-farm income accounts for significant and increasing share of total income of rural farm households in the developing countries, the implications on efficiency of household resource allocation, food supply and overall household welfare remain uncertain, and vary widely by locality and socio-economic environments. While for some, off-farm labour supply might be a response to remunerative wage employment and high return business opportunities, for many, working off-farm may be borne out of necessity to seek ways and means to relax credit constraints, raise supplementary income to complement what obtains from the farm most especially during lean seasons, or otherwise. Therefore, bearing in mind the likely presence of rigidity in rural off-farm labour market and/or reliance on joint technology for farm and off-farm activities among the rural-folks, more empirical evidences are required to assess the implications of off-farm labour supply on efficiency of resource allocation and household welfare. This study is an effort along this direction, with focus on resource poor farm households in rural Southwest Nigeria.

Conceptual framework

The conceptual framework for this study is based on a variant of agricultural household models developed in Chavas, *et al.* (2005). Reliance on agricultural household modelling framework is in recognition of the fact that the semi-commercial nature of smallholder agriculture in the developing countries makes it imperative that the production, consumption and labour allocation decisions of the farm households are interdependent. The framework is further justified given the well-documented evidences of labour market imperfection in the developing countries and/or the fact that farm households may rely on joint technology for their farm and off farm activities. The framework, as developed by Chavas, *et al.* (2005), may be summarised as follows:

Consider a farm household with m family members making production, consumption, and labour allocation decisions during a specific time period. Let $\mathbf{F} = (F_1, \dots F_m)$ and $\mathbf{L} = (L_1 , \dots, L_m)$ be the amount of labour supplied by the m family members in pursuit of the household farm and off-farm activities respectively; and *H*, the amount hired labour hired and used in conjunction with the non-labour inputs x (including land) and **F** to produce the vector of farm output, y in addition to the off-farm income, N earned from **L**. The technology facing the household is represented by the feasible set *X*, where $(x, \mathbf{F}, H, \mathrm{L}; \mathrm{y}, N) \in X$ means that inputs $(x, \mathbf{F}, H, \mathrm{L})$ can feasibly produce outputs (y, N), and farm and off-farm labour productivity is allowed to vary across family members.

If the total amount of time available to any family member over a time period is T; and the m family members can allocate their time between leisure activities $I = (l_1, \ldots, l_m)$, on-farm labour $\mathrm{F} = (F_1, \ldots, F_m)$, and off-farm employment $\mathrm{L} = (L_1, \ldots, L_m)$, the time constraint facing each family member can be written as:

$$I_i + F_i + L_i = T \tag{1}$$

If the farm-household consumes goods **z**, purchased at market prices **q**, and faces competitive markets for its products and inputs with **p** as the price vector for farm outputs **y**, **r** the price vector for non-labour inputs x, and w the wage rate for hired labour H, the household consumption decisions would be made subject to the budget constraint, which requires that consumer expenditure ($\boldsymbol{q}'\boldsymbol{z}$) cannot exceed the net farm income ($\mathbf{p}'\mathbf{y} - \boldsymbol{r}'x - wH$) plus the non-farm income (N). That is:

$$q'\boldsymbol{z} \leq (\mathbf{p}'\mathbf{y} - \boldsymbol{r}'x - wH) + N \tag{2}$$

Therefore, if it is assumed that household members make production, consumption, and labour allocation decisions under cooperative bargaining, and that members' preferences can be aggregated into a non-satiated and quasi-concave "social utility function" function $U(z, l)$ defined over $(\boldsymbol{z}, \boldsymbol{l}) \geq 0$, reflecting their relative bargaining power; then, the household decisions may be analysed based on the following optimisation problem:

$$\max_{x,\mathrm{F},\mathrm{H},\mathrm{L};\mathrm{y},\mathrm{N}} U(\boldsymbol{z}, \boldsymbol{l}) : l_i + F_i + L_i = T; \boldsymbol{q}'\boldsymbol{z} \leq \mathbf{p}'\mathbf{y} - \boldsymbol{r}'x - wH + N; (x, \mathbf{F}, H, \mathbf{L}; \mathbf{y}, N) \in X \tag{3}$$

Chavas, *et al.* (2005) asserted that under non-satiation of the utility function $U(\boldsymbol{z}, \boldsymbol{l})$, the budget constraint (2) is necessarily binding, and the optimisation problem (3) can be decomposed into two stages: first, choose $(x, \mathbf{F}, H, \mathbf{L}; \mathbf{y}, N)$; and second, choose $(\mathbf{z}, l)$.

The first stage optimisation with respect to $(x, \mathbf{F}, H, \mathbf{L}; \mathbf{y}, N)$ can be written as:

$$\pi(\mathbf{p}, \mathbf{r}, w, \mathbf{T} - \boldsymbol{l}) = \max_{x,\mathrm{F},\mathrm{H},\mathrm{L};\mathrm{y},\mathrm{N}} \{\mathbf{p}'\mathbf{y} - \boldsymbol{r}'x - wH + N : \ (x, \mathbf{F}, H, \mathbf{L}; \mathbf{y}, N) \in X;$$

$$F_i + L_i = T - l_i; i = 1, \ldots, m\} \tag{4}$$

where $(\mathbf{T} - \boldsymbol{l}) \equiv (T - l_1, \ldots, T - l_m)$ are the amounts of time the m family members spend working either on or off the farm. Equation (4) establishes profit maximization with respect to the household choice of $(x, \mathbf{F}, H, \mathbf{L}, \mathbf{y}, N)$, with $\pi(\mathbf{p}, \mathbf{r}, w, \mathbf{T} - \boldsymbol{l})$ being the indirect profit function conditional on $(\mathbf{T} - \boldsymbol{l})$.

Chavas, *et al.* (2005) drew attention to the fact that for a given amount of time allocated to work by household members $(\mathbf{T} - \boldsymbol{l})$, a failure to maximize profit would reduce household income, which would restrict consumer expenditure (from equation (2)), and which under non-satiation, would make the household worse-off. Thus, a failure to maximize profit would be inconsistent with household utility maximization.

Furthermore, considering that solution to (4) would yield the profit maximizing input and labour decisions, $x^*(p, \mathbf{r}, w, \mathbf{T} - \boldsymbol{l})$, $\mathrm{F}^*(p, \mathbf{r}, w, \mathbf{T} - \boldsymbol{l})$, $H^*(p, \mathbf{r}, w, \mathbf{T} - \boldsymbol{l})$, and $\mathbf{L}^*(p, \mathbf{r}, w, \mathbf{T} - \boldsymbol{l})$ as well as the profit maximizing outputs decisions, $\mathbf{y}^*(p, \mathbf{r}, w, \mathbf{T} - \boldsymbol{l})$ and $N^*(p, \mathbf{r}, w, \mathbf{T} - \boldsymbol{l})$ that together with the profit function $\pi(\mathbf{p}, \mathbf{r}, w, \mathbf{T} - \boldsymbol{l})$ do not depend on consumption levels $\mathbf{z}$, we find that production decisions are "separable" from consumption decisions. Hence, analysis of the production and consumption decisions of farm households can be undertaken separately as a two stage problem, starting with the profit maximisation problem (4) as a first stage optimisation.

The profit function $\pi(\mathbf{p}, \mathbf{r}, w, \mathbf{T} - \boldsymbol{l})$ and production decisions, $\mathbf{y}^*(\boldsymbol{p}, \mathrm{r}, w, \mathbf{T} - \boldsymbol{l})$ and $N^*(\boldsymbol{p}, \mathbf{r}, w, \mathbf{T} - \boldsymbol{l})$ are, however, jointly dependent on the amount of time available for work, $(\mathbf{T} - \mathit{1})$. Note also that equation (4) includes farm and non-farm activities, both in terms of labour allocation (**F** and **L**) and income (**p'y** and *N)* at the household level. It involves the general technology *X*, allowing for joint household decisions between farm and non-farm activities. Hence, decisions on labour allocation between farm and off-farm activities are dependent, and have to be jointly resolved within the profit maximisation problem (4). Chavas, *et al.* (2005) pointed out that examples of jointness in farm and off-farm activities include skills acquired in non-farm employment that improve farm management, and non-farm income that reduces the adverse effects of credit market imperfection on farm decisions.

Now, given that utility maximization (3) implies profit maximization (4) as a first stage optimisation, the second stage decisions with respect to $(\boldsymbol{z}, \boldsymbol{l})$ becomes:

$$\max_{\mathrm{z,l}} \; U(\boldsymbol{z}, \boldsymbol{l}) : \boldsymbol{q}'\boldsymbol{z} \leq \pi(\mathbf{p}, \mathbf{r}, w, \mathbf{T}-l)\} \tag{5}$$

Equation (5) is a standard utility maximization problem subject to the household budget constraint. Combining the two stages (4) and (5) is fully consistent with utility maximization (3). Chavas, *et al.* (2005) noted that the profit maximization (4) is the relevant framework to analyse production efficiency at the household level. They observed that in the presence of market imperfections and/or poor managerial skills, it is possible that households may not behave in a way consistent with (4) because they do not or cannot respond to economic incentives. They concluded thus, that economic analysis based on (4) could yield useful insights into the nature and causes of economic inefficiency. They stressed further, that the profit maximization problem (4) implies the following revenue maximization:

$$\max_{\mathrm{y,N}} \; R(\mathbf{p}, x, \mathbf{F}, H, \mathbf{L}, X) = \mathbf{p}'\mathbf{y} + N\text{: } (x, \mathbf{F}, H, \mathbf{L}; \mathbf{y}, N) \in X \tag{6}$$

where $R(p, x, \mathrm{F}, H, \mathrm{L}, X)$ is the revenue function, conditional on inputs $(x, \mathrm{F}, H, \mathrm{L})$. This suggests that analysis of production efficiency of farm households that exhibits significant off-farm labour market participation can be undertaken by focusing on output decisions, conditional on available inputs $(x, \mathrm{F}, H, \mathrm{L})$. Chavas, *et al.* (2005) pointed out that equation (6) assumes only well-functioning output markets. And, that this is important in the sense that analysis of farm household production efficiency, such as would be undertaken in this study, remains valid in the presence of factor market imperfections.

Given a representation of the production possibility frontier of a household involved in both farm and off-farm activities that is characterized by the use of inputs *(x,* F, *H,* L) in producing outputs (y, *N).* Chavas, *et* al. (2005) observe that the output based technical efficiency index, TE, is defined as:

$$TE(x, \mathbf{F}, H, \mathbf{L}, \mathbf{y}, N, X) = \min_{\theta}\{\theta : (x, \mathbf{F}, H, \mathbf{L}; \mathbf{y}/\theta, N/\theta) \in X, \theta > 0\} \quad (7)$$

Where $0 \leq \text{TE} \leq 1$, and $\text{TE} = 1$ when the household is producing on the production frontier and is said to be technically efficient, while $\text{TE} < 1$ shows the farm is not technically efficient.

Similarly, the allocative efficiency index, AE, with respect to farm outputs may be defined as:

$$AE\ (\mathbf{p}, x, \mathbf{F}, H, \mathbf{L}, X) = [\mathbf{p}'(\mathbf{y}/TE + N/TE]/R(\mathbf{p}, x, \mathbf{F}, H, \mathbf{L}, X) \quad (8)$$

where (y/TE, *N*/TE*)* is a technically efficient output vector. In general, $0 \leq \text{AE} \leq 1$, where $\text{AE} = 1$ represents a revenue maximizing firm that is allocatively efficient with respect to outputs, and $\text{AE} < 1$ shows that the farm is not allocatively efficient.

Methods

Data collection method

This study was based on primary data obtained in a cross-section survey of rural farm households in Ogun and Oyo states, in the Southwest rainforest zone of Nigeria. The respondent farm households were drawn in a three-stage sampling process that yielded a total 537 rural farm households. These were drawn across 80 randomly selected rural communities/villages in eight randomly selected Local Government Areas (LGAs) among those that are predominantly agrarian and rural in Ogun and Oyo states in Nigeria. The selected LGAs include Obafemi-Owode, Ijebu-North, Remo-North and Yewa South LGAs in Ogun state; and Oluyole, Ibarapa-East, Oyo-West and Saki-West LGAs in Oyo state.

It is instructive to note that in Nigeria, an area is considered urban if its human population is at least 20,000 and/or if it is located within a local government's or state's headquarter town or within the Federal Capital Territory (Okali, *et al.*, 2001), otherwise it is considered rural. However, areas designated as villages/farming communities on the maps of the various LGAs were typically with human population below 3000, and are mostly in remote locations from the urban centres. The study respondents were drawn from these "small" and remote communities. The mean distance (± standard error) of the sampled communities to the nearest town /LGA headquarters was 11.22 ± 0.41 kilometres; with only about half (54.2%) linked to the nearest town/LGA headquarter by a paved road, 52.7% linked to the national electricity grid and 43.7% having access to portable water either through the public tap (10.7%) or public borehole (32.6%) (See: Shittu, 2011).

Relevant data on socio-economic (community and household) characteristics, household labour and other resource use in both farm and non-farm activities during the 2005/2006 production season, and the associated costs, returns, outputs, and prices, among others were collected with the aid of a structured questionnaire/interview schedule. Information supplied by some of the respondents was considered inconsistent and/or incomplete, such that only data from 489 farm households were included in the final analysis.

Model specification

In this study, a stochastic conditional revenue frontier (9) was specified and estimated jointly with a production inefficiency equation (10). The estimated model may be stated as follows:

$$Y_i = f(X, \beta) + v_i - u_i \tag{9}$$

where,

Y_i is logarithm of total household income, including farm income and off-farm labour income (₦),

X is a vector of the logarithm of explanatory variables, including selected product prices, quantities of conventional inputs used, household labour and hired labour.

β is a vector of unknown parameters;

v_i is a random disturbance term, which is assumed to be identically, independently and normally distributed with mean 0 and variance σ_v^2, and independent of u_i;

u_i is a non-negative random variable assumed to account for production inefficiency across farm households, and is assumed to be independently distributed as truncations at zero of the $N(m_i, \sigma_u^2)$ distribution; where the means of u_i, m_i is defined as follows:

$$m_i = f\ (Z_1, Z_2, \ldots\ldots, Z_k) \tag{10}$$

Details of variables included in equations (9) and (10) as well as their descriptive statistics based on the study data are presented in Table 1.

A-priori expectations

A-priori, it is expected that coefficients of all the variables in the conditional revenue frontier will be positive, since increased use of factors and higher prices are expected to lead to increase in farm household income, *ceteris paribus*. Focusing on the inefficiency equation, however, it is expected that coefficients Z_8 and Z_9 would be positive, given that farm fragmentation (Z_8) and having to trek long distances to farm (Z_9) tends to raise production inefficiency. Coefficients associated with education (Z_3) and youth factor (Z_4) as well as those of asset income (Z_{10}), remittances received from migrants (Z_{11}) and total credit received (Z_{12}) are expected to be negative; since education, youthfulness and access to financial capital are expected to enhance productivity (production efficiency) in agriculture. Coefficients of other variables could take on either positive or negative signs depending on their influence on farm household production efficiency.

Results and Discussion

As background information, Table 2 provides key socio-economic information on the 489 sampled farm households included in the study. As shown on the Table 2, an average household in the sample was composed of six members, three (3) of which were economically active. These households are generally resource poor, with the members jointly cultivating an average of 1.83 hectares of land, and raising an

Table 1 Descriptive statistics of variables in the Stochastic Revenue Frontier and inefficiency equation

Variable	Definition of variables	Mean	Std. error
Variables in the conditional revenue function			
ADLAB	Number of economically active adults	2.88	0.23
CHLAB	Number of children (<18 years) supporting household income generating activities	1.44	0.10
HRLAB	Hired labour cost (N)	25,823.34	2,336.35
FSIZECRP	Size of arable crop farm (Ha)	1.27	0.02
FSIZETRE	Size of tree crop farm (Ha)	0.57	0.01
ANU	Size of livestock farm (Animal unit)	0.34	0.08
FERT	Quantify of fertilizer (50 kg Bags)	0.96	0.16
TRACTDAY	Tractor services used (Workdays)	0.45	0.05
MATCOS	Cost of other intermediate materials (N)	11,710.26	1,774.14
PGARRI	Price of garri (N/kg)	80.43	1.32
PMAIZE	Price of maize grain (N/kg)	80.07	1.13
PYAM	Price of yam (N per dozen tubers)	1,110.48	16.11
PEGGS	Price of eggs (N/tray)	345.51	2.24
POIL	Price of palm oil (N/litre)	91.18	1.12
Variables in the production inefficiency equation			
GENDER	Household head is a female (Yes =1, No = 0)	0.08	0.02
NATIVITY	Household head is non-native (Yes =1, No = 0)	0.19	0.03
HIGHEDU	Years of schooling by the most educated worker in the household	7.95	0.31
YOUTHFAC	Proportion of youths (30 years) in household	0.52	0.03
GENDERFA	Proportion of females in the household	0.41	0.03
PTREE	Proportion of farmland devoted to tree crops	0.30	0.02
PFREHOLD	Proportion of farmland on freehold	0.62	0.03
SIMPSON	Simpson index measure of farm fragmentation	0.39	0.02
DISTANCE	Average distance of farm parcels from home	3.18	0.15
ASSETINC	Asset income	39,083.86	4,141.02
REMITTA	Remittances: income received from rural out-migrants	35,401.28	3,198.20
LOANGOT	Amount of credits accessed	7,471.43	2,528.83
POFFFARM	Proportion of household income derived from off-farm activities	0.22	0.02

SOURCE: Computed from survey data, 2006.

average of 0.36 animal units of livestock. This is equivalent to about four (4) sheep/goats/pigs or 36 chickens.

Returns to household labour use in farm and off-farm sectors

Table 3 summarises the pattern of returns to farm and off-farm activities undertaken by members of the sampled farm households during the 2005/2006 farming season, while Table 4 present results of t-tests/F-tests of differences between the mean incomes across different mix of livelihood strategies employed.

As shown on Table 3, 47.1%of the sampled rural farm households derived some income from off-farm labour sources, with only 10.1% having at least one member engaged in formal employment. With regards to the relative contributions of the various labour and non-labour income sources to an average rural farm household's income,

Table 2 Distribution of sampled farm households by membership composition

Description	Ogun		Oyo		Total	
	No	%	No	%	No	%
Household size						
• 1 – 4	89	35.5	96	40.3	185	37.8
• 5 – 8	121	48.2	115	48.3	236	48.3
• 9 or more	41	16.3	27	11.3	68	13.9
Total	251	100	238	100	489	100
Mean household size	6.09		5.55		5.83	
Mean No. of economically active members	3.22		2.85		3.04	
Farm size (Hectares)						
• At most 1	91	36.3	113	47.5	204	43.9
• 1 – 3	126	50.2	101	42.4	227	44.9
• Above 3	34	13.5	24	10.1	58	11.2
Total	251	100	238	100	489	100
Mean farm size	1.92		1.73		1.83	
Tree crop share of farm land						
• None	84	33.5	78	32.8	162	33.1
• At most 50%	101	40.2	79	33.2	180	36.8
• Above 50%	66	26.1	81	35	147	30.1
Total	251	100	238	100	489	100
Livestock size (Animal units)						
• None	153	61	79	33.2	232	47.4
• Below 1	66	26.3	121	50.8	187	38.3
• At least 1	32	12.7	38	16	70	14.3
Total	251	100	238	100	489	100
Mean livestock size	0.26		0.46		0.36	

SOURCE: Field survey, 2006.

results on Table 3 shows that labour supplied to the non-farm sector by members of the sampled rural farm household accounted for 27.6% of the farm households labour based income (estimated at ₦282, 263.54 per annum) and 21.9% of gross income (estimated at ₦356, 748.68 per annual). Arable crop production accounted for the largest share of an average sampled farm household's annual gross income (30.6%) as well as total labour income (38.7%), followed by tree crops production. Overall, efforts put into the household farms yielded an average of ₦175, 638.21 per annum per household, accounting for 49.2% of an average farm household's annual gross income and 62.2% of total income from members work efforts.

As shown in Table 4, ANOVA results show that significant differences exists at $p < 0.01$ between the mean farm income of some categories of the sampled rural farm households when grouped across the kind of farm enterprise they were involved in during the 2005/2006 production season. Results of Duncan multiple range tests show that the sampled farm households may be classified into three homogeneous subsets, distinguished on the table by superscripts a, b and c. Households that cultivated both arable and tree crops recorded the highest mean farm income (₦230, 586.35) and fell in the highest income group while those that focused on tree crops and/or livestock only recorded the lowest income (₦61, 556.07) fell in the least farm income group.

Table 3 Descriptive statistics of contributions of labour and non-labour sources to farm households' income

Income source	Households deriving income from source				Income per average household in the entire sample			
	Number	%	Avg. receipt	Std. error	Mean	Std. error	% of GTA	% of TLS
Household farms								
Arable crop production	480	98.1	111,281.26	6,602.30	109,141.24	6,561.58	30.6	38.7
Tree crops production	263	53.8	86,767.81	5,185.84	57,984.26	4,477.04	16.3	20.5
Livestock production	229	46.8	14,755.36	1,767.86	8,512.71	1,137.22	2.4	3.0
Sub-total (Household farms)	**489**	**100.0**	**175,638.21**	**7,684.42**	**175,638.21**	**7,684.42**	**49.2**	**62.2**
Add: Natural resource collection	115	23.6	40,724.10	7,159.62	9,593.66	2,059.84	2.7	3.4
Agro-processing	82	16.8	86,154.40	11,807.07	14,497.13	2,978.50	4.1	5.1
Labour on others' farms	85	17.3	26,348.45	6,813.71	4,560.31	1,355.97	1.3	1.6
Sub total (Farm sector)	**489**	**100.0**	**204,289.30**	**8,574.58**	**204,289.31**	**8,574.57**	**57.3**	**72.4**
Artisanship & Craft	94	19.2	149,417.81	12,120.70	28,734.19	4,698.43	8.1	10.2
Trading	109	22.2	112,817.49	10,011.92	24,950.02	3,925.55	7.0	8.8
Formal employment	49	10.1	175,620.31	32,551.50	17,730.90	4,884.71	5.0	6.3
Other non-farm labour sources	37	7.6	85,268.60	12,615.72	6,559.12	1,838.80	1.8	2.3
Sub total (Non-farm sector)	**230**	**47.1**	**165,496.33**	**12,706.03**	**77,974.23**	**8,283.25**	**21.9**	**27.6**
Total Labour Sources (TLS)	**489**	**100.0**	**282,263.54**	**10,978.18**	**282,263.54**	**10,978.19**	**79.1**	**100.0**
Add: Remittances	320	65.4	54,143.14	4,058.35	35,401.28	3,198.20	9.9	
Income generating assets	334	68.3	57,249.60	5,431.39	39,083.86	4,141.02	11.0	
Grand Total, All Sources (GTA)	**489**	**100.0**	**356,748.68**	**12,389.87**	**356,748.68**	**12,389.87**	**100.0**	

Similar analysis in respect of off-farm labour supply mix also shows that significant differences exist at $p < 0.01$ between the mean off-farm income some groups of the sampled rural farm households. Farm households that had at least one members involved in paid employment in addition to trading and/or artisanship recorded the highest off-farm income (₦357, 269.22), and were placed in a separate (and the highest) off-farm income group, based on results of Duncan multiple range tests. Meanwhile, farm households whose members concentrated on only one line of off-farm activities except paid employment (i.e. artisanship only, trading only or services only) fell in the lowest off-farm income group. Furthermore, result of t-test conducted in respect of total labour income shows that farm households whose members participated in off-farm activities recorded, on the average, an annual labour income of ₦383, 162.75, which is significantly higher at $p < 0.01$ than the ₦192, 371.76 recorded by those whose members' efforts were concentrated on farming activities only.

ANOVA results also show that significant differences exist in the mean gross income of farm households that derived their income from different mix of labour use, income

Table 4 Results of F/t-tests of differences in mean income across investment and labour use patterns

Description	N	%	Mean	Std. error
Farm income by farm enterprise mix				
Arable crops only	82	16.8	142,215.80^{a,b}	17,590.50
Arable & tree crops	122	25.0	230,586.35^{c}	15,658.46
Arable crops & livestock	75	15.4	138,596.11^{a,b}	15,680.56
Arable crops, tree crops & livestock	200	40.9	175,099.07^{b,c}	11,727.26
Tree crops and/or livestock	9	1.9	61,556.07^{a}	22,466.14
F-value = 6.587; p < 0.01				
Off-farm labour Income by activity mix				
Artisanship	61	26.5	153,977.53^{a,b}	15,767.33
Trading	70	30.6	121,573.38^{a,b}	13,528.98
Paid Job	26	11.3	196,092.74^{b}	52,361.29
Other services	28	12.2	91,293.44^{a}	16,134.62
Artisanship, trading and/or other service	21	9.2	193,644.16^{b}	30,030.38
Paid job + trading/artisanship	23	10.2	357,269.22^{c}	57,770.53
F-value = 9.228; p < 0.01				
Total labour income by labour use mix				
Farming only	259	52.9	192,371.76^{a}	12,097.18
Farm and off-farm activities	230	47.1	383,162.75^{b}	16,937.51
t-value = 9.166; p < 0.01				
Gross income by income source mix				
Labour only	52	10.6	266,766.86^{a}	30,367.86
Labour & remittances	104	21.2	339,215.71^{a,b}	26,224.03
Labour & asset	120	24.5	356,064.14^{b}	27,998.20
Labour, asset & remittances	214	43.7	387,364.01^{b}	21,238.92
F-value = 2.538; p < 0.10				

NOTE: Mean incomes within each socio-economic group that carry the same superscript, a b or c are not significantly different, while those having different superscripts are significantly different at $p < 0.05$.

generating assets and remittances at 10% level of significance. Result of Duncan multiple range tests show that the mean income of those households that relied solely on their members labour, on the average, is significantly lower at $p < 0.05$ than the average income of those households that, in addition to their members labour, also owned some income generating assets. Evidence in respect of farm households that received some remittances, however, showed that such receipt did not significantly raise the affected farm households' gross income above what was obtained by other comparable households without such remittances.

Overall, results on Table 4 provide some important insight into appropriate strategies for raising income level among rural farm households in the study area. First, some level of household labour use diversification both within agriculture and into off-farm activities might be necessary, not only for income risk reduction, but also for significant increases in rural farm households' income. Note that apart from the seasonal nature of agricultural production in Nigeria, the fact that most farm households operates small landholdings make off-farm diversification almost inevitable if full employment is to be guaranteed.

Second, the fact that those households whose members were able to gain access to paid employment, recorded significantly higher income than their peers whose members could not suggests that creation of appropriate opportunities for the establishment of industries and other employment generating institutions in the rural areas may be a veritable strategy for raising rural household income. Furthermore, given that farm households that owned some income generating assets earned significantly higher income than their peers who do not, governments' may have to come up with policies and programmes that provide support for wealth creation among the rural folks as part of its overall poverty eradication strategies.

Econometric results

The central theme of this study has been to examine the implications of off-farm labour supply by members of farm household on household production efficiency. The results, based on specification and joint estimation of a conditional revenue function and a production (economic) inefficiency model using the parametric - Stochastic Frontier approach of Battesse and Coelli (1995) is presented Table 5.

Evidence from the generalised likelihood ratio test of the one-sided error term confirms that substantial inefficiency exists in the allocation of production resources, including labour, by farm households in the sample, and by extension the study area: The calculated Chi-square value was 235.4 as against the critical value of 32.00 at $p < 0.01$ and 16 degrees of freedom. This shows that the OLS version of the conditional revenue frontiers is not an adequate representation of the study data. The joint MLE of the revenue frontier and inefficiency model is therefore, a much more appropriate framework for analysing the production systems of farm household in the study area.

The coefficient of gamma ($\gamma = \sigma_U^2/(\sigma_V^2 + \sigma_U^2)$, which measures the proportion of the total variance ($\sigma^2 = \sigma_V^2 + \sigma_U^2$) that is due to inefficiency in the production system was found to be 0.99 and is significant at $p < 0.01$. This shows clearly that variation in income of farm households in the study area, conditional on their resource endowment, is predominantly due to inefficiency in their allocation of production resources.

Income response to resource use and prices

The top part of Table 5 presents coefficients of explanatory variables in the conditional revenue frontier of farm households in the study area; which are the partial income (revenue) elasticity with respect to changes the variables in the model. Virtually all the variables except LNPYAM, LNTLAND and LNCHLAB were associated with the a-priori expected positive signs, meaning that increase in their values is associated with increase in farm household income. Examination of t-ratio associated with the coefficients of variables in the MLE version of the conditional revenue frontiers shows that it is only the influence of five variables - LNPOIL, LNCLAND, LNANU, LNHRLAB and LNADLAB that were significant.

Only one out of the five variables depicting vector of prices, LNPOIL (price of palm oil, proxy for price of output from tree crops), had coefficient that was significant at $p < 0.05$. The non-significance of coefficient of other price variables may be because the study was based on a cross-section data, typically characterised by minimal level of variation in prices across observation points. The coefficient of LNPOIL is 0.43, which shows that 1% increase in the farm gate price of products from tree crops (specifically, oil palm: which

Table 5 Estimates of the Stochastic Revenue Frontier and inefficiency model

	OLS estimates		ML Estimates	
	Coefficient	t-ratio	Coefficient	t-ratio
Conditional revenue frontier				
Constant	1.6742	0.53	9.5649***	7.21
LNPCAS	0.5010***	2.78	0.1074	0.86
LNPMAIZE	0.3527*	1.79	0.2384	1.52
LNPYAM	−0.4067**	−2.16	−0.0534	−0.38
LNPEGGS	1.4744***	3.57	0.2035	0.68
LNPPOIL	0.1742	0.74	0.4268**	2.71
LNCLAND	0.3815***	9.42	0.7075***	15.92
LNTLAND	0.1795***	7.86	−0.0309	−0.79
LNANU	0.0595**	1.97	0.0836***	3.82
LNFERT	0.0103	0.37	0.0286	1.57
LNTRACT	0.0667*	1.85	0.0370	1.47
LNMATCOS	0.0145*	1.66	0.0050	0.77
LNHRLAB	0.0264***	2.98	0.0112**	2.05
LNADLAB	0.2326***	3.28	0.1128***	2.89
LNCHLAB	0.0272**	2.01	−0.0054	−0.48
Inefficiency equation				
Constant			0.2696***	4.31
GENDER			−0.0719	−0.98
NATIVITY			−0.0351	−0.85
HIGHEDU			−0.0012	−0.22
YOUTHFAC			0.0114	0.19
GENDERFA			−0.0482	−0.74
PTREE			−1.7265***	−7.43
PFREHOLD			0.1471***	2.78
SIMPSON			0.2182***	3.08
DISTANCE			0.0028	0.25
ASSETINC			−3.3e-7*	−1.66
REMITTA			0.4257	0.95
LOANGOT			−6.8e-7	−1.27
POFFFARM			−0.1595***	−15.24
Sigma squared	0.2169		0.0661***	9.43
Gamma			0.99***	17702
Log likelihood function	−129.73		−12.03	
LR test of one sided error			235.4	

NOTE: ***, ** and * imply the associated coefficients are significant at 1%, 5% and 10% levels respectively.

was the most widely cultivated tree crops by farm households in the study area) is associated with about 0.43% increase in income of farm households.

Two out of the three coefficients associated with variables on farm size, LNCLAND and LNANU representing size of land devoted to arable crops production and stock of livestock raised by the farm households respectively, were positive and significant at $p < 0.01$. This shows that increase in size of arable crops and livestock farms raised by the farm households are associated with significant increases in their income. One per cent increase in farmland devoted to arable crops production is associated with about 0.71% increase in farm household income. Similar increase in stock of livestock was found to be associated with about 0.08% increase in farm household income. The fact that coefficient of LNTLAND (size of farmland devoted to tree crops production) is negative and not significant suggests that further increase in size of tree crop farm being cultivated by an average farm household in the sample, ceteris paribus, may not results in improvement in her income.

Focusing on the use of modern inputs / capital items like fertilizer (LNFERT), tractors services (LNTRACT) and intermediate inputs like seed, feed, etc. (LNMATCOS), results on Table 5 shows that none of the associated coefficients was significant even at $p < 0.10$. This outcome is not unlikely to be because the use of these inputs by farm households in the study area was very limited, and probably not in line with recommendations. It is also worthy of note that majority of the rural farm households in the sample that raised livestock kept their animals on a free range system, with little or no feed provided. The results with respect to labour use shows that farm household income is significantly affected by hired labour use (HRLAB) and numbers of economically active adults in the household (LNADLAB). One per cent increase in hired labour use is associated with about 0.01% increase in income while similar increase in the number of economically active adults in the household is associated with about 0.11% increase in farm household income.

Table 6 presents selected indicators of production (income) response to changes in various factors. It shows that partial elasticity of output response was highest with respect to increase in arable crop land cultivated, followed the number of economically active adults in the households. The overall elasticity of output response (return to scale) was found to be 0.949; meaning that 1% increase in use of all factors will lead to about 0.95% increase in farm household income.

Comparing the value of marginal products of the various factors, evaluated at the geometric mean of the variables in the revenue frontier with the estimated unit cost of each input, evidence on Table 6 suggests excessive use of both household and hired labour in the production system. Note, for example, that at the margin, ₦1.00 spent in employing hired labour yielded a marginal income of just about ₦0.63, while an

Table 6 Selected indicators of production response to factor use

Production factor	Partial elasticity	Geometric mean	Value of marginal product	Estimated unit factor cost (N)
Arable crop land (Ha)	0.7075	0.8263	191,383.23	n/a
Tree crop land (Ha)	−0.0309	0.2333	−29,604.58	n/a
Size of livestock (ANU)	0.0836	0.2276	82,101.14	30,000
Fertilizer (50 kg-bag)	0.0286	0.2298	27,818.34	2,400
Tractor services (workdays)	0.0370	0.1849	44,728.05	15,000
Intermediate materials (N)	0.0050	1,056.18	1.06	1.00
Hired Labour cost	0.0112	3,995.07	0.63	1.00
Number of adult workers	0.1128	1.8547	13,594.10	n/a
Number of supporting child	−0.0054	0.2936	−4,111.05	n/a
Returns to scale	**0.9494**			

NOTE: Values of marginal products were evaluated at the geometric mean level of each factor and output (household labour income). The geometric mean output was found to be ₦223, 519.38.

increase in the number of economically active adult in the farm household by one person will only cause income to be raised, at the margin, by N13, 954 per year showing that labour use in their production system is excessive. Contrarily, marginal returns to the use of modern inputs like fertilizer, tractor services and intermediate materials were all greater than the respective unit factor costs, showing that capital items were underutilised.

The main implication of these results is that farm household would fare better if they employ more capital items and reduce the level of their labour use in their production system. It also points to the possibility that massive rural out-migration by youths in the study area might be a response to poor marginal returns to labour use in both farm and off-farm activities in the study area.

Production efficiency estimates

Table 7 presents the distribution of the efficiency indices [a] computed for each farm household in the sample. It shows that most (70.5%) of the farm households had production efficiency index that was below 0.2 with a mean of 0.18. It thus imply that an average farm household in the sample can have her income raised by more than five times the present level if the limiting factors are mitigated.

Comparing the production efficiency indices across farm household categories based on whether their members participated or did not participate in the rural non-farm sector, results on Table 7 shows that an average household whose members participated in off-farm activities was significantly ($p < 0.01$) more efficient (PE = 0.2114) than her counterpart whose members focused their attention only on activities in the farm sector (PE = 0.1374). It thus imply that access to off-farm employment opportunities can, *ceteris paribus,* help raise production efficiency of rural farm households in the study area. The efficiency gain can enable such farm households to regain as much as 7.4% of the frontier (potential) income or 53.9% of what would have been obtained if their members did not participate in off-farm activities.

Table 7 Distribution of sampled farm households by level of production efficiency and source of livelihood

Level of farm household production efficiency	Households that derived their labour based income from:				All farm households in the sample	
	Farm sector only		Farm and non-farm sectors			
	No	%	No	%	No	%
Below 0.10	121	46.7	47	20.5	168	34.3
0.10 - <0.20	90	34.8	87	37.8	177	36.2
0.20 - <0.30	28	10.8	49	21.3	77	15.7
0.30 - <0.40	10	3.9	20	8.7	30	6.2
0.40 - <0.50	5	2.0	18	7.9	23	4.8
0.50 & higher	5	1.9	9	3.9	14	2.9
Total	259	100.0	230	100	489	100
Mean	0.1374		0.2114		0.1782	
Std. error of mean	0.0073		0.0095		0.0061	

NOTE: The calculated t-value in a test of difference between the mean production efficiency of the two categories of farm households was 6.18 as against a critical t-value of 2.58 at $p < 0.01$ and 487 degree of freedom.

Determinants of inefficiency in the production system

The lower section of Table 5 presents coefficients of variables in the inefficiency equation and their associated t-ratios. It shows that coefficients associated with five out of the 14 variables - PTREE, PFREHOLD, SIMPSON, ASSETINC and POFFFAM were significant. Coefficient of PTREE is negative and significant at $p < 0.01$ level, which means that the higher the proportion of the cultivated farmland devoted to tree crops production the lower is the inefficiency in the production system adopted by farm households. This result may appear to be a sharp contrast to earlier evidence that coefficient of LNTLAND (area of land devoted to tree crops production) was negative and insignificant in the revenue frontier. However, the import of these results is that while raising areas of land devoted to tree crops production when all other things are held constant would bring no improvement to household income, converting some areas of land from arable to tree crops production would lead to significant improvement in household income. This evidence would suggest that the tendency to convert tree crop estates to arable crop farms in search of short-term funds, would amount to greater inefficiency in farm households' resource use. It is instructive to note that, until very recent times, the proportion of arable land devoted to permanent crops production in Nigeria had been on a steady decline, dropping from about 9.1% in 1980 steadily to about 8.36% in 1995 (FAO, 2014). However, the trend is being reversed in recent times, with total area devoted to permanent crops growing from about 3 million hectares in 1995 to 3.2 million hectares in 2011 (FAO, 2014) in response to policy reforms.

Focusing on other variables in the inefficiency model, results on Table 5 show that coefficient of PFREEHOLD as well as SIMPSON (Simpson index, measuring the extent of farm fragmentation) were positive and significant at $p < 0.01$. This shows that cultivation of fragmented landholdings by rural farm households and increase in rights on land are associated with significant increases in inefficiency in the production system. The results in respect of farm fragmentation is not unexpected: evidence in literature (e.g. Blarel *et al.,* 1992) have shown, among others, that cultivation of fragmented landholdings raises operation costs, leads to loss of time spent commuting between parcels, makes mechanisation (and therefore, deriving the associated economies of size) difficult.

The fact that farm households with higher level of rights / control on their farmland were associated with significantly higher level of inefficiency, however, goes against a-priori expectations. A-priori, it is expected that farm households that enjoys greater security on their farmland are more likely to invest on land development, which could make such households to operate at higher levels of production efficiency (Chavas *et al.,* 2005). This outcome may however, not be unconnected with the fact that most freely held land by farm households in the study area was acquired by inheritance. Meanwhile, land tenure system dominated by land ownership through inheritance has been a major cause of farm fragmentation in Africa (Blarel *et al.,* 1992).

Coefficients of ASSETINC and POFFFARM were significant and associated with negative signs indicating that an increase in each of these variables is associated with lower production inefficiency. In other words, raising the level of income that rural farm household derive from non-labour sources, particularly income yielding assets and off-farm activities, raises production efficiency of rural farm households. This outcome supports suggestions in literature (e.g. Chavas *et al.,* 2005; Hazell and Hojjati,

1995) that in the presence of poorly functioning capital markets, cash from off-farm earnings (including asset income and off-farm employment income) can help stimulate farm investments and improve agricultural productivity.

Meanwhile, coefficient of REMITTA (remittances received from members of rural farm families that have migrated to the urban centre) was not significant even at $p < 0.10$. This shows that while increase in off-farm earnings and asset income enhances production efficiency of rural farm households in the study area, increase in the quantum of remittances they receive from their members that have migrated away to the urban centre exercise insignificant influence on their production efficiency. It is also instructive to note that coefficient of LOANGOT (amount of credit secured by the rural farm households, which was predominantly from informal sources) is also not significant; although it is associated with the desirable negative sign that suggests that access to credits tends to lower production inefficiency.

Overall, analysis of production system of rural farm households in the study area shows that substantial inefficiency exists in their allocation of production resources. Labour, including household and hired labour, is over-utilised, while modern inputs like fertilizer, tractor services, etc. are underutilised. The level of the rural farm households' production efficiency is enhanced by adoption of a diversified portfolio of income sources vis-à-vis diversification of labour use to include farm and off-farm activities, raising proportion of tree crops on their farms and investment in income yielding assets. It is instructive to note that relative influence of within farm sector diversification (e.g. raising the shares of tree crops and livestock in the farm holdings) is much stronger than the influence of diversification into off-farm labour supply. A plausible explanation for this is the fact that rates of returns to household labour use in most of the available rural non-farm activities is much lower than what obtains from labour supply/investment in tree cropping (mostly cash crops) and livestock production (Shittu, 2011). Hence, observed diversification into rural non-farm economic activities in the area are largely out of necessity (distress-push) rather than as a response to remunerative wage employment and high return off-farm business opportunities (demand-pull).

Another important evidence is that production efficiency at the farm level is enhance if the farm households operates a much more consolidated landholdings, just as raising the proportion of leased land (as against freehold land) brings about greater production efficiency. Meanwhile, household composition and personal factors like education level, gender and nativity of household heads as well as gender/youth factors in the households exercise little or no influence on production efficiency of rural farm households in the study area.

Conclusions

This central theme of this study has been to examine the influence of off-farm labour supply among members of rural farm households on farm household production efficiency in Southwest Nigeria. The study was based on primary data collected from a cross-section of 489 rural farm households, drawn by multi-stage random sampling from Ogun and Oyo States, in the Southwest rain-forest zone of Nigeria. The data was analysed by descriptive and econometric techniques, including specification and estimation of a conditional revenue frontier and a production inefficiency equation.

The results showed that 47.1% of the rural farm households had some of their members involved in off-farm activities with a typical member devoting 34.3% of his/her work efforts to off farm activities, while off farm activities contributed 27.6% of a typical farm household's labour income (N282, 263.54 in 2005/2006). Increase in off-farm labour supply was found to be associated with significant ($p < 0.01$) reduction in production in-efficiency among the rural farm households. The study also found that farm household production efficiency is significantly enhanced by increasing the share of tree crops and livestock in farm household farming activities, having access to land by leasing and operating a much more consolidated landholdings. However, remittances from rural out-migrants do not significantly affect rural farm household production efficiency.

Two main conclusions may be drawn from evidences in the study. First, availability of urban-type employment opportunities and increased participation of farm household members in the rural non-farm sector contribute significantly to farm household income and significantly enhance production efficiency. The patterns and returns to these off-farm activities, however, suggest they are economically motivated response to inadequate access to production resources and poverty within the rural farm sector. This is because returns to labour allocated to off-farm activities in the rural sector was found to be lower, on the average, than what obtains in agriculture. This shows that more attention still need to be focused at understanding and developing the rural non-farm sector in Nigeria to make activities in the sector much more rewarding and attractive to the unemployed/under-employed youths and landless households.

Second, substantial inefficiency exists in the allocation of production resources by farm households in the study area. Labour, including household and hired labour, is over utilised, while modern inputs like fertilizer, tractor services, etc. are underutilised. The level of the rural farm households' production efficiency is enhanced by adoption of a diversified portfolio of income sources vis-à-vis diversification of labour use to include farm and off-farm activities, raising proportion of tree crops and livestock on their farms, investment in income yielding assets, and by operating a much more consolidated landholding.

Endnotes

[a]Efficiency estimates associated with the revenue frontier are estimates of overall economic efficiency. Meanwhile, the Battesse and Coelli (1995) version of the frontier model is such that allocative efficiency is imposed in an attempt to obtain an efficient and consistent estimates of the parameters of the model (See: Coelli, 1996: pg 6). The implication of this is that efficiency estimates obtained may also be interpreted in the realm of technical efficiency.

Abbreviations

ADLAB: Number of economically active adults; ANOVA: Analysis of Variance; ANU: Animal unit; ASSETINC: Asset income; CHLAB: Number of children (<18 years) supporting income generating activities; DFID: Department for International Development of the United Kingdom; FAO: Food and Agriculture Organisation of the United Nations; FOS: Federal Office of Statistics (FOS); FERT: Quantify of fertilizer; FSIZECRP: Size of arable crop farm; FSIZETRE: Size of tree crop farm; GENDERFA: Proportion of females in the household; HIGHEDU: Years of schooling by the most educated worker in the household; HRLAB: Hired labour cost; LOANGOT: Amount of credits accessed; MATCOS: Cost of other intermediate materials; MLE: Maximum Likelihood Estimator; NBS: National Bureau of Statistics (NBS); PE: Production (Technical) Efficiency; PEGGS: Price of eggs; PFREHOLD: Proportion of farmland on freehold; PGARRI: Price of garri; PMAIZE: Price of maize grain; POFFFARM: Proportion of household income derived from off-farm activities; POIL: Price of palm oil; PTREE: Proportion of farmland devoted to tree crops; PYAM: Price of yam; REMITTA: Remittances: income received from rural out-migrants; SIMPSON: Simpson index measure of farm fragmentation; TRACTDAY: Tractor services used; USAID: United State Agricultural and Industrial Development; UNDP: United Nations Development Programme (UNDP); YOUTHFAC: Proportion of youths (15 - 30 years) in the household.

Authors' contributions

The research from which this paper was written was conceived, designed and implemented by AM in pursuit of PhD (Agricultural Economics) at the Department of Agricultural Economics and Farm Management, Federal University of Agriculture, Abeokuta, Nigeria. He solely undertook the drafting and revision of the paper, and submitted it for publication. He read and approved the final manuscript.

Author's Information

Adebayo Musediku Shittu is an Associate Professor of Agricultural & Resource Economics in the Department of Agricultural Economics and Farm Management, Federal University of Agriculture, Abeokuta (FUNAAB), Nigeria. Degrees earned include B.Agric in Agricultural Economics and Farm Management from Ogun State University (Now Olabisi Onabanjo University), Ago-Iwoye (1992), MSc Agricultural Economics from the University of Ibadan, Nigeria (1995), and PhD Agricultural Economics from FUNAAB (2008). Research interests include applied microeconomic analysis, economics of natural resources, environment and development. He has to his credit over 30 articles published in local and international journals. He is a member of International Association of Agricultural Economists, Agricultural and Applied Economics Association, Association of Environmental and Resource Economists, Northeastern Agricultural and Resource Economics Association, and Farm Management Association of Nigeria. He has mentored over 100 undergraduate & postgraduate students, and currently functions as a Part-time Consultant to Lagos State Fadama III Project Coordinating Office.

Acknowledgements

The author wish to acknowledge the very useful comments and suggestions of members of AM's PhD thesis Supervisory Committee – Professors DAO Phillip, S Momoh and SO Apantaku, to the design and implementation of the study that served as basis for writing up this paper. The author was also an employee of Olabisi Onabanjo University, Ago-Iwoye while conducting the PhD research. The research was however, solely funded by the author. The author also wishes to acknowledge the very useful comments and suggestions of the anonymous reviewers as well as editorial comments of the Handling Editor. These contributed immensely to the quality of the final output of the paper.

References

Anderson D, Leiserson MW (1984) Rural Nonfarm Employment in Developing Countries. Econ Dev Cult Change 28:227–248

Babatunde RO, Qaim M (2010) Impact of off-farm income on food security and nutrition in Nigeria. Food Policy 35 (4):303–311. doi:10.1016/j.foodpol.2010.01.006

Babatunde RO, Olorunsanya EO, Adejola AD (2008) Assessment of Rural Household Poverty: Evidence from South-western Nigeria. American-Eurasian J Agric & Environ Sci 3(6):900–905

Battese GE, Coelli TJ (1995) A Model for Technical Inefficiency Effects in a Stochastic Frontier Function for Panel Data. Empir Econ 20:325–332

Blarel B, Hazell P, Place F, Quiggin J (1992) The Economics of Farm Fragmentation: Evidence from Ghana and Rwanda. The World Bank Econ Rev 6(2):233–254

Chang H, Mishra A (2008) Impact of off-farm labor supply on food expenditures of the farm household. Food Pol 33 (6):657–664. doi:10.1016/j.foodpol.2008.02.002

Chang H, Wen F (2011) Off-farm work, technical efficiency, and rice production risk in Taiwan. Agric Econ 42(2):269–278. doi:10.1111/j.15740862.2010.00513.x

Chavas J, Petrie R, Roth M (2005) Farm Household Production Efficiency: Evidence from the Gambia. Am J Agric Econ 87(1):160–179

Coelli TJ (1996) A Guide to Frontier 4.1: A Computer Program for Stochastic Frontier Production and Cost Function Estimation. In: CEPA Working Paper 96/07. Centre for Efficiency and Productivity Analysis (CEPA), University of New England, Australia

Department for International Development of the United Kingdom (DFID) (2004) Rural Urban Development Case Study – Nigeria. Oxford Policy Management, Oxford, United Kingdom

Escobal J (2001) The determinants of non-farm income diversification in rural Peru. World Dev 29(3):497–508

Federal Office of Statistics (FOS) (1999) Poverty and Welfare in Nigeria. FOS, Nigeria

Food and Agriculture Organisation (FAO) (1998) The State of Food and Agriculture. FAO Series No 31, FAO, Rome, Italy

Food and Agriculture Organisation (FAO) (2014) FAOSTAT data. FAOSTAT data. FAO, Rome, Italy. http://faostat3.fao.org/faostat-gateway/go/to/download/R/RL/E. Accessed 25 Mar 2014

Goldsmith PD, Gunjal K, Ndarishikanye B (2004) Rural–urban Migration and Agricultural Productivity: the case of Senegal. Agric Econ 31:33–45

Haggblade S, Hazell P, Reardon T (2010) The Rural Non-farm Economy: Prospects for Growth and Poverty Reduction. World Dev 38(10):1429–1441

Hazell P, Hojjati B (1995) Farm/Non-farm Growth Linkages in Zambia. Journal of African Economies 4(3):406–435

Holden S, Shiferaw B, Pender J (2004) Non-farm income, household welfare, and sustainable land management in a less-favoured area in the Ethiopian highlands. Food Pol 29(4):369–392. doi:10.1016/j.foodpol.2004.07.007

Huang J, Wu Y, Rozelle S (2009) Moving off the farm and intensifying agricultural production in Shandong: a case study of rural labor market linkages in China. Agric Econ 40:203–218

Jacoby HG (1993) Shadow Wages and Peasant Family Labor Supply: Application to the Peruvian Sierra. Rev Econ Stud 60:903–921

Lanjouw P (2000) Non-farm Employment and Poverty in Rural El Salvador. World Dev 23(3):529–547

Lanjouw J, Lanjouw P (2001) Rural Non-farm Employment: A Survey. Agric Econ 26:1–23

Lien G, Kumbhakar SC, Hardaker JB (2010) Determinants of off-farm work and its effects on farm performance: the case of Norwegian grain farmers. Agric Econ 41(6):577–586. doi:10.1111/j.1574-0862.2010.00473.x

Matshe I, Young T (2004) Off-farm Labour Allocation Decisions in Small-scale Rural Households in Zimbabwe. Agric Econ 30:175–186

National Bureau of Statistics (NBS) (2011) National average of the indicator. "Unemployment Rates by State" in the Data Category – "Socio-Economic Data", NBS, Nigeria. Available at: http://nigeria.prognoz.com/en/DataAnalysis

National Bureau of Statistics (NBS) (2012) Press Briefing by the Statistician-General of the Federation/Chief Executive Officer. In: National Bureau of Statistics. Dr. Yemi Kale held at the Conference Room, 5th Floor, NBS Headquarters, Central Business District, Abuja on Monday, 13th February, 2012. NBS, Abuja, Nigeria. Available at: http://www.proshareng.com/reports/4632

Newman JL, Gertler PJ (1994) Family Productivity, Labor Supply, and Welfare in a Low Income Country. J Hum Resour 29:989–1026

Ogwumike OF, Aromolaran AB (2000) Poverty Dynamics in Nigeria. World Bank, Rome

Okali D, Okpara E, Olawoye J (2001) Rural–urban Interactions and Livelihood Strategies: The case of Aba and its region, southeastern Nigeria. Working Paper Series on Rural–urban Interactions and Livelihood Strategies, Human Settlements Programme, IIED, October

Owusu V, Abdulahi A, Abdul-Rahman S (2011) Non-farm work and food security among farm households in Northern Ghana. Food Pol 36(2):108–118

Pfeiffer L, L´opez-Feldman A, Taylor JE (2009) Is off-farm income reforming the farm? Evidence from Mexico. Agric Econ 40:125–138

Phimister E, Roberts D (2006) The Effect of Off-farm Work on the Intensity of Agricultural Production. Environ Resour Econ 34(4):493–515

Schultz TW (1990) Restoring Economic Equilibrium. Basil Black Well, Cambridge

Shi X, Heerink N, Qu F (2011) Does off-farm employment contribute to agriculture-based environmental pollution? New insights from a village-level analysis in Jiangxi Province, China. China Econ Rev 22(4):524–533

Shittu AM (2011) Off-farm Labour Drift in the Sub-Saharan Africa: Patterns, Determinants and Production Efficiency Outcomes in Southwest. LAP Lambert Academic Publishing, Germany, Nigeria

Shittu AM, Ashaolu OF, Odusanya OS (2006) Off-farm Labour Participation and Farm Household Livelihood Strategy in Yewa Division, Ogun State, Nigeria. FAMAN Journal 8(1):64–70

Stampini M, Davis B (2009) Does nonagricultural labor relax farmers' credit constraints? Evidence from longitudinal data for Vietnam. Agric Econ 40:177–188

United Nations Development Programme (UNDP) (2010) The Real Wealth of Nations: Pathways to Human Development (Human Development Report 2010). UNDP, New York

United State Agricultural and Industrial Development (USAID) (2002) Urban Profile – Nigeria. USAID, USA

Wandschneider T (2003) Determinants of access to rural non-farm employment. In: Evidence from Africa, South Asia and transition economies. NRI Report No 2758, Natural Resource Institute, University of Greenwich, United Kingdom

World Bank (2000) World Development Report 2000: Attacking Poverty. World Bank, Washington, D.C

7

Farm types and their economic characterization in complex agro-ecosystems for informed extension intervention: study from coastal West Bengal, India

Rupak Goswami[1*], Soumitra Chatterjee[2] and Binoy Prasad[1]

* Correspondence: goswamirupak@rediffmail.com
[1]Integrated Rural Development and Management Faculty Centre, Ramakrishna Mission Vivekananda University, Narendrapur, Kolkata-700103, INDIA
Full list of author information is available at the end of the article

Abstract

Identification and characterization of farming systems simplify huge diversity of farm types in complex agro-ecosystems, which is of critical importance for precise technological intervention and informed policy support. Multivariate statistical techniques like Principal Component Analysis (PCA) and Cluster Analysis (CA) may be used for a wide variety of situations associated with farm typology delineation. The present study conducted in coastal saline India demonstratively established the usefulness of such methodology in identification of predominant farm types and their characterization. Data collected from 144 farm households through questionnaire survey could identify four predominant farm types with differential income sources and resource-base. The methodological perspective employed in the study may be used as a decision support tool by extension agencies. On other hand, a differentiated, holistic and broad-based extension intervention with suitable institutional arrangement will be needed to address the need of these identified farm types. This will lead to a reduced transaction cost of the agricultural research and extension systems in diverse ecosystems in India and many similar situations in the developing countries.

Keywords: Cluster analysis; Economic characterization; Extension support; Farm heterogeneity; Farm typology; New Institutional Economics; Principal component analysis

Background

Adoption of new technologies in agriculture is of central interest to both academicians and policy makers since this is directly related to the efficiency of an agricultural research and extension system (Bozeman 2000). Higher adoption ensures higher return on investment in research and development and creates a cycle of production-led impact on economy and rural livelihoods. Unfortunately, there are numerous examples of technologies with great potential that have not been accepted by farmers, especially the smallholders of the developing countries. Quite often, these technologies do not fit well into heterogeneous smallholder systems, which need specific technological solutions. Such inherent variability often influences farmers' response to various technologies that aim at improving farm productivity and natural resource management (Lal et al. 2001; Emtage and Suh 2005). Unfortunately, both in agricultural and social sciences, complexity and diversity has been under-perceived and undervalued resulting in their neglect, under-estimation and exclusion from government statistics and policy framework (Chambers et al. 1989). The archetypal Green Revolution technologies and 'transfer-of-technology' paradigm has also historically failed to cater to the needs of these diverse resource-poor agro-ecosystems in the developing countries (Pender and Hazell 2001). Extension offered blanket recommendation for wide geographical areas and was largely used as a deterministic 'dart gun' (Roling 1988), i.e. 'take the technology and transfer it to farmers'. The heterogeneity of the farming systems for which different technologies are needed, has been ruefully ignored.

Thus, study of farm typology is of practical interest for precise and effective technological interventions. Farm typology study recognizes that farmers are not a monolithic group and face differential constraints in their farming decisions depending on the resources available to them and their lifestyle (Soule 2001). Ellis (1993) observes that small farmers are always and everywhere typified by internal variations along many lines. Although every farm and farmer is unique in nature, they can be clustered into roughly homogeneous groups. Developing a typology constitutes an essential step in any realistic evaluation of constraints and opportunities that farmers face and helps forwarding appropriate technological solutions, policy interventions (Ganpat and Bekele 2001, Timothy 1994; Vanclay 2005), and comprehensive environmental assessment (Andersen et al. 2009). Moreover, typology studies are of paramount importance for understanding the factors that explain the adoption and/or rejection of new technologies (Kostrowicki 1977; Mahapatra and Mitchell 2001).

The heterogeneity of farming systems are created by a host of biophysical (e.g. climate, soil fertility, slope etc.) and socio-economic (e.g. preferences, prices, production objectives etc.) factors (Ojiem et al. 2006). Researchers have examined factors such as farm resources such as cash and labor (Tittonell et al. 2007), infrastructure such as agency and markets (Tittonell et al. 2006), management practices (Tittonell et al. 2005) and technological level (Gómez-Limón et al. 2007). Few others have used string of factors together to explain the heterogeneity of farming systems (Ojiem et al. 2006; Bidogeza et al. 2009; Guto et al. 2010; Tittonell et al. 2010; Rowe et al. 2006). The selection of factors that define farm typology varies greatly from study to study and may be governed by the purpose of research. For example, farm typologies were used to study appropriate fertilizer application (Tittonell et al. 2006), resource use efficiency (Zingore 2007; Tittonell et al. 2007), water use efficiency (Senthilkumar et al. 2009), or overall classification of farm types (Bidogeza et al. 2009), IPM (Leeson et al. 1999) and may sometimes be crop specific in nature (Mwijage et al. 2009; Reckling 2011). Most of the farm typology study has focused on socio-economic and agro-ecological factors for classification of farms. Economic factors have been less used, especially in small-scale studies, for classifying farms (USDA, ERS 2000; Briggeman et al. 2007; Andersen 2009). With economic characterization of farms as the objective at hand, the present study assumes that classification of farms based on economic returns from farm enterprises, along with other related non-economic factors, will give more effective insights into the farm type identification.

Indian farmers are heterogeneous in terms of agro-ecology and resource endowments and the transfer of appropriate technology requires careful targeting. This problem is exacerbated in complex-diverse-risk-prone coastal agro-ecosystems of the country covering approximately 10 Mha cultivated area, which is characterized by a host of bio-physical and socio-economic constraints coupled with diversity in livelihoods. At present, agricultural research is based on primarily 15 agro-climatic zones and 127 sub-regions based on comprehensive research review of states (Khanna 1989). Socio-economic and structural factors are not reflected in such classifications. Economists classify farmers based on landholdings, which is not comprehensive enough for classifying complex farming systems. Indian Council of Agricultural Research (ICAR) used several factors for classification of farming systems, but the bases were more discussion-based with limited statistical applications, perhaps due to functional purpose of the classification (Sanjeev et al. 2010).

At the micro level, classification of farmers might be of practical use for localized technological solutions and extension support. During the height of farming system research and extension paradigm, conceptualization of recommendation domain wanted to address this issue. A recommendation domain is a group of farmers whose circumstances are similar enough that they are eligible for the same recommendation (Harrington and Trip 1984). This led to informed decisions in part of technology managers and higher rate of technology integration in smallholder systems. Classification of farming situation by the farmers has also been suggested by some authors (Conroy and Sutherland 2004; Goswami et al. 2012). However, these were more project-based and were hardly mainstreamed in the national research and extension systems in the developing countries (Frankenberger et al. 1989), India being no exception. Participatory research is not mainstreamed in National Agricultural Research System in India and state-owned monolithic extension system is not prepared to deal with the need of small farms in diverse ecosystems (Glendenning et al. 2010). This leads to poor adoption of technology and large yield gap (Aggarwal et al. 2008). Extension support and policy intervention also suffers due to lack of informed decision by public extension. Although there has been experimentation with reorganized and decentralized systems of technology assessment and refinement and revitalized public extension systems, development of sound analytical tools for targeting extension has remained undermined till date. A sound methodology for profiling farm typologies will help in rapid transfer of appropriate technology, precise extension support and development of policy environment sensitive to immense diversity of smallholder farms in coastal saline India.

Macro-economic perspective on this systematic problem leads us to the examination of efficiency in agricultural research and extension system. Ruttan and Hayami (1984) cites the shortcomings of 'induced innovation' model (Hayami and Ruttan 1981) that considers factor prices and user demand to induce the development of appropriate technology and the widely cited 'diffusion of innovations' theory (Rogers 1962, 1995) that assumes diffusion to have taken place in an institutional vacuum. Unlike the neoclassical theory in economics, New Institutional Economics recognized the importance of institutions in explaining uneven performance of economies (North 1997) and suggested that this type of analysis can lead to the necessary institutional developments vis-à-vis technical developments generated by R&D systems (Kydd 2002). Within the extension science literature, responses could be traced in the form of Agricultural Knowledge and Information System (AKIS) concept (Engel 1995) that sees agricultural innovation as a function of multi-stakeholder process. Roling (1988) departed from Rogers' theory towards a precise technological targeting and drew on the marketing research tradition (Kotler 1986) for the same.

Increasing focus on the role of institutions in differential economic growth has now added new dimension to the study of technology transfer and appropriate technology (Hall and Yoganand 2004; Radulovic 2005). New Institutional Economics (NIE) that makes explicit considerations of institutional arrangement and institutional environment (North 1997; Williamson 1990), are now frequently drawn on in the literature on efficient extension mechanisms (that reduces transaction cost and enhances economic efficiency) (Birner and Anderson 2007; Kherallah and Kirsten 2002). However, the application of NIE was mostly focused on the budgetary constraints and institutional evolution, markets, and institutional development (Pal et al. 2003). Methodology for identification of farm types for precise technological intervention and policy support, if mainstreamed, may be beneficial to the institutional arrangement that reduces transaction cost in smallholder agricultural production by providing appropriate technology to the smallholders rapidly. This is even more important in the regime of open economy where smallholders will have to be served with less transaction cost for long-term economic and environmental sustainability.

The present study was undertaken to identify the predominant farm types in coastal agro-ecosystem of India and to characterize them by some important socio-economic indicators. The article demonstrates the methodology of farm typology study when farming systems are heterogeneous and in need of appropriate technology for agricultural sustainability.

Methods

Area of the study

West Bengal is an eastern state of India and one of the major players in agricultural production. It has a high population density (976 per sq km) and varied farming systems. Agriculture is the mainstay of rural livelihoods and nearly 90% of the cultivators are small and marginal farmers jointly holding 84% of the State's agricultural lands. Another 3 million landless families have earned the right to cultivate and grow crops on their own land after enactment of Operation Barga system (land reform movement) by the State Government (GoWB 2005). The high population density has resulted in unabated fragmentation of cultivable land and insufficient farm size to sustain livelihoods. Moreover, increase in the price of agricultural inputs, uncertain price of perishable agricultural produce, inadequate market infrastructure, distress sale of produce by small and marginal farmers are some of the problems that pose serious challenges to sustainability of farm sector in the state. The socio-economic condition of the farming community is declining in the absence of appropriate technical, social, financial and market interventions (NABCOM 2009).

Based on climate, soil and physiography, there are six agro-climatic Zones for West Bengal. Among these, South 24 Parganas district comes under Coastal Saline Zone (Figure 1) (Gajbhiye and Mandal 2008), one of the biggest Zones with high cropping intensity (143%) i.e. the ratio of gross and net cultivated area in percentage terms, and covering part or whole of 6 districts of West Bengal. Climate is tropical moist sub-humid with 1796.2 mm rainfall, and wide range of air temperature (maximum 35.0°C, minimum 15.6°C). Rice is the main crop grown over different land terrains and seasons. Aus (spring paddy), sesame and green gram in pre-kharif (early wet season), jute and *aman* rice in

Figure 1 Study locations in South 24 Parganas district of West Bengal, India (shown by red circles).

kharif (wet season) and wheat, different oilseeds and pulses, and potato in rabi (winter season) are important crops. South 24 Parganas (22°32′N to 22.53°N & 88°20′E to 88.33°E) has an area of 8165.05 km^2 with population of 81,53,176 (sixth highest in India), of which 74.39% stay in rural areas. Percentage of households below the poverty line is 37.21, much higher than the state and country average (26% and 29%). The district is one of the poor districts of West Bengal having large number of resource poor farmers. Natural resource is fragile and highly prone to degradation.

Joynagar II, Baruipur, Basanti and Gosaba, the sampled Community Development Blocks, are distributed across the middle and eastern part of the district (Figure 1). Percentage of cropped and irrigated area of the district was 393.47 thousand ha and 98.31 thousand ha in 2006-07. Net cropped area of the district is 3,72,290 hectares, decreasing slowly due to the conversion of agricultural land for other purposes. Production of food grain, pulse and oilseed, vegetables and fruits has registered a growth in recent years (GoWB 2005).

The district is characterized by complex, diverse and risk-prone ecosystem, which is perturbed by extreme climatic events, water logging in low lands for considerable part of the year and soil salinity in winter and summer months resulting in dearth of appropriate technology for the district. Population density has risen rapidly over the last decade, with 819 people per km^2 arable land (2011 Census) and at a growth rate of 18.05% between 2001 and 2011 (GoWB 2005). Sustainable use of natural resources in the face of high population density and ecosystem fragility is critically important for the region. Overexploitation of natural resources is overwhelming in this district. Farming has diversified greatly over recent years (Ghosh and Kuri 2005) and technological intervention with appropriate technology has become critically challenging to the extension system. That is why classification of farming systems is a pragmatic starting point for facilitating appropriate technology in this region.

Sampling[a]

Following the guidelines of Indian Agricultural Statistical Research Institute, data collection was organized using a multi-stage random sample survey method (Sukhatme et al. 1984). Four blocks (namely Joynagar II, Baruipur, Basanti and Gosaba) were randomly selected from 29 blocks of South 24 Parganas district. Then, three villages from each of the block and 12 farmers from each of the village were selected randomly. Care was taken that four farmers each were selected from marginal-small (2.50-7.50 bigha), medium (7.50-12.00 bigha) and large (>12.00 bigha) categories (1 bigha = 0.13 ha). A quick survey of these 144 households was conducted with a focus on socio-economic information and income of the farms from different farm enterprises. This data was used for identification of predominant farm types for the region. After the identification of predominant farm types, half of the members from each of the identified farm types were selected randomly and were interviewed in details for eliciting additional socio-economic and farm management related information (to be used for characterization). Thus, a total of 72 farmers were interviewed in details for characterization of farm types. This helped us to avoid detailed and time-consuming interviewing with all 144 respondents.

Data

Instead of conducting a conventional way of identifying and characterizing farming system based on the source of the highest gross income received (from crops, vegetables, fruits and orchard, livestock, dairy, fishery, and poultry), the study employed multivariate statistics for identification of farm types and their characterization. Data were collected in two phases as mentioned in Sampling section.

Based on an extensive literature survey (Dixon et al. 2001; Fliegel et al. 1968; Mohammad 1978; Singh et al. 1993), initial case explorations in the field and expert counsel, a structured interview schedule was developed for the study. Apart from socio-economic parameters of the households, farm size, infrastructural facilities, information on cost of cultivation and yield of different crops, and price received by the farmers were included in the data collection instrument. The draft interview schedule was then pre-tested on 20 non-sampled respondents to incorporate necessary modifications. Description and measurement of some of the variables used in the study is given in Table 1.

Data were collected during May-June 2012 through personal interview method from 144 sampled households for quick survey in the first phase followed by a detailed survey on 72 respondents in the second phase. The dataset was prepared and analyzed using SPSS 17 (SPSS Version 17, SPSS Inc., Chicago, Illinois) and XLSTAT 2012 software (Version 2012, Addinsoft, Inc., Brooklyn, NY, USA).

Table 1 Selected variables used in farm type identification and characterization

Variable	Description
Education Index	Weighted average of formal education received by the household members (illiterate = 0, primary = 1, middle = 2, secondary = 3, higher secondary = 4, graduate = 5 and post-graduate = 6).
Crop Diversification	Measured by Herfindahl Index; index varies from 0 to 1, '1' - complete specialization and '0' - perfect diversification (Malik and Singh 2002)
System Cost of Cultivation	Cost of all inputs used (in Indian Rupees), for all enterprises of the farming system including the cost of family labor
System Gross Return	Total revenue earned (in Indian Rupees) from all the components of the farming system
System Net Return	System Gross Return – System Cost of Cultivation
Cost Benefit Ratio	Gross return from sale of output/Total cost of input used

Statistical analysis

Both univariate descriptive statistics and multivariate statistical techniques were employed for the analysis of data. Multivariate statistical techniques have been widely used for farm typology and characterization study (Kobrich et al. 2003; Andersen et al. 2009; Guto et al. 2010). These statistical techniques allow us creating farm typologies, particularly when an in-depth database is available. The combination of principal component analysis for necessary reduction of the number of variables followed by cluster analysis to identify typical farm households was used. Similar examples may be drawn from Gebauer (1987) in Germany that employs cluster analysis; Jolly (1988) in Senegal that employs canonical-discriminant analysis for identification of recommendation domains; Orr and Jere (1999) in Malawi employing cluster analysis followed by profiling through discriminant analysis; Hardiman et al. (1990) in China using cluster analysis; Köbrich et al. (2003) in Chile and Pakistan and Usai et al. (2006) in Sardinia, both using principal component analysis and successive cluster analysis.

For the present study, farm household typologies were identified by using two sequential multivariate statistical techniques: principal component analysis (PCA) and cluster analysis (CA) (Ding and He 2004). PCA condenses the information from the interdependent variables to a smaller set of factors (Jolliffe 2002; Abdi 2007). We used both socio-economic variables and income from different crop enterprises in the PCA. Since the impact of off-farm income on technology adoption is well-reported, the present study also used off-farm income as a factor for classification (Savadogo et al. 1998; Nehring et al. 2005; Fernandez-Cornejo et al. 2005).

Factors were identified using orthogonal rotation (varimax method as in Kaiser 1970; Gorsuch 1983) so that a smaller number of highly-correlated variables might be put under each factor and interpretation becomes easier (Field 2005). In accordance with Kaiser's criterion, all factors exceeding an eigenvalue of one were retained (Kaiser 1970). Kaiser's criterion is accurate when the number of variables is less than 30 (Field 2005), which was the case for our data set.

The sampled farms were clustered based on the five principal components identified by PCA. Cluster analysis is a collective term covering a wide variety of techniques for defining natural groups or clusters in data sets (Anderberg 1973). These groups are relatively homogeneous within themselves and heterogeneous between each other based on a defined set of variables. In a word, it is the art of finding groups in data (Kaufman and Rousseeuw 1990). Entities within a certain group or cluster are expected to be very similar to each other, and entities belonging to different classes are expected to be dissimilar (Bidogeza et al. 2009).

To determine the number of clusters two steps were followed for clustering – the hierarchical method and K-means clustering method. For hierarchical clustering, Euclidian distance and Ward's computation method was considered. The number of clusters retained from Ward's method (four in our study) was used as starting values in the K-means method. Accordingly, the number of clusters that seemed most realistic and meaningful was chosen for the final solution.

In addition to CA, a one-way analysis of variance test was performed. The test allowed us to identify the differences in variance between clusters in terms of PC scores (Field 2005). Thus, the variables that explained the largest differences between clusters could be identified.

Results and discussion

Description of the farming systems

Four major types of farms, based on the source from which maximum gross income was earned by the farmers, were identified. These were – rice-based farming system (34 households), vegetable-based farming system (70 households), fishery-based farming system (10 households) and farming system based on off-farm income (20 households). The farm size wise distribution of these farm types in four study Blocks has been reported in Figure 2.

A general observation revealed that large proportion of farmers followed rice-based farming systems, perhaps to grow rice for self-consumption and not for sale. On the other hand, medium farmers followed vegetable-based farming systems and marginal-small farmers relied on off-farm income sources. This might be due to the pressing need for income from their small holdings. Medium farmers could allocate enough land for vegetables after growing rice for subsistence purposes. However, vegetable-based

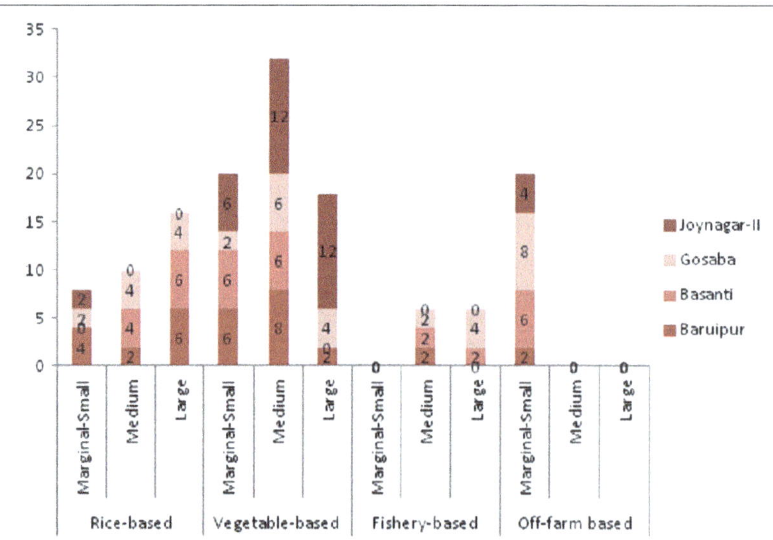

Figure 2 Farmer category wise sub-farming systems in four study blocks of South 24 Parganas district.

farming systems were found across farm sizes, perhaps for their high economic return. Chili, tomato, brinjal, ladies finger, and cole crops were common in this part of the state (Basu et al. 2009; Halder and Das 2012). Fishery-based farming systems were found among medium and large farmers only, who could afford to spare land for water bodies, in case of dug pond, and invest in commercial fishery. Spatially, most of the vegetable-based systems were found in Joynagar-II Block followed by Baruipur Block. In Basanti and Gosaba Block, all types of farm were found. A considerable proportion of these farms earned large off-farm income.

Had we used the conventional method for identifying farm types, i.e. classification based on farm size or income sources of the farm, the diversity of farming systems would have been grossly underestimated. For example, it is difficult to understand the importance of fishery (only 10 households) and animal husbandry (only four households, not featured in Figure 2) in different farming systems if we go by the numbers in Table 2 only.

In all the four study blocks, a large number of sub-farming systems were found. These sub-farming systems were determined by unique combination of farm enterprises (Table 2). The label of enterprises in a given sub-system was given according to their contributions to the gross farm income. For example, 'Rice + Jute + Vegetables' sub-system denoted that Rice contributed highest to the farm income, followed by Jute and Vegetables. Number of sub-farming systems in Joynagar-II, Baruipur, Basanti and Gosaba Blocks were 18, 16, 16 and 18 respectively. This huge diversity in sub-farming systems (68 in no.) was inherent to the farms of this region and they ideally required specific extension interventions. However, in practice, it was difficult to address the needs of all these 68 sub-farming systems.

Since these sub-systems are huge in number and farm size or enterprise-based classification is oversimplification of farm heterogeneity, we followed an alternative methodology for identification of farm types and their characterization by employing a combination of PCA and CA with the derived principal components. The results of these multivariate analyses are described below.

The principal component analysis

Table 3 shows rotated factor (Varimax) matrix of independent variables with factor loadings for each variable. The communality column shows the total amount of variance of each variable retained in the factors. For the interpretation of the PCs, variables with high factor loadings and high communality were considered from the rotated factor matrix (Harris 2001). In total, 17 variables were included in PCA, of which 5 principal components with eigenvalues greater than 1 were retained for further analysis. These five PCs explained 73.46% of total variability in the dataset. A closer look at each column of Table 3 helps us to define each component according to the strongly associated variables. The first component explains 27.78% variance and is correlated with the land holding, owned land, land fragment and income from rice crop. Thus the component represents land entitlement and income from rice crop. Principal component 2, 3, 4 and 5 explain 15.87%, 11.02%, 9.59% and 8.20% variance respectively. That means the first two factors have more importance in explaining the variance in dataset than the other three factors combined. PC 2 is correlated with crop

Table 2 Predominant sub-farm types under each farm types in the study blocks; sub-farm types are arranged according to the decreasing contribution of an enterprise towards gross farm income

Farming systems	Sub-farming systems			
	Joynagar-II	**Baruipur**	**Basanti**	**Gosaba**
Rice-based	Rice + Off-farm + Vegetables + Livestock + Pulses + Fishery + Fruit + Poultry	Rice + Vegetables + Off-farm + Livestock + Fishery + Poultry	Rice + Vegetables + Off-farm + Livestock + Fishery + Fruit + Goatery + Poultry	Rice + Vegetables + Off-farm + Livestock + Fishery + Goatery
		Rice + Vegetables + Fishery + Livestock + Plantation + Fruit + Poultry	Rice + Vegetables + Fishery + Livestock + Goatery + Plantation + Fruit	Rice + Vegetables + Off-farm + Fishery + Livestock + Fruit
		Rice + Poultry + Vegetables + Fishery + Plantation + Livestock	Rice + Vegetables + Off-farm + Fishery + Livestock + Fruit + Goatery	Rice + Fishery + Vegetables + Livestock + Goatery + Fruit + Poultry
		Rice + Vegetables + Off-farm + Livestock + Fishery + Tree + Poultry	Rice + Vegetables + Off-farm + Fishery + Livestock + Fruit	Rice + Off-farm + Fishery + Livestock + Goatery + Poultry
		Rice + Fruit + Off-farm + Vegetables + Livestock + Fishery + Plantation	Rice + Vegetables + Off-farm + Fishery + Livestock	Rice + Livestock + Vegetables + Fishery + Farm Machinery + Goatery + Poultry
		Rice + Vegetables + Off-farm + Fruit + Livestock + Fishery		
Vegetables-based	Vegetables + Rice + Livestock + Pulses + Fishery + Goatery + Tree + Poultry	Vegetables + Rice + Off-farm + Fishery + Livestock + Oilseed	Vegetables + Rice + Fishery + Off-farm + Livestock + Fruit + Goatery + Poultry	Vegetables + Rice + Livestock + Fishery + Goatery + Fruit crop
	Vegetables + Rice + Pulses + Fruit + Fishery + Livestock + Poultry	Vegetables + Rice + Fruit + Livestock + Fishery + Goatery	Vegetables + Rice + Off-farm + Fishery + Fruit + Livestock + Plantation + Goatery	Vegetables + Rice + Livestock + Fishery + Fruit + Goatery + Poultry
	Vegetables + Rice + Fruit crop + Pulses + Livestock + Oilseed + Fishery + Plantation + Poultry	Vegetables + Rice + Fruit crop + Fishery + Livestock	Vegetables + Fishery + Rice + Livestock + Poultry + Fruit	Vegetables + Rice + Fishery + Fruit crop + Fishery + Livestock
	Vegetables + Rice + Fruit + Fishery + Goatery + Pulses + livestock	Vegetables + Off-farm + Rice + Livestock + Goatery + Poultry	Vegetables + Rice + Fishery + Plantation + Goatery + Fruit + Poultry	Vegetables + Rice + Fishery + Livestock + Tree + Poultry
	Vegetables + Rice + Fruit + Pulses + Fishery + Livestock + Oilseed + Plantation	Vegetables + Poultry + Fruit + Rice + Livestock + Fishery + Pulses	Vegetables + Rice + Off-farm + Fishery + Livestock + Poultry	Vegetables + Rice + Off-farm + Livestock + Fishery + Goatery

Table 2 Predominant sub-farm types under each farm types in the study blocks; sub-farm types are arranged according to the decreasing contribution of an enterprise towards gross farm income *(Continued)*

	Vegetables + Rice + Pulses + Fruit + Fishery + Livestock + Tree + Poultry	Vegetables + Rice + Fishery + Goatery + livestock	Vegetables + Off-farm + Rice + Fishery + Livestock + Poultry	Vegetables + Rice + Fishery + livestock + Poultry + Fruit crop
	Vegetables + Rice + Fruit + Plantation + Pulses + Fishery + Livestock	Vegetables + Rice + Fruit + Fishery + Livestock + Goatery		
	Vegetables + Rice + Oilseed + Goatery + Fruit + Fishery + Livestock + Poultry	Vegetables + Rice + Off-farm + Goatery + Fishery + Tree + Fruit + Poultry		
	Vegetables + Rice + Fruit + Fishery + Pulses + Plantation + Livestock			
	Vegetables + Rice + Fishery + Goatery + Pulses + Plantation			
	Vegetables + Rice + Livestock + Pulses + Fishery + Tree + Fruit crop			
	Vegetables + Rice + Pulses + Fruit + Livestock + Plantation + Fishery + Oilseed			
	Vegetables + Rice + Livestock + Fishery + Tree + Goatery + Poultry			
	Vegetables + Fishery + Rice + Fruit + Pulses + Plantation + Livestock + Cotton			
	Vegetables + Rice + Off-farm + Goatery + Fishery + Livestock + Pulses + Plantation			

Table 2 Predominant sub-farm types under each farm types in the study blocks; sub-farm types are arranged according to the decreasing contribution of an enterprise towards gross farm income *(Continued)*

Fishery-based		Fishery + Vegetables + Cereal + Fruit + Livestock + Oilseed	Fishery + Cereal + Livestock + Goatery + Poultry	Fishery + Vegetables + Rice + Livestock + Goatery + Fruit + Poultry
			Fishery + Cereal + Vegetables + Livestock + Poultry	Fishery + Rice + fruit crop + Livestock + Poultry
				Fishery + Rice + Vegetables + Off-farm + Fruit crop + Livestock + Farm Machinery + Tree + Poultry
Off-farm based	Off-farm + Cereal + Vegetables Livestock + Fruit + Poultry + Fishery + Tree	Off-farm + Cereal + Livestock + Fishery + Poultry	Off-farm + Rice + Livestock + Plantation + Tree + Fishery + Fruit	Off-farm + Rice + Fishery + Livestock + Poultry
	Off-farm + Vegetables + Cereal + Livestock + Fishery + Tree + Poultry		Off-farm + Rice + Livestock + Fishery + Fruit + Poultry	Off-farm + Rice + Livestock + Poultry + Goatery + Fruit
			Off-farm + Rice + Livestock + Vegetables + Fishery + Poultry + Fruit	Off-farm + Vegetables + Rice + Livestock + Poultry + Fishery + Tree
				Off-farm + Rice + Livestock + Vegetables + Fishery + Poultry

Table 3 Four principal components derived by principal components analysis with loadings for individual crop enterprises and percent cumulative variance explained

	PC 1	PC 2	PC 3	PC 4	PC 5	Communality
Family Size	0.33	0.25	0.55	0.34	0.16	0.62
Land Holding	0.89	0.15	0.19	0.14	0.10	0.88
Own Land	0.69	0.16	0.14	0.19	0.17	0.60
Land Fragment	0.80	0.04	0.02	0.05	0.04	0.65
Crop diversification Index	0.22	−0.69	−0.07	0.26	0.05	0.61
Income from Rice	0.84	−0.12	0.09	−0.15	0.18	0.79
Income from Pulses	0.10	0.78	−0.16	0.34	−0.09	0.77
Income from Oilseed	−0.03	0.59	0.19	−0.06	0.33	0.54
Income from Vegetables	0.46	0.25	−0.21	0.47	0.20	0.56
Income from Fruits	0.37	0.75	0.14	0.04	−0.17	0.74
Income from Tree	−0.27	−0.10	−0.09	0.10	−0.47	0.53
Income from Plantation	0.28	0.63	−0.08	0.30	0.06	0.58
Income from Cattle	0.25	0.12	−0.07	0.10	−0.55	0.63
Income from Poultry	0.09	−0.04	0.92	0.00	0.01	0.86
Income from Fishery	0.41	−0.15	0.65	−0.02	−0.14	0.70
Income from Goatery	0.01	−0.15	−0.02	0.19	0.85	0.77
Off-farm income	0.07	−0.13	−0.19	−0.83	−0.08	0.79
Eigen values	4.72	2.36	1.36	1.29	1.05	
Cumulative explained variance	27.78	43.66	54.68	63.26	73.46	

diversification index and income from pulse, oilseed, fruits and plantation, representing crop and income diversification. PC 3 is correlated with income from poultry, fishery and family size, embodying the income from poultry and fishery enterprises driven by family labor. PC 4 represents off-farm income and PC 5, correlated with income from goatery and cattle, represents livestock enterprises. Thus the five principal components could be names as 'land ownership and income from staple food' (PC 1),'farm diversification' (PC 2), 'income from poultry and fishery' (PC 3), 'off-farm income' (PC 4), and 'income from livestock' (PC 5).

The cluster analysis

The first five factors from the principal component analysis were used for hierarchical clustering using Euclidean Distance as distance measure and Ward's technique as agglomerative clustering. The agglomeration schedule, resulting from this analysis, illustrated the sequence of analysis and produced coefficients. The objective of generating such schedule is to arrive at an appropriate number of clusters that best fit the data set. A check of the agglomeration schedule and Scree Diagram suggested that the number of clusters should be four. It was then checked that the number of retained clusters is realistic with respect to the field observation in order to be accepted as a meaningful classification. Thus, four clusters were found to be appropriate, as these seemed to be most representative of farm households within the study blocks. The k-means clustering method was employed with 4 numbers of clusters. The analysis produced final cluster centres (Table 4), and distribution of farms within a cluster among study blocks (Table 5) and size of farms (Table 6).

Table 4 Final cluster centres for all four clusters identified through K-means clustering method

Principal components	Clusters			
	1	2	3	4
PC 1 (Land ownership and income from cereals)	0.84	0.23	−0.31	0.53
PC 2 (Farm diversification)	−0.09	0.87	−0.17	0.53
PC 3 (Income from poultry and fishery)	3.58	0.10	−0.19	0.50
PC 4 (Off-farm income)	−0.03	−0.20	−0.72	0.35
PC 5 (Income from livestock)	−0.00	−0.50	−0.07	2.14

Final cluster centers interpret what is typical for a particular cluster. Cluster 1 is characterized by relatively higher income from poultry and fishery enterprises, has large holdings, and earns high from rice crop. Cluster 2 is characterized by higher diversification of farm income and Cluster 3 is characterized by smaller farms with higher off-farm income. Cluster 4 represents farms with higher income from livestock and other diverse sources such as rice, vegetables, poultry and fishery (Table 4). In all clusters, the contribution of several important crops and off-farm income could be observed.

Figures 3 & 4 (A, B) represents the observations and variables simultaneously in the two dimensional space. The individual farms have been labeled with study blocks (Figure 3A & B) and size category of farms (Figure 4A & B). The A series (3A & 4A) depicts the distribution of observations in a space defined by two principal components, after varimax rotation. The B series (3B & 4B) depicts the same along with the initial variables plotted in the form of vectors. Farms clustered (circled) together represent a farm type. A visual check suggests one group at the right and lower quadrant of the figure (Figure 3A), another clustered at the left lower quadrant, which is closer to the axis. The third one is also located at the left lower quadrant, but away from vertical axis. The scattered farms of upper right and left quadrants form the fourth group. A check of these Figures (3 & 4) in combination with Tables 5 and 6 helps us to understand this clustering better.

Noticeably, type of farms is not specific to a study block. The labels '1', '2', '3' and '4' are scattered throughout the plot. However, in some places similar labels are clustered signifying similar farms within a Block. The findings from Figure 4A are even more interesting. Small and marginal farms labeled as '1' are clustered together signifying their similarity in terms of a host of factors. Large farms labeled as '2' are also found closely clustered. Instead, medium farms are scattered throughout the plot representing

Table 5 Distribution of farm types, identified by cluster analysis, in four study blocks of South 24 Parganas districts

	Farm types				Total
	Cluster 1	Cluster 2	Cluster 3	Cluster 4	
Joynagar-II	0 (0.00, 0.00)*	18 (36.00, 50.00)	10 (14.29, 27.78)	8 (57.14, 22.22)	36
Baruipur	8 (100.0, 22.22)	8 (16.00, 22.22)	18 (25.71, 50.00)	2 (14.29, 5.56)	36
Basanti	0 (0.00, 0.00)	8 (16.00, 23.53)	22 (31.43, 64.71)	4 (28.57, 11.76)	34
Gosaba	0 (0.00, 0.00)	16 (32.00, 44.44)	20 (28.57, 55.56)	0 (0.00, 0.00)	36
Total	8	50	70	14	142
Likelihood ratio	46.13	Sig. = .000			

*Figures in parentheses are percentages to column and row marginal respectively.

Table 6 Distribution of farm types, identified by cluster analysis, among three farmer categories in South 24 Parganas districts

	Farm types				Total
	Cluster 1	Cluster 2	Cluster 3	Cluster 4	
Marginal-small	0 (0.00, 0.00)*	4 (8.00, 8.33)	44 (82.86, 91.67)	0 (0.00, 0.00)	48
Medium	2 (25.00, 4.35)	26 (52.00, 56.52)	14 (20.00, 30.43)	4 (28.57, 8.70)	46
Large	6 (75.00, 12.5)	20 (40.00, 41.67)	12 (17.14, 25.00)	10 (71.43, 20.83)	48
Total	8	50	70	14	142
Likelihood ratio	67.089	Sig. = .000			

*Figures in parentheses are percentages to column marginal and row marginal respectively.

higher instrinsic variability of medium farms. The nearness of blue dots (individual farms) to the red dots (initial variables) shows the characteristics of those farms in terms of the variables (represented by red dots). A close look of both 3A and 4A provides better understanding of farm classification in the study areas.

It is observed that all the Cluster 1 farms (100%) are situated in Baruipur Bock (Table 5). Cluster 2 farms are predominant in Joynagar II (36.00%) and Gosaba (32.00%), Cluster 3 farms are distributed across the four study blocks and Cluster 4 farms are predominant (57.14%) in Joynagar II Block. The data suggest that Joynagar II Block is predominated by diversified farms, Baruipur and Basanti Blocks by smaller farms with higher off-farm income and Gosaba Block with diversified and smaller farms with off-farm income. The distribution of clusters across the study blocks differed significantly (Likelihood ratio sig. value = .00) indicating heterogeneous nature of farm types in these blocks.

From Table 6, it may be observed that marginal-small farms are mostly Cluster 3 farms (91.67%), i.e. smaller farms with higher off-farm income. Medium farms are Cluster 2 (56.52%) and Cluster 3 (30.43%) farms i.e. diversified farms and small farms

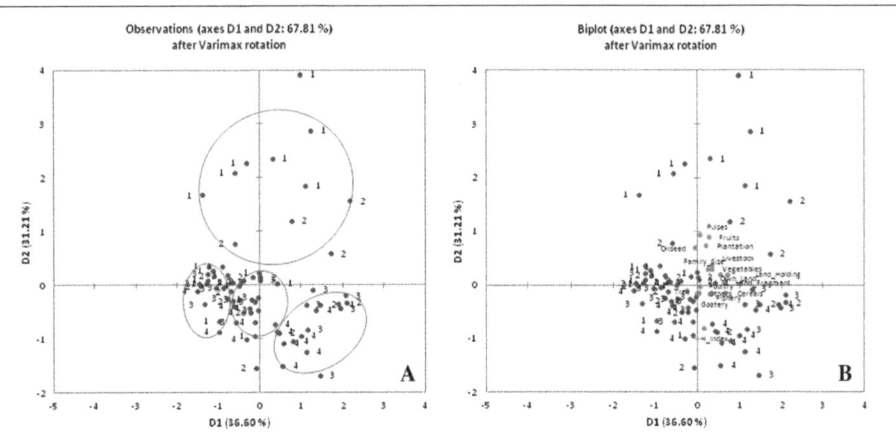

Figure 3 Distinct cluster of farms delineated by PCA axis and labeled by study blocks. A. Principal component analysis observations plot diagram of farming systems based on PCA of variables after varimax rotation. The numbers 1-4 on the plane indicate the community development blocks namely Joynagar II - 1, Baruipur - 2, Basanti - 3 and Gosaba - 4. The distance between the positions of the farms approximates the dissimilarity of their variables measured by Euclidean distance. Four circles/ovals represent four farm types. **B.** Ordination bi-plot diagram of farming system characterization based on PCA of variables. The numbers 1-4 on the plane indicate the community development blocks namely Joynagar II - 1, Baruipur - 2, Basanti - 3 and Gosaba - 4. Red dots represent variables used in PCA. The distance between the positions of the farms approximates the dissimilarity of their variables measured by Euclidean distance. Samples close to the origin have average values of a particular variable in a study sample.

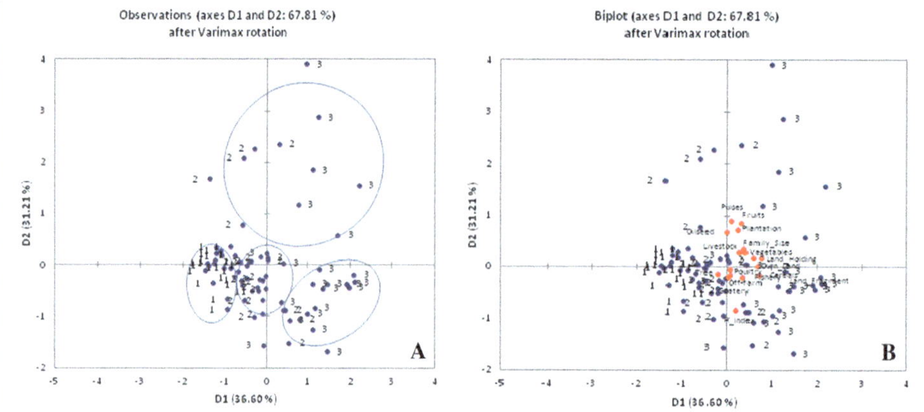

Figure 4 Distinct cluster of farms delineated by PCA axis and labeled by size category of farms. **A**. Principal component analysis observations plot diagram of farming systems based on PCA of variables after varimax rotation. The numbers 1-3 on the plane indicate the different farm sizes – Marginal-Small - 1, Medium – 2, and Large - 3. The distance between the positions of the farms approximates the dissimilarity of their variables measured by Euclidean distance. Four circles/ovals represent four farm types. **B**. Ordination bi-plot diagram of farming system characterization based on PCA of variables. The numbers 1-3 on the plane indicate the different farm sizes – Marginal-Small - 1, Medium – 2, and Large - 3. Red dots represent variables used in PCA. The distance between the positions of the farms approximates the dissimilarity of their variables measured by Euclidean distance. Samples close to the origin have average values of a particular variable in a study sample.

with higher off-farm income. Noticeably, large farms are distributed across all clusters, perhaps due to their higher flexibility in allocation of land resources among different crops/enterprises. Alternatively, Cluster 1 farms are mostly (75.00%) large farms, Cluster II farms are largely Medium (56.52%) and Large (41.67%) farms, Cluster 3 farms are mostly Marginal-small farms (82.86%) and Cluster 4 farms are Large farms (71.43%). The distribution of clusters across the farm size differed significantly (Likelihood ratio sig. value = .00) indicating heterogeneous nature of farm types across farm sizes.

In some parts of West Bengal including South 24 Parganas district, departure from traditional cropping pattern has been a trend for the last one and half decades (Goswami 2007; Bhattacharyya 2008) for enhancing income from fragmented land resources, which are challenged by abiotic stresses (e.g. salinity). The number of smallholders has in fact increased in West Bengal (De 2003) and there is evidence of increased crop diversity on fragmented lands in the state (De 2002; Bagchi et al. 2012). In many places, farmers have changed land shapes to grow vegetables in low lands (Basu et al. 2009). Introduction of farm-ponds and land-shaping has also resulted in high cropping and income diversity among small farmers (Bahu and Mishra 2001). Off-farm income has also become a burgeoning reality of rural India especially for the smallholders and this has often become the largest source of rural farm income. This also contributes to the investment in farming activity and asset creation (Mehta 2009).

To examine the authenticity of the cluster analysis, a one-way analysis of variance for each principal component (equality of group means) was performed with the four clusters (Table 7). The more distinct a PC value is among groups, the lower is the p-value. Judging from the p-values, all the factors namely 'land ownership and income from rice crop', 'income from poultry and fishery', 'off-farm income' and 'income from livestock' found to be significant in differentiating the clusters (p = 0.00 for all factors). However, factor 2 characterized by 'farm diversification' was not significantly differing across identified clusters, although the p value was not far from significant (p = 0.09).

Table 7 One-way analysis of variance of factor scores across four identified clusters

Factors	Cluster		Error		F	Sig.
	Mean square	df	Mean square	df		
PC 1 (land ownership and income from rice crop)	4.68	3	.92	138	5.10	0.00
PC 2 (farm diversification)	2.14	3	.98	138	2.20	0.09
PC 3 (income from poultry and fishery)	26.06	3	.46	138	57.25	0.00
PC 4 (off-farm income)	36.30	3	.23	138	156.12	0.00
PC 5 (income from livestock)	25.58	3	.47	138	54.95	0.00

Characterization of identified farm types

Table 8 helps us to understand the characteristics of the clusters in terms of background variables and economic performance indicators of the farming systems.

Cluster I accounts for 5.63% of farm households. The cluster is comprised of households having large land holding, large family size (representing joint extended families), relatively higher family education, and relatively higher crop diversification (Col. 1, Table 8). The households both lease out and lease in land, land are highly fragmented and situated at a high distance from the dwellings. In terms of economic performance indicators, this cluster is characterized by relatively high system gross return, higher cost of cultivation and system net return and relatively higher cost-benefit ratio. These households, on an average, secured high income from rice crop, poultry and vegetables followed by fruits and fishery. They depended wholly on agriculture with no off-farm income. These farms may be supported for technically sound intensification of agriculture with assured input and advisory services. Since these groups pursue a capital intensive diversified farming, access to credit is important for them. The farmers may be assisted in commercial production techniques with provision for building requisite capacity in farm and non-farm based entrepreneurship (Folmer et al. 2010). Export-oriented crops may be tried with these farmers with suitable incentive (Singh 2002). Since these farms are capital and management intensive, they are prone to become environmentally unsustainable in the long run. An environmental monitoring mechanism is needed to be in place in addition to an incentivized system in favour of sustainable farming (Dhiman and Thambi 2009).

Cluster II comprised of 35.21% of the farm households. The cluster members had relatively smaller land, smaller family size, relatively poor family education, and high crop diversification (Col 2, Table 8). These farms lease-in land and their fragmented lands are relatively nearer to their settlement. This cluster is characterized by low system gross return, low cost of cultivation, moderate system net return, but relatively better cost-benefit ratio. These households secured large proportion of income from vegetables, fishery, livestock and rice crop and may be grouped as diversified farms. These farms are management intensive, presumably supported by human labor, and pursue diversification in enterprises. Integrated farming with livestock and fishery will help in higher food and livelihood security without undermining farm resources (Prein 2002). This will require specialized extension support with a farming system approach, and institutional convergence for sound planning (Nabi 2008). Cash crops may be promoted for enhancing farm income with assured provision for credit and insurance (Fafchamps 1992). On account of the large proportion of such farms, collectivization might be a viable strategy for their market integration (Hazell 2005).

Table 8 Characteristics of identified clusters of farm households and p-value of one way analysis of variance (equality of group mean) for variables used in characterization of farm types

	Cluster 1 (N = 4)	Cluster 2 (N = 25)	Cluster 3 (N = 35)	Cluster 4 (N = 7)	P-value
	(1)	(2)	(3)	(4)	
Age (yrs)	50.50^{ab**}	51.68^{a}	46.03^{b}	53.86^{a}	0.00
Family size	9.50	6.72	5.60	7.29	0.00
Farm size (bigha)*	10.38^{a}	7.71^{bc}	5.28^{c}	9.65^{ab}	0.00
Own land (bigha)	6.88^{ab}	6.55^{b}	4.41^{b}	9.65^{a}	0.00
Leased in land (bigha)	3.50^{a}	1.16^{b}	0.87^{bc}	0.00^{c}	0.00
Leased out land (bigha)	0.28^{a}	0.12^{b}	0.07^{b}	0.36^{a}	0.02
Land fragment (no.)	4.25^{ab}	3.96^{b}	3.22^{c}	4.43^{a}	0.01
Distance from home	2.0^{a}	1.08^{b}	1.08^{b}	1.28^{b}	0.00
Education Index	2.27	1,92	1.92	1.97	0.20
Area of Rice (bigha)	8.00^{a}	6.16^{ab}	4.76^{b}	7.79^{a}	0.00
Area of Vegetables (bigha)	0.75^{b}	1.40^{a}	0.70^{b}	1.84^{a}	0.00
Area of oilseeds (bigha)	0.23^{b}	0.86^{ab}	0.09^{b}	1.43^{a}	0.00
Crop Diversification Index	0.29^{ab}	0.33^{a}	0.28^{b}	0.28^{b}	0.01
System Gross return (Rs.)	139640^{ab}	110980^{b}	66508^{c}	149550^{a}	0.00
System Cost of Cultivation (Rs.)	60391^{ab}	47228^{b}	30330^{c}	64218^{a}	0.00
System Net Return (Rs.)	79250^{a}	63755^{a}	34576^{b}	79449^{a}	0.00
Cost-Benefit Ratio	2.42	2.34	2.39	2.28	0.88
Income from Rice (Rs.)	37380^{a}	23165^{b}	20734^{b}	34563^{a}	0.00
Income from pulses (Rs.)	576^{c}	3538^{ab}	310^{bc}	4071^{a}	0.00
Income from Oilseed (Rs.)	1050^{b}	296^{b}	40^{b}	2657^{a}	0.00
Income from Vegetables (Rs.)	23610^{b}	42148^{a}	16817^{b}	46093^{a}	0.00
Income from Fruits (Rs.)	9700^{a}	6584ab	2986^{2}	5185^{b}	0.01
Income from plantation (Rs.)	1750^{b}	1856$^{\mathbf{b}}$	332^{c}	3614^{a}	0.00
Income from Livestock (Rs.)	5600^{ab}	6059^{a}	4738^{b}	6558^{a}	0.01
Income from poultry (Rs.)	50125^{a}	503^{b}	701^{b}	212^{b}	0.00
Income from Fishery (Rs.)	8250^{b}	18816^{a}	4995	15800^{a}	0.00
Income from goatery (Rs.)	1600^{b}	650^{c}	1145^{b}	6485^{a}	0.00
Off-farm Income (Rs.)	0^{c}	800^{c}	15171^{a}	2857^{b}	0.00

*1 bigha = 0.133 ha.
**Difference in mean among clusters denoted by different letters as found from Duncan Multiple Range Test. Highest mean value is denoted by a, followed by b and c when significant mean difference was found between clusters.

Cluster III comprises of 49.30% of the sampled farms. This cluster is characterized by small land holding, relatively poor family education, smaller family size and relatively low crop diversification (Col. 3, Table 8). These farms have started shifting to off-farm income to offset low income from farming. In terms of economic performance, this cluster is characterized by low system gross return, low cost of cultivation or low investment in farming, poor net return from farming, but higher benefit-cost ratio (since they invest less). The farm households earn most of their income from rice crop, vegetables and from off-farm sources. These are typical subsistence farms that have started to shift from traditional farming to non-farming vocations, often in the form of seasonal migration. Maintaining food security is a key policy goal for these farms, which may be achieved through sustainable and risk-averting cultivation of food grains (e.g. climate resilient agriculture) coupled with assured coverage under public distribution system for food grains (Thomas 2006). This might be supplemented with a comprehensive capacity building programme on non-farm enterprises, since these people are poor in human resources and exploited in the labor markets in nearer towns (Deb et al. 2002). Collectivization of small enterprises may follow this capacity building programme for their market integration.

Cluster IV comprises of 9.86% of the farm households. The cluster members had relatively larger land holding (owner cultivator), relatively larger family size, moderate family education, and moderate crop diversification (Col. 4, Table 8). They do not lease-in land, i.e. these farms represent owner cultivator who lease-out land occasionally. Land is highly fragmented. These households secure high income from rice crop, vegetables, and fishery and are characterized by high system gross return, high cost of cultivation, high system net return and relatively low benefit-cost ratio. The cluster members had little off-farm income. These farms are capital intensive, developed to meet growing food demand in the nearest town markets. Unlike Cluster I, these farms earn higher income from vegetables and fishery. These may be supported for intensive farming with improved technology. Since they specialize is in food grain and vegetables, assured marketing support and crop insurance are critical for these farms. Due to their available family labors, sustainable agriculture and agro-based enterprises also seem to be promising propositions (Bhattacharjee and Saravanan 2012).

This farm typology classification offers clear advantages over classifications based on farm size or agro-ecological characteristics. The farm types delineated are manageable in number and represent both socio-economic, resource ownership and management orientation of the farms. Farm size-based classifications undermine the huge diversity among size classes and agro-ecological classification and ignore socio-economic realities of the farms. At the same time, the farm classes are based on sound statistical procedures instead of size-based classifications and, hence, more acceptable to policymakers.

The four farm types differed among themselves in terms of all studied parameters including a string of background variables, income sources and economic indicators, showing the efficiency of the classification methodology employed. The farm types, however, did not differ in terms of education index and cost-benefit ratio. Increasing dependence on costly external inputs has perhaps rendered an increase in cost of cultivation, relative to system net return, irrespective of farm types. Education was not an important means to classify farms, as access and control over natural resource was more important in differentiating farm types.

Conclusions and implications

The implications of the study may be illustrated in practical and academic terms. While the practical focus is on institutional arrangement and policy interventions, the academic focus has been on methodological insights gained and its possible upscaling.

The methodological implication and scope for modification and upscaling

The study departs from conventional methodology of economic characterization of predominant farming systems (classification based on land holding or discussion with stakeholders with little statistical rigor) in terms of at least two dimensions. First, it uses multivariate statistics for identification of farm types, and second, it uses a string of economic and non-economic variables for such classification. This methodology has reduced the time of data collection significantly, as quick survey was employed for farm type identification and only half of the households were interviewed in detail for their economic characterization. In-depth and comprehensive classification of farms was possible resulting in a manageable number of farm types (four) that could be taken up for appropriate policy and technological intervention.

The methodology employed in this study can be modified under different situations of farm typology study. The variables used for PCA may be different, based on the nature of agro-ecosystem, nature of agriculture practiced, objective of classification (nutrient management, irrigation intervention, credit support) etc. Variables may also vary when the focus of farm characterization is different from economic characterization only (e.g. these may be energy efficiency, ecological sustainability). The implication would also reflect in the development of data collection instruments to be used in quick and detailed survey. The process of variable selection may involve stakeholder participation, thus promoting popular participation in the study. The multivariate approach (algorithm) followed for the farm type delineation may also be used for the development of software or web-based decision support system. Separate modules in the software may be incorporated with flexible set of indicators under each of them. This may be used by research and extension agencies after rigorous multi-locational field testing.

Proposed institutional environment for technology transfer

The results from the analysis suggested four major farm types in the study areas – (i) capital-intensive large diversified farms with high economic efficiency, (ii) management-intensive, medium, diversified farms with moderate economic efficiency, (iii) small, subsistence farms with high off-farm income and high economic efficiency, and (iv) large diversified owner-cultivated farms with relatively lower economic efficiency. Smaller farms have inclined towards non-farm income more than the medium and large farms. On the other hand, medium and large farms have diversified more than smaller farms with a departure from grain-based systems to vegetable-based farming systems. This asks for a holistic farm planning and extension intervention than confining efforts to technology transfer alone. India, along with a large number of countries having agrarian society, has entered the open economy regime and need to establish economic efficiency for their smallholder system (Birner et al. 2009). The extension system must precisely target agricultural inputs, advisory services, credit access and critical information for identified farm. The selection of beneficiaries for many public extension programs may also be guided by such farm typology.

To respond to the need of such comprehensive farm types, institutional innovation (development and/or convergence) often becomes necessary. In India, a nationwide revitalized extension system has just started functioning. It explicitly encourages such convergence in planning and execution of technology transfer programs. Linkage with existing donor supported and Ministry funded efforts might also lead to improved institutional environment. India is now experiencing a handful of flagship rural development programs that builds on constitutional provisions for basic services to its citizens and development of rural institutions (mostly in the form of rural collectives), and these may be efficiently tapped for effective transfer of agricultural technologies.

Extrapolation of the study outcome

The study does not necessarily claim to extrapolate the identified farm types to larger context. But, since the study locations are representative of coastal saline zone of India, this may be generalized for at least 10 Mha of cultivable lands with nearly 900 million people operating under this system. However, it is the methodology that is subject to generalization for even larger regions. India has almost 200 districts under complex agro-ecological systems covering a population of nearly 400 million. Integration of the methodology in the formal policy or in the form of web-based decision support tool may reach these people effectively. Need not to say, the same may be explored anywhere in the world using flexible and participatory indicator selection and their subsequent validation.

Endnote

[a]Logistic support during sampling and data collection was received from All India Coordinated Research Project on Integrated Farming System and academic research programme of Ramakrishna Mission Vivekananda University.

Competing interests

The authors declare that they have no competing interests.

Authors' contributions

RG designed research, analysed and interpreted data and prepared the manuscript, SC conceptualized and designed the research work, and BP collected data and drafted the manuscript with RG. All authors read and approved the final manuscript.

Acknowledgement

Logistic supports received from the All India Coordinated Research Project on Integrated Farming System and Academic Research Programme of Ramakrishna Mission Vivekananda University is duly acknowledged. Authors are also thankful to all the farmers who assisted and participated in the survey.

Author details

[1]Integrated Rural Development and Management Faculty Centre, Ramakrishna Mission Vivekananda University, Narendrapur, Kolkata-700103, INDIA. [2]Department of Agricultural Economics, Bidhan Chandra Krishi Viswavidyalaya, Mohanpur, Nadia-741252, West Bengal, INDIA.

References

Abdi H (2007) Discriminant Correspondence Analysis. In: Salkind NJ (ed) Encyclopedia of Measurement and Statistics. Sage, Thousand Oaks, CA

Aggarwal PK, Hebbar KB, Venugopalan MV, Rani S, Bala A, Biswal A, Wani SP (2008) Quantification of yield gaps in rainfed rice, wheat, cotton and mustard in india, Global theme on agro-ecosystems report # 43. ICRISAT, Hyderabad

Anderberg MR (1973) Cluster Analysis for Applications. Academic, New York

Andersen E (2009) Regional Typologies Of Farming Systems Contexts, seamless Integrated Project, Eu 6th Framework Programme. Available via http://www.seamless-ip.org/Reports/Report_53_PD4.4.3.pdf. Accessed 10 Apr 2012

Andersen E, Elbersen B, Godeschalk F, Verhoog D (2009) Farm management indicators and farm typologies as a basis for assessments in a changing policy environment. J Env Manage 82:353–362

Bagchi BD, Roy SB, Jaim WMH, Hossain M (2012) Diversity, spatial distribution, and the process of adoption of improved rice varieties in West Bengal. In: Hossain M, Jaim WMH, Paris TR, Hardy B (eds) Adoption and diffusion of modern rice varieties in Bangladesh and eastern India. IRRI, Philippines, pp 31–44

Bahu RR, Mishra P (2001) Farm ponds in Coastal Saline Zone of West Bengal: a techno-economic feasibility study. J Agric Eng 38:17–22

Basu D, Banerjee S, Goswami R (2009) Vegetable cultivation on ail (bund): a successful farmer-led technology transfer. Acta Hort 809:155–160

Bhattacharyya R (2008) Crop Diversification: A Search For An Alternative Income Of The Farmers In The State Of West Bengal In India. Available via http://kastoria.teikoz.gr/icoae2/wordpress/wp-content/uploads/articles/2011/10/012-2008.pdf Accessed 10 May 2013

Bhattacharjee S, Saravanan R (2012) Poverty In Natural Prosperity: Can Agriculture Bring The Renaissance In North-East India? In: Kashyap SK, Pathak A, Papnai G (eds) Saving Humanity Saving Humanity, Swami Vivekananda Perspective. Vivekanand Swadhyay Mandal. GBPUAT, Pantnagar, India, pp 260–274

Bidogeza JC, Berentsen PBM, de Graaff J, Lansink AGJMO (2009) A typology of farm households for the Umutara Province in Rwanda. Food Sec 1:321–335

Birner R, Anderson JR (2007) How To Make Agricultural Extension Demand-Driven? The case of india's agricultural extension, Discussion Paper 729. International Food Policy Research Institute, Washington DC

Birner R, Davis K, Pender J, Nkonya E, Anandajayasekeram P, Ekboir J, Mbabu A, Spielman D, Horna D, Benin S, Kisamba-Mugerwa W (2009) From best practice to best fit: a framework for designing and analyzing pluralistic agricultural advisory services worldwide. J Agric Edn Ext 15:341–355

Bozeman B (2000) Technology transfer and public policy: a review of research and theory. Res Policy 29:627–655

Briggeman BC, Gray AW, Morehart MJ, Baker TG, Wilson CA (2007) A new U.S. farm household typology: Implications for agricultural policy. Rev Agric Econ 29:765–782

Chambers R, Pacey A, Thrupp LA (1989) Farmer First: Farmer Innovation And Agricultural Research. Intermediate Technology Publications, London

Conroy C, Sutherland A (2004) Participatory technology development with resource-poor farmers: maximising impact through the use of recommendation domains, AgREN Network Paper no. 133. ODI, London

De UK (2002) Economics Of Crop Diversification. Akansha Pub, House, New Delhi

De UK (2003) Diversification of crop in West Bengal: a spatio-temporal analysis. Artha Vijnana 42:170–182

Deb UK, Rao GN, Rao YM, Slater R (2002) Diversification And Livelihood Options: A Study Of Two Villages In Andhra Pradesh, India. Department for International Development (DFID), United Kingdom

Dhiman SC, Thambi DS (2009) Ground water management in coastal areas. Bhu-Jal 24:69–73

Ding C, He X (2004) K-Means Clustering Via Principal Component Analysis. In: Proceedings Of The Twenty-First International Conference On Machine Learning. Banff, Canada. Available via http://ranger.uta.edu/~chqding/papers/KmeansPCA1.pdf. Accessed 16 Aug 2013

Dixon J, Gulliver A, Gibbon D (2001) Farming Systems and Poverty: Improving Farmers' Livelihoods In A Changing World. FAO, Rome

Ellis F (1993) Peasant Economics: Farm Households and Agrarian Development, 2nd edn. Cambridge University Press, Cambridge

Emtage N, Suh J (2005) Variations in socioeconomic characteristics, farming assets and livelihood systems of Leyte rural households. Ann Trop Res 27:35–54

Engel PGH (1995) Facilitating Innovation: An Action-Oriented Approach and Participatory Methodology To Improve Innovative Social Practice In Agriculture. Wageningen Agricultural University, Dissertation

Fafchamps M (1992) Cash crop production, food price volatility, and rural market integration in the third world. Am J Agric Econ 74:90–99

Fernandez-Cornejo J, Hendricks C, Mishra A (2005) Technology adoption and off-farm household income: the case of herbicide-tolerant soybeans. J Agric Appl Econ 37:549–63

Field A (2005) Discovering Statistics Using Spss For Windows, 2nd edn. Sage, London

Fliegel F, Kivlin JE, Sekon GS (1968) A cross-national comparison of farmers' perceptions of innovations as related to adoption behavior. Rural Sociol 33:437–449

Folmer H, Dutta S, Oud H (2010) Determinants of rural industrial entrepreneurship of farmers in West Bengal: a structural equations approach. Int Region Sc Rev 33(4):367–396

Frankenberger TR, Dewalt BR, McArthur HJ, Hudgens RE, Mitawa G, Rerkasem K, Finan T, Butler Flora C, Young N (1989) Identification of results of farming systems research and extension activities: a synthesis. FSRE Newsl 1:8–10

Gajbhiye KS, Mandal C (2008) Agro-Ecological Zones, Their Soil Resource and Cropping Systems. Available via http://www.agricoop.nic.in/Farm%20Mech%20PDF/05024-01.pdf. Accessed 29 Nov 2012

Ganpat W, Bekele I (2001) Looking For The Trees In The Forest: Farm Typology As A Useful Tool In Defining Targets For Extension. In: Lindner JR (ed) Proceedings of the 17th Annual Conference of the Association for International Agricultural and Extension Education. Baton Rouge, Louisiana. Available via http://agris.fao.org/agris-search/search.do?f=2002/TT/TT02001.xml;TT2002001006. Accessed 22 February 2014

Gebauer RH (1987) Socio-economic classification of farm household - conceptual, methodical and empirical considerations. Eur Rev Agric Econ 14:261–283

Ghosh BK, Kuri PK (2005) Changes in cropping pattern in West Bengal during 1970-71 to 2000-2001. IASSI Quart 24:39–56

Glendenning CJ, Babu S, Asenso-Okyere K (2010) Review Of Agricultural Extension In India. Are Farmers Information Needs Being Met? IFPRI Discussion Paper 01048. IFPRI, Washington DC

Gómez-Limón JA, Berbel J, Arriaza M (2007) MCDM Farm System Analysis For Public Management Of Irrigated Agriculture. In: Weintraub A, Bjorndal T, Epstein R, Romero YC (eds) Handbook On Operations Research In Natural Resources In International Series In Operations Research And Management Science. Páginas, Kluwer, New York, pp 93–114

Gorsuch RL (1983) Factor Analysis. Erlbaum, Hillsdale, NJ

Goswami R (2007) Understanding Farmer-To-Farmer Communication Within The Sustainable Rural Livelihood Framework. Dissertation, Bidhan Chandra Krishi Viswavidyalaya

Goswami R, Biswas MS, Basu D (2012) Validation of participatory farming situation identification: a case of rainfed rice cultivation in selected area of West Bengal, India. Ind J Tradition Knowl 11:471–479

Go WB (2005) District Statistical Handbook, South 24 Parganas. Bureau of Applied Economics & Statistics, Government of West Bengal, India

Guto SN, Pypers P, Vanlauwe B, de Ridder N, Giller KE (2010) Socio-ecological niches for minimum tillage and crop-residue retention in continuous maize cropping systems in smallholder farms of Central Kenya. Agron J 104:188–198

Halder JC, Das P (2012) Present status and futuristic view of horticulture in West Bengal. Geo-Analyst 2(1), Available via http://gswb.in/wp-content/uploads/2012/08/v2n1jully2012_1.pdf. Accessed 10 May 2013

Hall AJ, Yoganand B (2004) New Institutional Arrangements In Agricultural Research And Development In Africa: Concepts And Case Studies. In: Hall AJ, Sulaiman RV, Raina RS, Shambu Prasad C, Naik GC, Clark NG (eds) Innovations In Innovation: Reflections On Partnership, Institutions And Learning. International Crops Research Institute for the Semi-Arid Tropics/National Centre for Agricultural Economics and Policy Research, Andhra Pradesh/New Delhi, pp 105–134
Hardiman RT, Lacey R, Yang MY (1990) Use of cluster analysis for identification and classification of farming systems in Qingyang County, Central North China. Agric Syst 33:115–125
Harrington LH, Tripp R (1984) Recommendation domains: a framework for on-farm research. CIMMYT Economics Working Paper 02/84. Series: CIMMYT Economics Working Paper, CIMMYT, Mexico. Available via http://repository.cimmyt.org/xmlui/bitstream/handle/10883/1001/7013.pdf?sequence=1. Accessed 17 September 2013
Harris RJ (2001) A Primer Of Multivariate Statistics, 2nd edn. Academic, New York
Hayami Y, Ruttan VW (1981) Agricultural Development: An International Perspective. Johns Hopkins University Press, Baltimore, USA
Hazell PB (2005) Is there a future for small farms? Agric Econ 32:93–101
Jolliffe IT (2002) Principal Component Analysis. Springer Verlag, New York
Jolly CM (1988) The use of action variables in determining recommendation domains: senegalese farmers for research and Agric Admin Ext 30:253–267
Kaiser HF (1970) A second generation Little-Jiffy. Psychomelrika 35:401–415
Kaufman L, Rousseeuw PJ (1990) Finding groups in data: an introduction to cluster analysis. Wiley, New York
Khanna SS (1989) The Agro-Climatic Approach. In: Survey of Indian Agriculture. The Hindu, Madras, India, pp 28–35
Kherallah M, Kirsten JF (2002) The New Institutional Economics: applications for agricultural policy research in developing countries. Agrekon 41:111–134
Kobrich C, Rehman T, Khan M (2003) Typification of farming systems for constructing representative farm models: two illustrations of the application of multi-variate analyses in Chile and Pakistan. Agric Syst 76:141–157
Kostrowicki J (1977) Agricultural typology concept and method. Agric Syst 2:33–45
Kotler P (1986) Principles of Marketing, 3rd edn. Prentice-Hall, Englewood Cliffs, NJ
Kydd J (2002) Agriculture And Rural Livelihoods: Is Globalisation Opening Or Blocking Paths Out Of Rural Poverty? Agricultural Research and Extension Network (AgREN) paper no.122. Overseas Development Institute (ODI), London, UK
Lal P, Lim-Applegate H, Scoccimarro M (2001) The adaptive decision making process as a tool for integrated natural resource management: focus, attitudes and approach. Conserv Ecol 5:11
Leeson JY, Sheard JW, Thomas AG (1999) Multivariate classification of farming systems for use in integrated pest management studies. Can J Plant Sci 79:647–654
Mahapatra AK, Mitchell CP (2001) Classifying tree planters and non planters in a subsistence farming system using a discriminant analytical approach. Agroforest Syst 52:41–52
Malik DP, Singh IJ (2002) Crop diversification – an economic analysis. Ind J Agric Res 36:61–64
Mehta R (2009) Rural Livelihood Diversification and Its Measurement Issues: Focus India. Wye City Group on Statistics On Rural Development And Agriculture Household Income, Second Meeting, at FAO HQ, Italy, Rome, 11-12 June 2009. Available via www.fao.org/fileadmin/templates/ess/pages/rural/wye_city_group/2009/Paper_1_a1_Mehta-Wye-Grp-RurDiv-Final.doc Accessed 21 Jul 2012
Mohammad A (1978) Situation Of Agriculture, Food, And Nutrition In Rural India. Concept, New Delhi
Mwijage NA, de Ridder N, Baijukya F, Pacini C, Giller EK (2009) Exploring the variability among smallholder farms in the banana-based farming systems in Bukoba district, Northwest Tanzania. Afr J Agric Res 4:1410–1426
NABCOM (2009) State Agricultural Plan of Government of West Bengal. NABARD Consultancy Service Private Limited. Kolkata, India. Available via http://rkvy.nic.in/sap/wb/wb.pdf. Accessed 9 May 2013
Nabi R (2008) Constraints to the adoption of rice-fish farming by smallholders in Bangladesh: a farming systems analysis. Aquacult Econ Manage 12:145–153
Nehring R, Fernandez-Cornejo J, Banker D (2005) Off-farm labour and the structure of U.S. Agriculture: the case of Corn/Soybean farms. Appl Econ 10:633–650
North DC (1997) Prologue In The Frontiers Of New Institutional Economics. In: Drobak JN, Nye VCJ (eds) The Frontiers of The New Institutional Economics. Academic, New York
Ojiem J, Ridder N, Vanlauwe B, Giller KE (2006) Socio-ecological niche: a conceptual framework for integration of legumes in smallholder farming systems. Int J Agric Sust 4:79–93
Orr A, Jere P (1999) Identifying smallholder target groups for IPM in southern Malawi. Int JPest Manage 45:179–187
Pal S, Mruthyunjaya J, Joshi PK, Saxena R (2003) Institutional change in Indian agriculture. National Centre for Agricultural Economics and Policy Research. Indian Council of Agricultural Research, New Delhi
Pender J, Hazell P (2001) Promoting Sustainable Development In Less-Favored Areas, IFPRI Policy Brief, Focus 4. IFPRI, New York
Prein M (2002) Integration of aquaculture into crop–animal systems in Asia. Agric Syst 71:127–146
Radulovic V (2005) Are new institutional economics enough? Promoting photovoltaics in India's agricultural sector. En Policy 33:1883–1899
Reckling M (2011) Characterisation of Bean Farming System Across Farm Types in Northern and Eastern Rwanda.: Identification Of Potential Niches For Legume Technologies. Dissertation, Wageningen University, The Netherlands
Rogers EM (1962) Diffusion of Innovations, 1st edn. The Free Press, New York
Rogers EM (1995) Diffusion of innovation, 4th edn. The Free Press, New York
Roling N (1988) Extension Science: Information Systems In Agricultural Development. Cambridge University Press, Cambridge
Rowe EC, van Wijk MT, de Ridder N, Giller KE (2006) Nutrient allocation strategies across a simplified heterogeneous African smallholder farm. Agric Ecosyst Environ 116:60–71
Ruttan VW, Hayami Y (1984) Toward a theory of induced institutional innovation. J Dev Stud 20:203–223
Sanjeev MV, Venkatasubramanian V, Singha AK (2010) Farming Dystems Of North East India: Research And Development Strategies for KVKs. Zonal Project Directorate, Zone – III. Indian Council of Agricultural Research, Umiam, Meghalaya, India

Savadogo K, Reardon T, Pietola K (1998) Adoption of improved land-use technologies to increase food security in Burkina Faso: relating animal traction, productivity, and nonfarm income. Agric Syst 58:441–464

Senthilkumar K, Bindraban PS, de Boer W, de Ridder N, Thiyagarajan TM, Giller KE (2009) Characterising rice-based farming systems to identify opportunities for adopting water efficient cultivation methods in Tamil Nadu, India. Agric Water Manage 96:1851–1860

Singh S (2002) Contracting out solutions: political economy of contract farming in Indian Punjab. World Dev 30:1621–1638

Singh AK, Singh S, Venkatasubramanian V (1993) Relationship between farmers. Socio- personal traits and adoption of dairy innovations. Charion 22:84–87

Soule MJ (2001) Soil management and the farm typology: do small family farms manage soil and nutrient resources differently than large family farms? Agric Resour Econ Rev 30:179–188

Sukhatme PV, Sukhatme BV, Sukhatme S, Asok C (1984) Theory Of Sample Surveys With Applications. ISAS, IASRI, New Delhi

Thomas HC (ed) (2006) Trade Reforms And Food Security: Country Case Studies And Synthesis. Food and Agriculture Organisation of UN, Rome

Timothy WO (1994) Identifying target groups for livestock improvement research: the classification of sedentary livestock producers in western Niger. Agric Syst 46:227–237

Tittonell P, Vanlauwe B, Leffelaar PA, Shepherd KD, Giller KE (2005) Exploring diversity in soil fertility management of smallholder farms in western Kenya. II. Within-farm variability in resource allocation, nutrient flows and soil fertility status. Agric Ecosyst Environ 110:166–184

Tittonell P, Leffelaar PA, Vanlauwe B, van Wijk MT, Giller KE (2006) Exploring diversity of crop and soil management within smallholder African farms: a dynamic model for simulation of N balances and use efficiencies at field scale. Agric Syst 91:71–101

Tittonell P, Vanlauwe B, de Ridder N, Giller KE (2007) Heterogeneity of crop productivity and resource use efficiency within smallholder Kenyan farms: soil fertility gradients or management intensity gradients. Agric Syst 94:376–390

Tittonell P, Muriukid A, Shepherde KD, Mugendif D, Kaizzig KC, Okeyoa J, Verchote L, Coee R, Vanlauwea B (2010) The diversity of rural livelihoods and their influence on soil fertility in agricultural systems of East Africa – A typology of smallholder farms. Agric Syst 103:83–97

Usai MG, Casu S, Molle G, Decandia M, Ligios S, Carta A (2006) Using cluster analysis to characterize the goat farming system in Sardinia. Livest Sci 104:63–76

USDA, ERS (2000) ERS Farm Typology For A Diverse Agricultural Sector, Agriculture Information Bulletin Number 759. USDA, US. Accessed via http://ageconsearch.umn.edu/bitstream/33657/1/ai000759.pdf. Accessed 12 August 2012

Vanclay JK (2005) Using a typology of tree-growers to guide forestry extension. Ann Trop Res 27:97–103

Williamson OE (1990) A comparison of alternative approaches to economic organization. J InstTheor Econ 146:61–71

Zingore S, Murwiraa HK, Delvea RJ, Gillerb KE (2007) Soil type, management history and current resource allocation: Three dimensions regulating variability in crop productivity on African smallholder farms. Field Crops Res 101(3):296–305

8

Changes in the technical and scale efficiency of rice production activities in the Mekong delta, Vietnam

Diep Thanh Tung[1,2]

Correspondence: Thanh.T.Diep@agrar.uni-giessen.de
[1]Faculty of Economics, Laws and Foreign Languages, Tra Vinh University, 126, National Road 53, Ward 5, Tra Vinh, Vietnam
[2]Institute of Project and Regional Planning, Justus-Liebig-University, Giessen, Germany

Abstract

I used bootstrapping data envelopment analysis (DEA) to measure changes in technical and scale efficiency in rice production in the Mekong delta region, Vietnam. The data include sample production sets from 1998 and bi-annual updates from 2002 to 2010. Technical efficiency changed significantly over this period, showing an upward trend. On the other hand, increasing return to scale is dominant trend, which reflects the need to increase the rice production scale generally and expand land use particularly. Meanwhile, the government has been trying to prevent land consolidation activities and restrict production scales using quotas and agricultural land use duration rules.

Keywords: Data envelopment analysis; Technical efficiency; Bootstrap; Returns to scale; Rice production; Mekong delta

Background

The Mekong delta region is popularly considered the rice bowl of Vietnam, as it accounts for over 50% of domestic rice production and 90% of rice exports. Over the last decade, economic growth has led to dramatic changes in the rural features and rice production activities in the Mekong delta. During the period 1995 to 2011, rice production in the Mekong delta grew by 3.89%, increasing by 1.6 and 2.22% per annum in terms of area and yield growth respectively (own calculation based on GSO (2012)) (Figure 1).

In the same time period, the number of agricultural households in Vietnam has declined significantly, from 10.15 million households in 2001, to 9.15 and then 8.89 million households in 2006 and 2011 respectively (GSO, General Statistics Office of Vietnam 2001, GSO, General Statistics Office of Vietnam 2006a, GSO, General Statistics Office of Vietnam 2011). In the Mekong delta in 2011, the number of agricultural households was about 1.83 million, and these covered more than four million hectares of rice land (GSO, General Statistics Office of Vietnam 2012). The number of agricultural households declined consistently, and the amount of agricultural land available remained quite limited.

The Vietnamese government has tried to reduce the rate of land consolidation in the country - a key cause of landlessness - by imposing quotas on agricultural land use

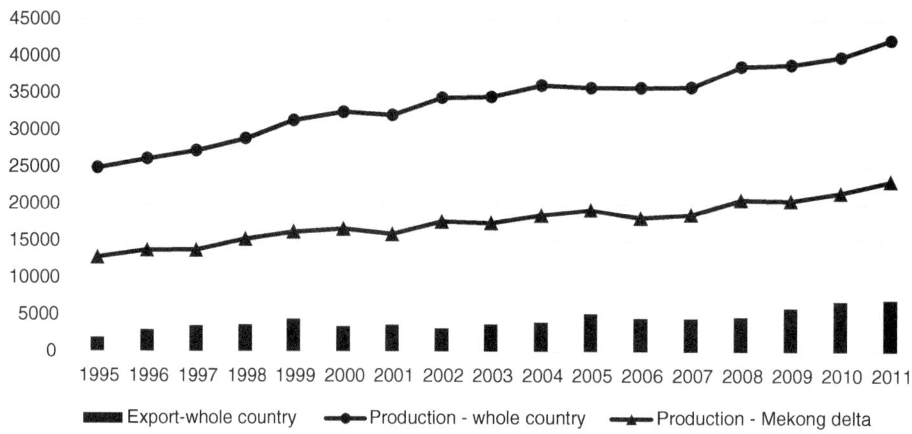

Figure 1 Rice production and exports (1000 tons). Source: Own summary from GSO, General Statistics Office of Vietnam (2012).

and limiting land use durations since the 1993 Land Law (Government of Vietnam 1993, 2003) was introduced. Article 70 of the Land Law stipulates that "*The quota on allocation to each family household or individual of land for planting annual crops, land for aquaculture and land for salt production shall be no more than three hectares of each type of land*", and specifies a twenty-year period for agricultural land use in the Mekong delta region (Government of Vietnam 2003). In 2007, this quota was increased to six hectares by Resolution No. 1126/2007/NQ-UBTVQH11 (The Standing Committee of National Assembly 2007), without any change being made to land use duration.

In practice, in 1994 11.71% of agricultural households had more than one hectare of agricultural land, but the figure increased to 15.08% by 2001 and 17.8% by 2006 (GSO, General Statistics Office of Vietnam 2001, GSO, General Statistics Office of Vietnam 2006a, GSO, General Statistics Office of Vietnam 2011). Akram-Lodhi (2001) showed that there were already 113,700 farms in excess of five hectares and 1,900 farms exceeding ten hectares by 1995. Even in some southern provinces and in the Mekong delta, it is still possible to find private farms exceeding of 1,000 hectares in size[a]. The upcoming issue is that the government's policies are limiting production scales in terms of both the area of land being used and its duration of use, and this may consequently affect the technical efficiency of rice production in the area.

Studies have analyzed technical efficiency in rice production on a nationwide scale, such as those by Minh and Long (2008), Khai and Yabe (2011), and Linh (2012). In addition, other studies used cross-section data to analyze the technical efficiency in the Mekong delta. For example, Huy (2009) analyzed the technical efficiency of 261 households during the 2006 winter-spring rice crop season, while Tuong (2010) used data from 200 households interviewed in 2010 to analyze growth rates in the area, as well as yield and production levels, and those factors affecting rice production. Thong et al. (2011) compared the economic efficiency of the summer-autumn and autumn-winter rice crop activities based on data taken from 479 households across four provinces of the Mekong delta. These studies were carried out in different years using different methods and a variety of sample sizes, making results hard to compare. Furthermore, estimates might be inconsistent due to sample size issues (Alirezaee et al. 1998; Andor & Hesse 2011; Staat 2001; Zhang & Bartels 1998).

Using a different approach, this study looked at the technical and scale efficiencies of rice production in the Mekong Delta over the period 1998 to 2010, using the same sample sizes and same study methods. Such a consistent analysis helped me to draw general conclusions and allowed me to measure changes across years in the Mekong delta's rice production activities. As a consequence, the results of this study may have policy implications for agricultural land management activities in the delta region.

Methods

Measuring technical efficiency and scale efficiency

The concept of technical efficiency was first proposed by Farrell (1957) based on the works of Debreu (1951) and Koopmans (1951). In his study, Farrell identified two forms of efficiency: *technical efficiency* and *allocative efficiency*. Technical efficiency reflects the ability of a firm to obtain maximum output based on a given set of inputs. Meanwhile, allocative efficiency measures the ability to use the optimal input set based on available prices and production techniques. Allocative efficiency is therefore also called *price efficiency*. Technical efficiency and allocative efficiency can be combined to measure the *economic efficiency* (or overall cost efficiency) of a firm.

Farrell illustrated the concepts above using the example of a firm whose work process has two inputs (x_1, x_2) and one output (q), and based on an assumption of constant returns to scale (CRS). The CRS assumption states that a given and proportionate increase in all inputs in the long run will result in an increase in outputs in the same proportion. Figure 2 shows technical efficiency based on an *input-oriented measure* which compares the actual production set (point P) and the fully-efficient production set (point Q), Q being a point which lies on the isoquant SS'. Meanwhile, in the study technical efficiency (TE) is measured using the ratio: $TE = 0Q/0P$, with the resulting TE values falling between zero and one. The distance QP represents technical inefficiencies, and inputs can be reduced without changing the outputs. If point P is able to move to point Q on the isoquant, the firm is said technical efficient and its TE score is unity.

In the case of the available price information, the isocost line AA' can be added to measure allocative efficiency (AE), which is calculated using the ratio: $AE = 0R/0Q$. By combining technical efficiency and allocative efficiency, one can generate the *overall cost efficiency*, using the following equation: $CE = TE \text{ x } AE = (0Q/0P) \times (0R/0Q) = 0R/0P$.

The analysis above illustrates Farrell's input-oriented measure of the ability of a firm to reduce inputs without changing outputs. An alternative approach, known as the *output-oriented measure*, measures the ability of a firm to increase outputs without changing inputs. In practice, the choice between input-oriented and output-oriented measures depends on whether the objective is input minimization (in which case use the input-oriented measure) or output maximization (use the output-oriented measure), without changing any of the other elements (FAO, Food and Agriculture Organization of the United Nations 2003).

The concepts mentioned above are based on an assumption of constant returns to scale. Although a firm may achieve both technical and allocative efficiency, it may not operate optimally in terms of *scale efficiency*. Using variable

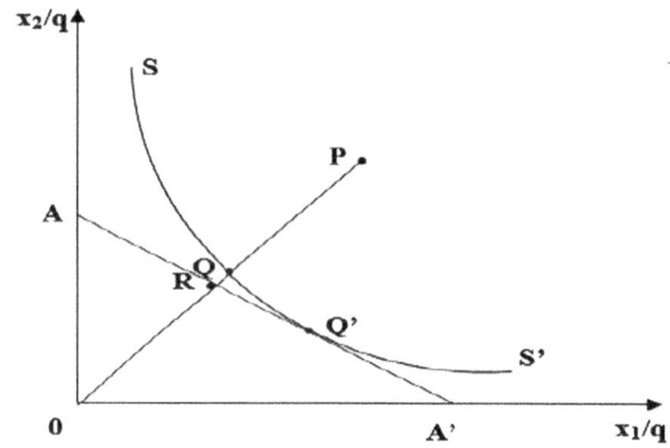

Figure 2 Technical and allocative efficiencies. Source: Coelli et al. (1998).

returns to scale (VRS) technology, the operational scale of a firm may be too small and so well within the limit of increasing returns to scale (IRS). On the other hand, a firm may be too big and operate with decreasing returns to scale (DRS), partly in the production function. If a firm operates within the IRS or DRS limit, its efficiency might be improved by changing its scale of operation (Coelli et al. 1998).

Data envelopment analysis (DEA)

Technical efficiency and scale efficiency can be measured using parametric or non-parametric approaches. In this study, a non-parametric *data envelopment analysis* (DEA) model was employed. DEA was described by Fare (1985) as creating a non-parametric piece-wise surface (or frontier) over the data (Coelli et al. 1998).

The input-oriented CRS DEA and VRS DEA models are described in Table 1. In addition to these models, I also employed an input-oriented non-increasing return to scale (NIRS) DEA model to examine the different types of returns to scale, these being CRS, IRS and DRS.

sing the DEA model, scale efficiency (SE) is given by the ratio between two corresponding efficiency scores, estimated using CRS and VRS technology, as follows:

Table 1 A summary of input-oriented DEA models

Input-oriented constant returns to scale DEA (CRS): min $_{\theta,\lambda}\theta$, subject to	Input-oriented variable returns to scale DEA (VRS): min $_{\theta,\lambda}\theta$, subject to	Input-oriented non-increasing returns to scale DEA (NIRS): min $_{\theta,\lambda}\theta$, subject to
$-q_i + Q\lambda \geq 0$	$-q_i + Q\lambda \geq 0$	$-q_i + Q\lambda \geq 0$
$\theta x_i - X\lambda \geq 0$	$\theta x_i - X\lambda \geq 0$	$\theta x_i - X\lambda \geq 0$
$\lambda \geq 0$	$N1'\lambda = 1$	$N1'\lambda \leq 1$
	$\lambda \geq 0$	$\lambda \geq 0$

Where θ is the total technical efficiency score of the ith firm and λ is a Nx1 vector of constants. The value of θ must satisfy the restriction: $0 \leq \theta \leq 1$. If $\theta = 1$; it indicates that the firm is on the production frontier and is technically efficient. When $\theta \leq 1$, the firm is technically inefficient.

In the VRS DEA model, $N1'\lambda = 1$ is added to show a convexity constraint which ensures that an inefficient firm is only benchmarked against firms of a similar size. On the other hand, in the NIRS DEA model, $N1'\lambda = 1$ is replaced by $N1'\lambda \leq 1$ to ensure that the ith firm is not "benchmarked" against firms that are substantially larger than it, but may be compared with firms smaller than it.

Source: Summarized from Coelli et al. (1998).

$$SE = TE_{CRS}/TE_{VRS}$$

To determine whether a firm operates under IRS or DRS, an additional DEA problem with NIRS should be employed (Coelli et al. 1998), so:

- If $TE_{CRS} = TE_{VRS}$: the firm is operating under CRS
- If $TE_{NIRS} = TE_{VRS}$: the firm is operating under DRS, and
- If $TE_{CRS} \neq TE_{VRS}$: the firm is operating under IRS

Bootstrapping DEA

Many studies use the two-stage approach to measure efficiency. For the first stage, the DEA method is employed to calculate a technical efficiency scores, then this score is used as a dependent variable to regress with exogenous variables during the second stage, using Tobit regression or OLS. Unfortunately, Simar and Wilson (2007) has shown that DEA efficiency estimates were serially correlated in these cases and that standard approaches to inference are, therefore, invalid. As an alternative, Simar and Wilson (2007) suggested using a bootstrap method to estimate bias-corrected efficiency scores during the first step.

For the second step, regression analysis determines the influence of environmental variables on the bias-corrected efficiency scores. Based on comparative analysis, Simar and Wilson (2007) proposed a truncated regression instead of a Tobit regression, or a choice between a single and double bootstrap to measure efficiency.

Data and analysis

Sampling

This study was based a database from the Vietnam General Statistics Office (GSO), and included data from the Vietnam Living Standard Survey (VLSS) 1998 and the Vietnam Household Living Standards Surveys (VHLSS) carried out in 2002, 2004, 2006, 2008 and 2010. Since 2002, the VHLSS have been conducted every two years on a nationwide scale. The sample was composed of 9,000 households on a nationwide scale, except in 1988 when the sample size was 6,000 households and 2002 when the sample size was 30,000.

The databases used for these surveys did not include panel data, meaning a household which appears in one survey may or may not have appeared in the others. An effort could have been made to create panel data across these databases based on the demographic characteristics of the households; however, this might have proved impossible or the results might have involved a too small sample size for the panel period of 1998 to 2010 in the Mekong delta region. Despite the use of panel data, in this study I was only concerned with average technical and scale efficiency scores taken from the 1,000 household sample. As a result of taking this approach, the results could be compared across years to identify changes during the study period (Table 2).

In addition, in the DEA analysis, two issues may be debated regarding those factors affecting the efficiency scores, these being:

(1) DEA is sensitive to outliners. Since DEA relies on identifying best practice reference units, it can be sensitive to extreme points, known as outliners, especially data contaminated by measurement error as described by Kuosmanen and Post (1999) and Nam et al. (2008).

Table 2 Sample size of VHLSSs (No. of observations)

	2010	2008	2006	2004	2002	1998
Vietnam	9,399	9,189	9,198	9,198	30,000	6,000
Mekong	1,905	1,863	1,863	1,863	6,298	1,112
Random sample			1,000			

Source: Own summary from (GSO, General Statistics Office of Vietnam 1998, GSO, General Statistics Office of Vietnam 2002, GSO, General Statistics Office of Vietnam 2004, GSO, General Statistics Office of Vietnam 2006b, GSO, General Statistics Office of Vietnam 2008, GSO, General Statistics Office of Vietnam 2010).

(2) DEA efficiency scores may be affected by sample size. A study by Zhang and Bartels (1998), comparing the structural inefficiency of different sized samples, found that this can lead to biased results. In addition, Staat (2001) supported Zhang and Bartels' results, saying it not only applies to studies explicitly comparing efficiency scores derived using samples of different sizes, but also to some types of DEA models such as the FDH, hierarchical DEA and DEA models for non-discretionary variables. Specifically, Alirezaee et al. (1998) argued that when the number of decision-making units is small, the number of dominant units or efficient sets will be relatively large and the average efficiency generally high.

To reduce the probability of such factors, the sampling process in this study was carried out over two stages: (1) First, I extracted observations of the Mekong delta from the national scale data and removed observations suspected of being outliners, (2) I randomly selected 1000 observations from each year.

Analysis

In this study, a single-bootstrapping DEA model was chosen to analyze technical efficiency under the input-orientation model, the aim being to measure the ability of the sample households to minimize inputs without changing outputs (Simar and Wilson (1998). During the first stage, bias-corrected technical efficiencies from the original technical efficiencies were estimated based on the same production set across all years, including five inputs and one output recoded (see Additional file 1).

The software package FEAR 1.0 of Wilson (2008) was used to estimate the TE_{CRS}, TE_{VRS} and TE_{NIRS} scores using the *DEA* function. After that, bias-corrected TE_{VRS} values were calculated by replicating the bootstrap 2,000 times using the *boot.sw98* function with an alpha value of 0.05 to estimate the statistical sizes of the confidence intervals. However, the *boot.sw98* function resulted in a homogeneous bootstrap for the Shephard distance functions, so Farrell efficiencies were re-calculated using the reciprocals of the Shephard efficiencies.

This study follows the scale efficiency estimation using DEA-based statistics instead of bootstrap-based statistics to estimate the returns to scale and scale efficiency[b]. To explore technical inefficiency factors, during the next stage, bias-corrected TE_{VRS} was used as a dependent variable, combined with exogenous independent variables in the truncated regression models. These potential factors were basically similar by years, though some additional variables from specific years were used to expand the models' levels of significance. Variables used in the truncated regression models included one dependent variable and eight independent variables recoded (see Additional file 1).

Results and discussion

Technical efficiencies

Values for the bias-corrected technical efficiencies (VRS) were found to be in the low range, at about 3% to 5% when compared to the original technical efficiencies (VRS). For instance, the original technical efficiency score in 2010 was 0.650, while the respective bias-corrected value was only 0.606. A natural characteristic of bootstrapping processes, efficiency distribution scores became "smoother" than the original values, as shown in Figure A1 (see Additional file 1).

The mean bias-correct technical efficiency (variable returns to scale) of rice production activities in the Mekong delta increased, rising from 0.484 in 1998 to 0.606 in 2010, despite a slight downward trend after 2008 (Table 3). The proportion of higher technical efficiency scores also significantly changed over the years. In 1998, less than a half of the farms (45.6%) reached a technical efficiency level between 50 and 100%, but nearly three quarters (74.2%) had reached this level of technical efficiency by 2010.

Compared to some recent studies, including Linh (2012), Khai and Yabe (2011), and Huy (2009), the technical efficiency scores found in this study were relatively low. One of the main reasons for this may have been the difference in sample sizes. Sample size, as the studies of Smith (1993), Alirezaee et al. (1998), Zhang and Bartels (1998), and Staat (2001) showed, can have a significant impact on efficiencies when those efficiencies are evaluated using the non-parametric DEA approach. In such cases, when the number of decision making unit (DMU) is small, the average efficiency will generally be high. To examine this trend, this study used data from 2010 and randomly chose a reduced sample of 200 DMUs to calculate the technical efficiencies. The results of a Welch t-test for unequal variance samples proved that the average technical efficiency score (VRS) of the sample of 200 households was significantly different to that of the sample of 1,000; the scores coming out at 0.745 and 0.650 respectively (Additional file 1). This result tends to support the findings of Linh (2012), which used about 82 DMUs from the Mekong delta region.

Determinants of technical inefficiency variation

Basically, inefficiency factors varied year on year. However, some popular inefficiency factors could be frequently found, including household demographic characteristics such as the age, gender, ethnicity and marriage status of the household head, household size, economic status of the household and the proportion of income from growing rice as a proportion of total income.

Table 3 Summary of efficiency scores and percentage forms of returns to scale

Mean scores		1998	2002	2004	2006	2008	2010
Original technical efficiency (VRS)		0.534	0.547	0.631	0.699	0.677	0.650
Corrected technical efficiency (VRS)		0.484	0.497	0.593	0.667	0.645	0.606
Scale efficiency		0.925	0.895	0.927	0.935	0.942	0.902
Percentage	*Constant returns to scale*	*1.2*	*1.2*	*1.3*	*1.1*	*2.7*	*0.9*
	Increasing returns to scale	*52.7*	*73.9*	*69.6*	*32.6*	*75.4*	*69.3*
	Decreasing returns to scale	*46.1*	*24.9*	*29.1*	*66.3*	*21.9*	*29.8*

Source: Own calculations based on (GSO, General Statistics Office of Vietnam 1998, GSO, General Statistics Office of Vietnam 2002, GSO, General Statistics Office of Vietnam 2004, GSO, General Statistics Office of Vietnam 2006b, GSO, General Statistics Office of Vietnam 2008, GSO, General Statistics Office of Vietnam 2010).

The proportion of rice income as compared to total income affected technical efficiency in studied years. Except for a very small negative impact in 2006, those households with a higher proportion of rice income usually reached higher technical efficiency levels. Having rice as a main income source implies that household resources may be focused on rice production and, therefore, they become more efficient.

While the education level and age of the household head had no or very little impact, the ethnicity of the head reflected the higher technical efficiency of the Kinh people when compared to the other ethnic minority groups. The Kinh, the ethnic majority group in Vietnam, basically have advantages over the others. During the period 1993 to 2006, government efforts to reduce poverty lead positive but unequal effects across the ethnic groups. The general poverty rate of the Kinh ethnic group went down by 71%, while this rate fell by only 42% among the other ethnic group between 1993 and 2006 (Dang 2010). This trend, in turn, may have had a negative impact on the capacity of minority group households to adopt new technology and improve their technical efficiency. Similar evidence from this study shows that households classified as poor by the government tended to attain lower technical efficiency scores when compared to non-poor households.

The gender of the household head, seemed to impact upon technical efficiency in a variety of ways. Female household heads attained higher efficiency scores in terms of rice production in 2006 and 2008, yet lower scores in 2002 and 2010 - when compared to the male household heads. Similarly, households with improved lives in general had higher technical efficiency scores, however, the impact of this did not manifest itself very often, in this case only in 2008. On the other hand, households with both a wife and husband present had significantly higher technical efficiencies scores when compared to other married statuses.

Scale efficiency

As mentioned above, scale efficiency is calculated by the ratio between technical efficiency under constant returns to scale and technical efficiency under variable returns to scale, which indicates how optimal a farm's scale is. In this study scale efficiencies were found to be relatively stable - at around 90% over the period 1998 to 2010. The mean scale efficiency scores, which is in the range of 50 and 100%, changed slightly from 0.925 in 1998 to 0.902 in 2010. In this case, the average efficiency scale value of 90% implies that the observed rice farms in Mekong delta could have further increased their output by about 10% if they had reached an optimal scale.

However, one should ask: which scale is optimal? Efficiency scales have a relationship to the different forms of returns to scale. The results here show that increasing returns to scale was a dominant characteristic in most periods, reflecting the need to expand production scales in future years in order to attain greater efficiency. In 2010, the proportion of increasing returns to scale was 69.3%, which is much higher than the proportion of decreasing returns to scale (29.8%), while the optimal scale accounts for only a small proportion, at 0.9%.

The need to increase returns to scale would seem to reflect an economic incentive, which works in opposition to government efforts to impose land use quotas. In fact, some of the farms were found to be over-sized when compared to these limits (Additional file 1).

Being a major competitor of Vietnam in the rice export market, Thailand may be a typical example on land management policies. There have been recently changes in number of Thai farms and average farm size, whereas the number of farms has increased from 5.15 million farms in 1992 to 5.70 million farms in 2001 (ALRO, Agricultural Land Reform Office 2006). The average farm size has trended to be smaller, from 5.6 hectares per household in 1980s (LePoer 1989), to approximately 4 hectares per household in 1992 and to 3.68 hectares per household in 2001 for the whole country (ALRO, Agricultural Land Reform Office 2006). On the other hand, land is relatively concentrated. The holdings with only one parcel takes account for approximately half of total number in 2003 (NSO, National Statistical Office of Thailand 2003). Although population pressure on the land has led increasing of smaller farms, this average farm size is still much higher than the average number of farms in the Mekong delta, Vietnam.

Conclusion

In this paper, a bootstrapping DEA model with an input orientation was employed across 1,000 observations in 1998 and bi-annual updates from 2002 to 2010. Using the same method throughout in terms of sample size and production sets, allowed me to reduce the risk of bias and to demonstrate changes in technical efficiencies and scale efficiencies with respect to rice production activities in the Mekong delta.

The bias-corrected efficiency scores were "smoother" than the original scores, while being lower than some other recent studies carried out in the region. This may have been due to the difference in sample sizes used, as a small number of DMUs may result in a upward bias of the efficiency estimates.

The level of technical efficiency of the rice production activities tended to increase over the study years, while optimal scale efficiency was only achieved by a small number of farms. Scale efficiency scores changed approximately by 90%, and the proportion of those achieving increasing returns to scale was high. Using a truncated regression, the main inefficiency factors could be linked to the demographic characteristics of the households, as well as the proportion of income from rice as a proportion of total household income.

In the Mekong delta region in particular and in Vietnam in general, the government has imposed agricultural land quotas per household and set a limit on the duration of agricultural land use. The empirical results challenge for the government's land management policies, as the farmers' essential need to expand their production scale to meet increased efficiency rice production targets would seem to be in opposition to the current, more restrictive land management policies to be found in the Mekong delta.

To achieve a deeper understanding of rice technical efficiency of the Mekong delta, there is still the need to conduct more comparative studies on rice production between this region and other rice production regions in other countries which focus on the relation between technical efficiency and farm size as well as their consequence – landlessness problem. In fact, Thailand is biggest rice exporter in the world and may be considered as a good experience for policy changes on land management in the Mekong delta, Vietnam.

Endnotes

[a]The informal numbers should be verified, it however shows an upward trend in land consolidation.

[b]There are two reasons to follow the DEA-based statistics instead of the bootstrap-based statistics: (1) Based on the experiment method, Banker et al. (2009)) argued that "There is no need to use the Bootstrap-based test procedures since they yield comparable results to DEA-based procedures", and (2) both bootstrap-based estimations of returns to scale and double-bootstrapping DEA require a high performance computer to be used to carry out a large number of calculations, and this would seem to be a big challenge for an average computer, particularly when using a dataset holding 1,000 DMUs (5 inputs, 1 output), plus carrying out 2,000 replications and repeated calculations over 6 different years.

Additional file

Additional file 1: Table A1. Variables used in the estimation of technical efficiency and technical inefficiency factors. **Figure A1**. TE score distributions under VRS; comparing initial (kdensityini_vrs) and bias-corrected (kdensitycor_vrs) scores over various years. **Table A2.** Test differences between various sample sizes. Two-sample t test with unequal variances. **Table A3.** Percentage of over-sized farms per sample - compared to the Land Law 1993 and Land Law 2003 quotas. **Table A4.** Statistical results of Technical Efficiency (TE_{VRS}) and Scale Efficiency (SE) by input-oriented DEA. **Table A5.** Statistics summary of Truncated regression models. Table A5. Statistics summary of Truncated regression models.

Competing interests

The authors declare that they have no competing interests.

References

Akram-Lodhi AH (2001) Vietnam's agriculture: is there an inverse relationship? Int Inst Soc Stud Erasmus Univ (ISS) 348:1–36

Alirezaee MR, Howland M, Van de Panne C (1998) Sampling size and efficiency bias in Data envelopment analysis. J Appl Math Decis Sci 2(1):51–64

ALRO, Agricultural Land Reform Office (2006) National Report on Agrarian Reform and Rural Development in Thailand. Paper presented at the The International Conference on Agrarian Reform and Rural Development (ICARRD), Brazil

Andor M, Hesse F (2011) A Monte Carlo Simulation comparing DEA, SFA and two simple approaches to combine efficiency estimates. CAWM Discussion Paper No. 51, University of Münster

Banker RD, Chang H, Chang SC (2009) Statistical Tests of Returns to Scale Using DEA. Paper presented at the The seventh International Conference on Data Envelopment Analysis Philadelphia, USA. http://astro.temple.edu/~banker/dea2009/paper/Fri350746B.ppt

Coelli T, Prasada Rao DS, Battese GE (1998) An introduction to efficiency and productivity analysis. Kluwer Academic Publishers, Boston

Dang HA (2010) Indigenous Peoples, Poverty and Development: A Widening Poverty Gap for Ethnic Minorities. discussion paper. World Bank, Washington

Debreu G (1951) The coefficient of resource utilization. Econometr J Econometr Soc 19(3):273–292

FAO, Food and Agriculture Organization of the United Nations (2003) Measuring and assessing capacity infisheries. 2. Issues and methods. FAO Fisheries Technical Paper. No. 433/2. FAO, Rome, p 130

Fare R (1985) The measurement of efficiency of production / by Rolf Fare, Shawna Grosskopf, and C.A. Knox Lovell. Hingham, MA, USA: Kluwer-Nijhoff Pub; Distributors for North America, Kluwer Academic Publishers, Boston

Farrell MJ (1957) The measurement of productive efficiency. J Royal Stat Soc Series A (General) 120(3):253–290

Government of Vietnam (1993) Law on Land. Hong Duc Publishing House, Hanoi

Government of Vietnam (2003) Law on Land. Hong Duc Publishing House, Hanoi

GSO, General Statistics Office of Vietnam (1998) The Viet Nam Living Standard Survey. Statistical Publishing House, Hanoi

GSO, General Statistics Office of Vietnam (2001) The Rural, Agricultural and Fishery Census in Vietnam. Statistical Publishing House, Hanoi

GSO, General Statistics Office of Vietnam (2002) The Vietnam Household Living Standards Survey. Statistical Publishing House, Hanoi

GSO, General Statistics Office of Vietnam (2004) The Vietnam Household Living Standards Survey. Statistical Publishing House, Hanoi

GSO, General Statistics Office of Vietnam (2006a) The Rural, Agricultural and Fishery Census in Vietnam. Statistical Publishing House, Hanoi

GSO, General Statistics Office of Vietnam (2006b) The Vietnam Household Living Standards Survey. Statistical Publishing House, Hanoi
GSO, General Statistics Office of Vietnam (2008) The Vietnam Household Living Standards Survey. Statistical Publishing House, Hanoi
GSO, General Statistics Office of Vietnam (2010) The Vietnam Household Living Standards Survey. Statistical Publishing House, Hanoi
GSO, General Statistics Office of Vietnam (2011) The Rural, Agricultural and Fishery Census in Vietnam. Statistical Publishing House, Hanoi
GSO, General Statistics Office of Vietnam (2012). Retrieved 12 April 2013 http://gso.gov.vn/default.aspx?tabid=717
Huy HT (2009) Technical efficiency of rice producing households in the Mekong delta of Vietnam. Asian J Agricult Dev 6(2):35–50
Khai HV, Yabe M (2011) Technical efficiency analysis of rice production in Vietnam. J. ISSAAS 17(1):135–146
Koopmans TC (1951) An analysis of production as an efficient combination of activities. In: Koopmans TC (ed) Activity Analysis of Production and Allocation. John Wiley and Sons, Inc, New York
Kuosmanen T, Post T (1999) Dealing with outliners in Data Envelopment Analysis. Rotterdam Institute for Business Economic Studies (RIBES)
LePoer BL (1989) Thailand a Country Study. Federal Research Division, Library of Congress, Washington
Linh VH (2012) Efficiency of rice farming households in Vietnam. Int J Dev Issues 11(1):60–73
Minh NK, Long GT (2008) Measuring agricultural production efficiency in Vietnam: An application of data envelopment analysis (DEA). Publishing House of Social Labour, Hanoi
Nam TA, Shively G, Preckel P (2008) A new method for detecting outliners in Data Development Analysis. Appl Econ Lett 2008:1–4
NSO, National Statistical Office of Thailand (2003) Agricultural Census. Retrieved 4 September 2013 http://service.nso.go.th/nso/nso_center/project/search_center/23project-en.htm
Simar L, Wilson PW (1998) Sensitivity analysis of efficiency scores: how to bootstrap in nonparametric frontier models. Manag Sci 44(11):49–61
Simar L, Wilson PW (2007) Estimation and inference in two-sage semi-parametric models of production processes. J Econ 136:31–64
Smith P (1993) Misspecification Bias in Data Envelopment Analysis. Discussion Paper, Department of Economics and the Related Studies. University of York, York
Staat M (2001) The effect of sample size on the mean efficiency in DEA: comment. J Prod Anal 15:129–137
The Standing Committee of National Assembly (2007) Resolution No. 1126/2007/NQ-UBTVQH11, Specifying limits for receipt of agricultural land use rights transferred to households or individuals for agricultural purpose
Thong PL, Xuan HTD, Duyen TTT (2011) So sánh hiệu quả kinh tế của vụ lúa hè thu và thu đông ở Đồng bằng sông Cửu Long (Vietnamese). Sci J Can Tho Univ 18a:267–276
Tuong DM (2010) Economic analysis on production of high quality rice in Cuulong delta, Vietnam. Omonrice 17:174–178
Wilson PW (2008) FEAR 1.0: A Software Package for Frontier Efficiency Analysis with R. Socioecon Plann Sci 42:247–254
Zhang Y, Bartels R (1998) The effect of sample size on mean efficiency in DEA with an application to electricity distribution in Australia, Sweden and New Zealand. J Prod Anal 9:187–204

9

Factors influencing willingness to participate in multi-stakeholder platform by smallholder farmers in Northern Ghana: implication for research and development

Edward Martey*, Prince M Etwire, Alexander N Wiredu and Wilson Dogbe

* Correspondence:
marteywayo2@yahoo.com
CSIR-Savanna Agricultural Research Institute, Nyankpala, P.O. Box TL 52, Tamale, Ghana

Abstract

Innovation platforms are established to facilitate open communication and collaboration among various actors usually within a value chain to promote collective resource management. The concept of innovation platform as a strategy of entry point for operationalization, validation and dissemination of agricultural technology have received much attention in recent times among researchers in Ghana. However, very limited information exists on the factors that influence willingness of farmers' participation. The present study identified the factors influencing willingness to participate in innovation platform and also tested the level of agreement among the identified constraints associated with participation using the Probit model and Kendall's coefficient of concordance respectively. Data for the study was collected from a cross sectional primary observations of 250 smallholder rice farmers in northern Ghana. The study observes, inter alia, that age of household head, household size and household income significantly influenced willingness to participate on the platform. The Kendall's coefficient of concordance value indicated that there was 21 percent agreement between the rankings of the participation constraints faced by the sampled rice farmers. Income-earning and younger household heads must be targeted coupled with situation of the platforms close to smallholder farmers to encourage participation.

Keywords: Multi-stakeholder platform; Constraint; Probit; Kendalls coefficient of concordance; Northern Ghana

Background

Development research in Africa has centered mainly on smallholder commercialization as a way of ensuring food security and economic growth. Agriculture in most African economies is on smallholder basis. In sub-Saharan Africa, Asia and the Pacific, the agriculture-dependent population is over 60%, while in Latin America and high-income economies the proportions are estimated at 18% and 4%, respectively (World Bank 2006). Most of the smallholder farmers in sub-Saharan Africa are confronted with multifaceted challenges. These problems require several interventions such as institutional reforms that facilitate efficient rural service delivery, development of markets, creation of physical infrastructure, and government policies that are supportive while ensuring a stable and conducive political environment. Smallholder farmers require systems that are responsive to their needs: access to markets, market information, market intelligence and effective farmer organization as the agricultural sector in developing countries transforms towards commercialization.

According to Pingali et al. (2005), smallholder farmers in most developing economies find it difficult to participate in markets because of numerous constraints and barriers. These are mostly reflected in the hidden costs that make it difficult to access input and output markets. Transaction costs are the embodiment of access barriers to market participation for most resource-poor smallholders (Delgado 1999; Holloway et al. 2000). A fundamental transaction cost these farmers face is the cost of obtaining information (Shepherd 1997). Though neoclassical economists essentially assume that information is costless, this assumption does not match reality, especially in developing countries (Stiglitz 1988). The fact that information is not costless has important implications for contracts and transactions, as has been pointed out in work pioneered by Coase (1937) and later expanded in Coase (1960). Commercialization studies such as Goetz (1992), Key et al. (2000) and Makhura et al. (2001) have identified high transaction costs as one of the key reasons for smallholder farmers' failure to participate in markets. Most are located in remote areas with poor transport, market infrastructure and lack of reliable information on markets and potential exchange partners. In some instances, these transaction costs are so high that markets can be said to be 'missing' (Omamo 1998; Key et al. 2000). The adoption and use of multi-stakeholder platform (MSP) in most recent times has the potential to address the high transaction cost incurred by smallholder farmers with regard to market information.

The three most targeted regions in Ghana for most agricultural development projects are Northern, Upper East and Upper West Regions. The regions are inundated with high levels of food insecurity and poverty due to over-reliance on rain-fed agriculture under low farm input conditions. About half of the population of the area face annual food deficit and are net buyers of food (GSS 2008). Nearly 80 percent of the population depends on subsistence agriculture with very low productivity and low farm income (MoFA 2010). Several programmes have been designed over the years especially in the rice sector to address the food insecurity and poverty levels in northern Ghana. One of the major interventions was the Multinational NERICA Rice Dissemination Project which sought to contribute to poverty reduction and food security in Ghana through capacity building, technology transfer and production support.

The Commercial Development of Farmer Based Organization (CDFO) of the Millennium Challenge Account (MCA) Programme was an intervention in addressing one of the objectives of the Food and Agriculture Sector Development Policy II (FASDEP II). The objective was to enhance increase competitiveness and integration of farmers into domestic and international markets. To overcome the market constraints of producers, farmers were organized into Farmer-Based Organization (FBO) to access credit and engage in bulk purchase and group marketing to lower the transaction cost they incur in marketing. Agricultural Business Centres (ABCs) were established to enable farmers to access market information on prices, standards and advisory services. The other component of the MCA program sought to strengthen input delivery systems and financial institutions that advances credit to farmers. The intervention however failed to use the integrated approach such as innovation platform to address the marketing problems of the farmers. Most of these interventions fail the test of sustainability after the completion of the entire project.

In an attempt to address the weak linkages among the rice value chain actors as well as increase in rice production in Ghana, the Africa Rice Center (AfricaRice) in collaboration with the Savanna Agricultural Research Institute of the Council for Scientific and Industrial Research (CSIR-SARI) established two Multi-Stakeholder Platforms (MSP) in two rice development hubs[a] in northern Ghana. The Multi-Stakeholder Platform establishment was used to facilitate open communication and collaboration among various actors within the value chain to promote collective resource management. The platform serves as a point of entry, operationalization, validation and dissemination of agricultural technologies. Sustainable development of the rice sector and the rice value chain through an MSP approach prompts key stakeholders to identify and prioritize their needs, concerns, constraints, and vision for key activities and interventions. They also act as a governing entity that coordinates decision-making at local, regional, and national levels regarding rice sector development in the long run. The MSP was also intended to assist farmers overcome their production, marketing and post-production challenges in the regions.

In northern Ghana, the MSP for the rice sector consists of researchers, producers, processors, traders, financial institutions, input dealers, tractor operators and policy makers. The platform brings together all the necessary stakeholders of the rice sector to dialogue on ways of increasing rice production to enhance food security with Savanna Agricultural Research Institute providing the backstopping. Introduction, operationalization, validation and dissemination of all rice development programmes in the region occurs on the platform. All the various actors play a significant role in ensuring a boost in the rice sector as well as strengthening of the rice value chain. Most established platforms are confronted with some challenges which include meeting and sustaining participants' interest, effective participation and collaboration, conflict resolution and collective action.

Problem statement and justification

Despite the benefits of the platform especially in transforming the rice value chain system, it is critical to identify the potential factors that are likely to influence farmers' willingness to participate in such a platform as well as the constraints associated with participation. Currently, there is no documented evidence of the factors and constraints limiting farmers' willingness to participate on innovative platforms especially in the study area. It is against this backdrop that the present study attempts to fill the gap in knowledge. The documented factors and constraints will serve as the basis for addressing the sustainability issues regarding the intervention. The study contributes significantly to the scanty literature on MSP and also serve as a guide for future implementation of similar platforms in other parts of Africa. Finally, facilitators of MSP and rice development programmes will benefit from the knowledge generated by the study.

Literature review

Concept and empirical evidence of multi-stakeholder platforms

The term multi-stakeholder platform (MSPs) describes processes which aim to bring together all major stakeholders in a new form of communication, decision-finding (and possibly decision-making) on a particular issue. They are also based on recognition of the importance of achieving equity and accountability in communication between stakeholders, involving equitable representation of three or more stakeholder groups and their views. They are based on democratic principles of transparency and participation, and aim to develop partnerships and strengthened networks between stakeholders. MSPs cover a wide spectrum of structures and levels of engagement. They can comprise dialogues on policy or grow into consensus-building, decision-making and implementation of practical solutions. The exact nature of any such process will depend on the issues, its objectives, participants, scope, time lines, etc. (www.earthsummit2002.org).

Multi-stakeholder platforms were first proposed in the context of natural resource management, where stakeholders share a common-pool resource, such as access to water in a river basin, and the platform contributes to the collective management of the resource (Rölling et al. 2002). A platform has value for stakeholders, because they are, or may become, interdependent. Interdependence can create tension, conflict, maneuvering to seek advantage and even group displacement. But it also opens opportunities for mutual understanding, building confidence, social learning and joint action (Rölling et al. 2002). The platform makes possible actions that none of the members could have achieved on their own. Because of its complex membership and potential for conflict, a platform is likely to require facilitation and may have a lengthy initial phase of mutual learning and role definition, before it can get down to business (Thiele et al. 2005).

Most value chain does not frequently use the approach of platforms. Several studies including a recent overview of collective action for small farmer market access considered small farmer organizations without mentioning platforms (Markelova et al. 2009). An exception is the study by Vellema et al. (2009), which analyses the oil seed subsectoral platform in Uganda. A platform can perform three different, but interlinked, functions in a value chain. First, it can create a space for learning and joint innovation, as an innovation intermediary or broker. Second, it can perform a governance function within the value chain to improve coordination of business activities by actors and reduce transaction costs. Finally, a platform can perform advocacy functions to secure policy change or influence.

Value chain governance may be provided by: (a) market mechanisms, (b) hierarchical non-market mechanisms and (c) by non-market-based voluntary coordination between actors of a collective action type (Markelova et al. 2009). Dorward et al. (2009), writing from a New Institutional Economics perspective, noted that coordination provided through different non-market mechanisms can help market actors reduce transaction costs and escape the low-level equilibrium trap associated with underdeveloped economies, as a weak institutional environment and high transaction risks limit investment opportunities.

Countries that have witnessed success in overcoming marketing constraints by using the MSP include Peru, Bolivia and Ecuador. The platform Cadinas Agricolas Productivas de Calidad (CAPAC) in Peru which focused on native potatoes grown largely by poorer farmers primarily for home consumption or local markets has been instrumental in providing transparent information on price and volumes. The CAPAC has also assumed an important role in market governance by linking farmers' native potato production to Frito-Lay and Ayllin Papa through intermediary NGOs (Thiele et al. 2011).

The platform Andino Boliviana in Bolivia played an important facilitating and coordinating role so as to link farmers' organizations with exporting companies, contributing to the supply of quality chun˜o for the export market (Thiele et al. 2011). In Ecuador, Chimborazo platform played the most extensive role in market governance. The platform developed and monitored production plans with farmer quotas by market catchment areas to smooth the supply of potatoes to clients. In addition, the platform empowered farmers' organizations and associations to assume a greater leadership role which led to the creation of CONPAPA (Consorcio de Pequen˜os Productores de Papa, Consortium of Small Potato Producers), which from late 2006 took over the technical assistance functions, production planning, bulking up and marketing functions that the Chimborazo platform had previously performed (Thiele et al. 2011).

Evidence from the Innovation Platform for Technology Adoption (IPTA) in Burkina Faso indicates that maize producers and processors are making profit on their investments and processors are making a difference of only CFA 1.00 over and above the profit margin of producers. Farmers are also increasing the area under maize cultivation due to a guaranteed market as provided by the platform.

Most of the literature reviewed on MSP is descriptive in presentation with little information on specific quantifiable factors that influence willingness of participation. However, the current study used a regression model to analyze farmers' willingness to participate in MSP by drawing lessons from the reviewed literature. Future studies can look at the extent of participation in MSP on household income and food security.

Methods

Study area

Northern Ghana consists of Northern, Upper East and Upper West Regions. The vegetation is a typical Guinea Savannah type characterized by drought-resistant grasses and trees. The area plays an important role in agriculture and is normally referred to as the grain basket of the country. It shares borders with Republic of Togo to the east, Ivory Coast to the west and Burkina Faso to the north. Within the country, the northern Ghana is bordered by Volta region to south east and Brong-Ahafo region to the south east. Geographically, the three regions are between longitude 8°46'01.88" N and 10° 58'34" N and latitude 2°45'45.40" W and 0°32'59.95" E and covers a total land area of 97,666 km^2 with an estimated population of 3,317,478. The major crops cultivated in the area are rice, maize, sorghum, millet, groundnut, beans and Shea butter. Livestock and poultry production are also important ventures for most households. The study area consists of two main rice development hubs namely the Navrongo and Savelegu hubs.

Data and sampling technique

The study was conducted between August and September, 2012. The basic information for the analysis was obtained from primary data collected with the aid of a structured questionnaire. A total of 250 smallholder rice producers were systematically selected and interviewed. The selection of the rice producers followed a multi-stage systematic random sampling technique. In the first stage, two (2) rice hubs in Navrongo and Savelugu districts were purposively selected. Secondly, five (5) communities each were randomly selected from a list of rice producing communities in each of the rice hubs. Within the selected communities in each of the hubs, 26 and 24 rice producers were further selected from a list of rice producers in the Navrongo and Savelugu selected communities respectively (Table 1). In addition to the survey, key informants interviews and focus group discussion were conducted to augment the household survey.

Probit model

The probit regression model was employed to quantify the factors that determine the willingness of farmers to participate in multi-stakeholder platform due to the dichotomous nature of the dependent variable. The justification for the use of the probit model over the logit model is as a result of its ability to constrain the utility value of the decision to join variable to lie within 0 and 1, and its ability to resolve the problem of heteroscedasticity (Asante et al. 2011). Willingness to participate in MSP (Y) was captured as a dummy variable with the value of 1 assigned to a farmer who is willing to participate and 0 for otherwise. Following from Greene (2003), the binary probit for the two choice models can be written as;

$$Y_i^* = \begin{cases} 1 \; if \; Y_i^* > Y \\ 0 \; if \; Y_i^* \leq 0 \end{cases} \tag{1}$$

The probit model is given by:

$$P\left(Y = \frac{1}{X}\right) = F(XB) = \frac{1}{\sqrt{2\pi}} \int_{-\infty}^{XB} e^{\frac{-(XB)^2}{2}} dx \tag{2}$$

where:

$$X = (1, x_{1i}, x_{2i}, \ldots\ldots, x_{ki}) \tag{3}$$

$$\beta' = (\beta_0, \beta_1, \ldots\ldots\ldots\ldots, \beta_k) \tag{4}$$

Specifically, the empirical model for determining the willingness to participate in in MSP is specified in equation as follows (5).

$$Y_i = \beta_0 + \beta_1 Age + \beta_2 Gend + \beta_3 Mari + \beta_4 HSize(5) + \beta_5 Edu + \beta_6 Assoc + \beta_7 Lab$$
$$+ \beta_8 FarmDec + \beta_9 Dist + \beta_{10} Inc + \mu_i$$

Where Y_i and μ_i represents willingness to participate in MSP and the error term respectively. The set of potential explanatory variables, definitions and their *a priori* expectations are presented in Table 2.

The marginal effect of the variables is calculated using the formula:

$$Marginal\ effects = B_i \varnothing(Z)\ where\ B_i\ are\ the\ coefficient\ of\ the\ variable \tag{6}$$

Table 1 Sampling frame

Rice hub	Communities	Households
Navrongo	5	130
Savelegu	5	120
Total	10	250

Table 2 Explanatory variables for the probit model

Variable	Definition	Expected sign
Gender (Gend)	1 if male and 0 otherwise	+/–
Age (Age)	Age of household head in years	-
Marital status (Mar)	1 if married and 0 otherwise	+
Education (Edu)	Number of years of formal education	+
Household size (HSize.)	Number of household members that assist with farming	+
Land availability (Lab)	Total land size cultivated in 2012	+
Total income (Inc.)	Total income earned from rice sales in 2012	+
Membership of association (Assoc.)	1 if farmer belongs to FBO and 0 otherwise	+
Distance (Dist.)	Distance from farmers house to the meeting place in km	-
Major farming decision (FarmDec)	1 if household head makes decision alone and 0 otherwise	+/–

Where B_i represents the estimated coefficients and the $\varnothing$ (Z) is the cumulative normal distribution value associated with the mean dependent variable from the Probit estimation.

Table 2 shows the set of potential explanatory variables that are expected to explain variation in willingness to participate in MSP and their *a priori* expectations.

Description of explanatory variables

Gender is expected to have a significant effect on participation in MSP. Males are expected to have higher willingness of participation in MSP relative to females. Female farmers usually lack access to agricultural resource that enhances their participation in social and innovation platforms. Age is expected to influence willingness to participate in MSP negatively. It is a proxy measure of experience and availability of resources. According to Etwire et al. (2013), younger farmers are usually innovative, risk loving and may want to try new concepts. Marital status allows an individual to have access to information and resource. It is expected that household head that are married are more willing to participate on the platform. Nnadi and Akwiwu (2008) noted that marriage increases a farmer's concern for household welfare and food security which is therefore likely to have a positive effect on their decision to participate in an agricultural project.

Education is posited to have a positive effect on willingness to participate on the platform. Education enables an individual to make independent choices and to act on the basis of the decision, as well as increase the tendency to co-operate with other people and participate in group activities (Enete and Igbokwe 2009; Southworth and Johnston 1967; Schultz 1945 and Ofori 1973). It is also possible that education could increase the chances of the household head earning non-farm income. This could reduce the household dependency on agriculture and thus participation on the platform. Household size is expected to positively influence farmers' willingness to participate on the platform. Household size serves as a form of family labour and compliments the effort of the household heads on the farm (Martey et al. 2013). The availability of family labour provides the household head the opportunity to share responsibility and save time for other useful activities like the MSP. Also, larger households spend more on food and other household needs. The higher expenditure associated with larger households sizes tend to make them more resource constrained and hence the need for external support (Martey et al. 2013).

Land availability has an indirect positive effect on farmers' willingness to participate. Household heads with more land are able to produce more *ceteris paribus* and earn higher income to overcome the transportation constraint as well as financial demands of the platform. Most researchers have found a positive relationship between farm size and decision to join or adopt (Adimado 2001; Kheralla et al. 2001; Langyintuo and Mekuria 2005). Other studies such as Mussei et al. (2001), and Gockowski and Ndoumbe (2004) found a negative relationship between farm size and decision to join or adopt. The coefficient of this variable can be either positive or negative. A household head that earns higher income from rice sales will be more willing to participate on the platform. Higher income household heads are able to meet the financial demands of any group they belong to relative to lower income households. Household head membership of an association/group increases access to information which is important to production and marketing decisions (Olwande and Mathenge 2010). Most farmer groups engage in group marketing, bulk purchasing of inputs and credit provision for its members. It is therefore expected that household head membership of association/group will positively affect willingness to participate on the platform. Distance serves as a major cost constraint to willingness to participate on the platform. Household heads residing farther away from the meeting place may find it difficult to effectively participate especially in situations where they do not have means of transport.

The Kendall's coefficient of concordance

The Kendall's concordance analysis was used to test for the agreement between the ranked constraints of willingness to participate on the MSP. It establishes the extent of disagreements and agreements among responses. The Kendall's coefficient of concordance (W) is the measure of the degree of agreement among m set of n ranks. W is an index that measures the ratio of the observed variance of the sum of ranks to the maximum possible variance of sum of ranks. If the rankings are in perfect agreement, the variability among sums will be a maximum (Mattson 1986). Computing the total rank score for each constraint and objective, the constraint and objective with the least score is ranked as the most pressing whilst the one with the highest score is ranked as the least pressing. The total rank score computed is then used to calculate for the coefficient of concordance (W), to measure the degree of agreement in the rankings (Edwards 1964). The formula for the coefficient of concordance W is then given by:

$$W = \frac{\left(\sum T^2 - \left(\sum T\right)^2 / n\right) / n}{m^2(n^2-1)/12} \tag{7}$$

The formula is further simplified as follows:

$$W = \frac{12\left[\sum T^2 - \left(\sum T\right)^2 / n\right]}{nm^2(n^2-1)} \tag{8}$$

Where;

T = sum of ranks for each thing being ranked.
m = number of rankings (smallholder farmers) and
n = number of things (factors) being ranked

The following hypothesis was tested for the potential constraints as follows:

H_0: There is no agreement among the constraints faced by smallholder rice farmers' willingness to participate on the MSP.
H_1: There is agreement among the constraints faced by smallholder rice farmers' willingness to participate on the MSP.

The Coefficient of concordance *W* was tested for significance using the F distribution.

Results and discussion

Multi-stakeholder platform participation by smallholder farmers

Participation in multi-stakeholder platform based on gender revealed that 59% of the sampled male-headed households whereas 43% of the female-headed households were members on the platform (Table 3). The higher representation by males relative to females on the platform was as a result of unlimited access to agricultural resources that enhances participation in social platforms. Easy access to agricultural resource by male-headed households makes them more adaptive to new innovations and technology compared to female-headed households.

Willingness to participate in MSP based on gender revealed that 82% (120) of male-headed households were willing to participate on the platform. Female headed households were more willing to participate on the platform as represented by 92% of the sampled farmers (Table 4). In northern Ghana, it is mostly common to see women group flourishing. Female-headed households are more likely to adopt a technology or innovation that will enhance their productivity and income. This finding implies that gender must be critically considered in the selection of the participants. Marketing of agricultural produce is mostly dominated by women who are also faced with myriads of marketing constraints such as lower market price and higher transaction cost. The opportunity offered by the platform as a way of overcoming these constraints as well as knowledge sharing could explain for the higher willingness of participation on the part of female producers.

Table 5 shows that 52% (92) of the sampled households who are members of the MSP are within the age bracket of less than or equal to 40 years (≤40) whereas 46%

Table 3 Membership of MSP by gender

Membership of MSP	Gender of household head					
	Male	Percentage	Female	Percentage	Total	Percentage
No	61	41.50	59	57.30	120	48.00
Yes	86	58.50	44	42.70	130	52.00
Total	147	100.00	103	100.00	250	100.00

Source: Author's Household Survey Data (2012).

Table 4 Willingness to participate in MSP by gender

MSP participation	Gender of household head					
	Male	Percentage	Female	Percentage	Total	Percentage
Yes	120	81.60	95	92.20	215	86.00
No	27	18.40	8	7.80	35	14.00
Total	147	100.00	103	100.00	250	100.00

Source: Author's Household Survey Data (2012).

and 73% of household head falls within the age category of 41 to 60 years and greater than 60 years (>60 years) respectively. In absolute terms majority of the members of the MSP are within the economically active age group of the country. The result has implication on technology adoption and sustainability. Relatively younger household heads should be targeted for the innovation platform approach to enhancing food security and reduction in poverty. Younger household heads are more likely to adopt technology and innovations relative to older household heads (Enete and Igbokwe 2009).

The results in Table 6 also show high level of willingness of participation across all the age categories. The willingness of participation response is relatively higher among the economically active age group (41–60 years) and older age category (>60 years) relative to the younger age group category (≤40 years) (Table 6). Willingness to participate on the MSP was highly correlated with age where the older age category (>60 years) recorded the highest percentage. The result implies that there is opportunity to reach out to other smallholder farmers irrespective of the age. For sustainability of the platform, it is more expedient for the facilitators to target both the young and the economically active age group category.

Fifty-five (55) percent of the household heads belonging to a Farmer-Based Organization (FBO) were members of the MSP whilst 50% of non-members of FBO household head were also members of the platform (Table 7). Farmer Based Organization is one of the main channels for reaching out to numerous farmers for collective actions. It also serves as a platform for capacity building, marketing, bulk purchasing and advocacy. Increase in MSP participation by members of FBO can only be realized through consistent education of farmers on the concept of the MSP.

Members and non-members of FBO were more willing to participate in the MSP (Table 8). About 87% of household heads belonging to a FBO were willing to participate in the MSP whilst 86% of non-members of FBO were also willing to join the platform. Previous benefits from FBOs and the need for social networks may have influenced the willingness decision of household heads in northern Ghana.

Table 5 Membership of MSP by age

Membership of MSP	Household head age (Years)							
	≤ 40	%	41-60	%	> 60	%	Total	%
No	84	47.70	32	54.20	4	26.70	120	48.00
Yes	92	52.30	27	45.80	11	73.30	130	52.00
Total	176	100.00	59	100.00	15	100.00	250	100.00

Source: Author's Household Survey Data (2012).

Table 6 Willingness of participation based on age

MSP participation	Household head age (Years)							
	≤ 40	%	41-60	%	> 60	%	Total	%
Yes	149	84.70	52	88.10	14	93.30	215	86.00
No	27	15.30	7	11.90	1	6.70	35	14.00
Total	176	100.00	59	100.00	15	100.00	250	100.00

Source: Author's Household Survey Data (2012).

Determinants of multi-stakeholder platform participation

The probit model was used to estimate the parameters of the determinants of willingness to participate on the multi-stakeholder platform by smallholder farmers in northern Ghana. The E-views software was used to estimate these parameters as well as the marginal effects. The McFadden R-squared value indicates that 36 percent of the variation in the willingness to participate on the platform was explained by the independent variables. The significant likelihood ratio (LR) of 73.43 with 10 degrees of freedom indicates that at least one of the variables in the model had a significant effect on farmers' willingness to participate on the platform and that the explanatory variables jointly influence the farmers' willingness of participation on the platform (Table 9). Farmers' willingness to participate on the multi-stakeholder platform was significantly determined by age of the household head, household size and household income. Numerically and statistically, age was the most influential determinant of willingness to participate in MSP. Details of the regression result is presented in Appendix 1.

Age of the household head was significantly associated with a lower probability of willingness to participate on the platform. The probability of willingness to participate in MSP decreases by 7.6 percent for every additional year added to the age of the household head (Table 9). It can be inferred from the result that younger household heads were more willing to participate on the platform than older household heads. Older household heads are normally experienced in farming and may have more social networks which enhance their farming business thus will be more reluctant in joining the platform. Secondly, older household heads may not have the motivation to join such a platform as the level of willingness to adopt innovation declines with age. The result is consistent with Ayamga (2006) who found that as age increases, the probability of a farmer to participate in microcredit programmes in northern Ghana, decreases. However, the finding is contrary to Asante et al. (2011). They established a positive relationship between age and farmers' decision to join farmer based organization in Ghana.

Household size was significantly associated with a higher probability of farmer's willingness to participate on the platform. The probability of farmer's willingness to participate increases by 3.3 percent for every additional member added to the household

Table 7 Membership of MSP by membership of FBO

Membership of MSP	Membership of FBO					
	Yes	Percentage	No	Percentage	Total	Percentage
No	51	45.50	69	50.00	120	48.00
Yes	61	54.50	69	50.00	130	52.00
Total	112	100.00	138	100.00	250	100.00

Source: Author's Household Survey Data (2012).

Table 8 Willingness of participation in MSP by membership of FBO

MSP participation	Membership of FBO					
	Yes	Percentage	No	Percentage	Total	Percentage
Yes	97	86.60	118	85.50	215	86.00
No	15	13.40	20	14.50	35	14.00
Total	112	100.00	138	100.00	250	100.00

Source: Author's Household Survey Data (2012)

(Table 9). Household size represents the supply of family labour for production activities. Farming in the tropics is mostly rain-fed and labour-intensive. A household head with large household size will be more willing to participate on the platform because of excess or additional labour to work on the farm whilst absent from the farm. The demand of economically inactive household members coerces household heads to search for innovative ways to improve upon their farming business.

Willingness to participate on the platform was significantly influenced by household income. The likelihood of farmers' willingness to participate on the platform increases by 0.04 percent for every additional increase in household income (Table 9). Transportation cost is one of the major constraints in MSP participation. Household head with higher income are able to overcome this cost and also make financial contributions in the form of dues and levy as demanded by the platform to ensure its sustainability. Household heads with higher income may necessarily participate on the platform to widen their social networks. Asante et al. (2011) also established a positive relationship between farmers' income and willingness to join Farmer Based Organizations. According to their findings, increasing farmers' income by one Ghana cedi increases the likelihood of joining FBOs by 0.026%.

Constraints in multi-stakeholder platform participation

The constraints of smallholder rice farmers identified during the survey are presented in Table 10. The Kendall's 'W' was found to be 0.205 and significant at 1% level. The

Table 9 Determinants of willingness of MSP participation

Variable	Estimated result of probit model		
	Coefficient	Std error	Marginal effect
Age of household head	−0.0688	0.0106	−0.0757***
Gender	0.0001	0.2592	0.0002
Marital status of household	0.0185	0.6966	0.0203
Household size	0.0302	0.0174	0.0332*
Years of education of household head	0.0616	0.0463	0.0678
Membership of association	0.3354	0.2533	0.3689
Land availability	−0.0802	0.0561	−0.0882
Major farming decision	−0.0628	0.3570	−0.0691
Distance to meeting place	−0.0456	0.0663	−0.0501
Household income	0.0004	0.0002	0.0004**
Constant	3.5952	0.9925	3.9547
Number of Observations	250	Log likelihood	−64.5260
Mean dependent var.	0.8600	McFadden R-squared	0.3626

Source: Regression Estimation from Author's Household Survey Data (2012) ***$p < 0.01$, **$p < 0.05$ and *$p < 0.10$.

Table 10 Constraints in multi-stakeholder platform participation

Identified constraints	Mean rank
Distance to meeting place	3.14
Confliction of meeting days with market days	3.18
Information dissemination regarding meeting days	3.70
Risk	3.77
Weather condition	3.82
Means of transport	4.38
Farming season for holding meeting	6.00
Number of observation	250
Kendall's W	0.205
Chi-square	307.110
Df.	6
Assymp. Sig.	0.000

Source: Estimation from Author's Household Survey Data (2012).

null hypothesis was rejected in favour of the alternate hypothesis which stated that there is agreement among the constraints faced by smallholder farmers in MSP participation. The Kendall's 'W' of 0.205 indicates that there was 21 percent agreement between the respondents in the ranking of the constraints faced by smallholder rice farmers with respect to participation on the platform (Table 10).

Among the identified ranked constraints, distance to the meeting place, confliction of meeting days with market days, poor dissemination of information regarding meeting days and risk were the top four most constraining factors of farmers' willingness to the MSP participation. Distance to the meeting place was found to be the most constraining factor of MSP participation. The success of the MSP is highly dependent on the location of the meeting place. The situation becomes aggravated where there is no access to transport services. Conflicting meeting days with market days was another major constraint to a successful willingness to participate on the platform. Most of the smallholder rice farmers are engaged in marketing activities either as sellers or buyers. Marketing of farm produce generates income for the household. Timely dissemination of information about meeting days was crucial to ensuring maximum participation on the platform. Season for holding meeting was the least identified constraint according to the respondents which indicated that the MSP meeting can be held during both the rainy and dry season.

Conclusions and recommendations

The study described multi-stakeholder platform participation based on key demographic indicators. There was high willingness of participation on the platform among smallholder rice farmers in Northern Ghana. The Probit regression analysis revealed that age of household head, household size and income were the main determinants of farmers' willingness to the MSP participation. Distance to meeting place was the most important constraint faced by the smallholder rice farmers in relation to their willingness to participate on the platform. Smallholder farmers residing at distant place from the meeting grounds were less likely to participate on the platform. The facilitators of the MSP must target and educate younger household heads within the region to ensure maximum participation and sustainability of the platform. Secondly, in designing of agricultural programmes, policy makers must incorporate activities that will impact positively on the income of farmers. Finally, the MSP must be situated at a central location coupled with improvements in road infrastructure in order to lessen the transportation burden of participants and consequently increase willingness of participation.

Endnotes

[a]Hubs represent key rice ecologies and different market opportunities across African countries, linked to major national or regional rice-development efforts to facilitate broader uptake of rice knowledge and technologies. It involves large groups of farmers and other value chain actors such as rice millers, input dealers and traders.

Abbreviations

ABCs: Agricultural Business Centres; AfricaRice: Africa Rice Center; CAPAC: Cadinas Agricolas Productivas de Calidad; CDFO: Commercial Development of Farmer Based Organization; CFA: Financial Cooperation in Central Africa; CIRAD: Center for International Research in Agricultural Development; CONPAPA: Consorcio de Pequeños Productores de Papa, Consortium of Small Potato Producers; CSIR: Council for Scientific and Industrial Research; Df: Degrees of freedom; FASDEP: Food and Agriculture Sector Development Policy; FBO: Farmer-Based Organization; IPTA: Innovation platform for technology adoption; LR: Likelihood ratio; MCA: Millennium challenge account; MoFA: Ministry of food and agriculture; MSP: Multi-stakeholder platform; NGOs: Non-Government Organizations; SARI: Savanna Agricultural Research Institute; SRID: Statistical research and information directorate.

Competing interests

The authors declare that they have no competing interests both financial and non-financial.

Authors' contributions

This work was carried out in with the contribution from all listed authors. EM designed the study, performed the statistical analysis, and wrote the first draft of the manuscript. PME was in charge of the collection and management of the data. ANW managed and streamlined the analyses of the study and author WD managed the literature searches and edited the manuscript. All authors read and approved the final manuscript.

Authors' information

Edward Martey is an agricultural economist at the Savanna Agricultural Research Institute of the Council for Scientific and Industrial Research (CSIR-SARI), Nyankpala, Ghana.

Acknowledgements

The authors greatly acknowledged financial support by USAID and the smallholder rice farmers that voluntarily participated in the survey.

References

Adimado S (2001) Willingness to pay for research findings: a case study of pineapple farmers in Ghana. University of Ghana, Legon, Accra Ghana, pp 38–50, Unpublished MPhil Thesis Submitted, Department of Agricultural Economics and Agribusiness

Asante BO, Afari-Sefa V, Sarpong DB (2011) Determinants of small-scale farmers' decision to join farmer based organizations in Ghana. Afr J Agric Res 6(10):2273–2279

Ayamga M (2006) Factors influencing the decision to participate in microcredit programmes: an illustration for Northern Ghana. Ghana J Dev Stud 3(2):57–65

Coase RH (1937) The nature of the firm. Economica 4:386–405

Coase RH (1960) The problem of social cost. J Law Econ 3:1–44

Delgado C (1999) Sources of growth in smallholder agriculture in sub-Saharan Africa: the role of vertical integration of smallholders with processors and marketers of high value-added items. Agrekon 38(S1):165–89

Dorward A, Kydd J, Poulton C, Bezemer D (2009) Coordination risk and cost impacts on economic development in poor rural areas'. J Dev Stud 45(1):1–20

Edwards AL (1964) Statistical methods for the behavioural sciences. Holt, Rmehart and Wintson, New York, pp 402–410

Enete AA, Igbokwe EM (2009) Cassava market participation decision of household in Africa. Tropicultura 27(3):129–136

Etwire PM, Dogbe W, Wiredu AN, Martey E, Etwire E, Owusu RK, Wahaga E (2013) Factors influencing farmer's participation in agricultural projects: the case of the agricultural value chain mentorship project in the Northern Region of Ghana. J Econ Sustain Dev 4(No.10):36–43

Ghana Statistical Service, GSS (2008) Ghana living standards survey report of the fifth round (GLSS 5), Accra, Ghana. http://www.statsghana.gov.gh/docfiles/glss5_report.pdf

Gockowski J, Ndoumbe M (2004) The adoption of intensive monocrop horticulture in Southern Cameroon. J Agric Econ 30:195–202

Goetz S (1992) A selectivity model of household food marketing behavior in sub-Saharan Africa. Am J Agr Econ 74:444–52

Greene HW (2003) Econometric analysis, 5th edn. Pearson Education, Inc., Upper Saddle River, New Jersey, USA
Holloway G, Nicholson C, Delgado C, Staal S, Ehui S (2000) Agro-industrialization through institutional innovation, transaction costs, cooperatives and milk market development in the East African Highlands. Agr Econ 23(3):279–88
Key N, Sadoulet E, de Janvry A (2000) Transaction costs and agricultural household supply response. Am J Agr Econ 82(1):245–59
Kheralla M, Minet N, Kachule R, Souce BG, Berry P (2001) Impact of Agricultural Market Reforms on Smallholder Farmers in Benin and Malawi, Research Report, Vol. 2. IFPRI
Langyintuo AS, Mekuria M (2005) Accounting for neighborhood influence in estimating factors determining the adoption of improved agricultural technologies. A paper at American Agricultural Economics Association annual meeting, providence. Rode Island, pp 1–28
Makhura M, Kirsten J, Delgado C (2001) Transaction costs and smallholder participation in the maize market in the Northern Province of South Africa, Seventh Eastern and Southern Africa Regional Maize Conference, February, pp. 463-467. Agriculture in Ghana: Facts and Figures, Accra, Ghana
Markelova H, Meinzen-Dick R, Hellin J, Dohrn S (2009) Collective action for smallholder market access'. Food Pol 34:1–7
Martey E, Wiredu AN, Asante BO, Annin K, Dogbe W, Attoh C, Al-Hassan RM (2013) Factors influencing participation in rice development projects: the case of smallholder rice farmers in Northern Ghana. Int J Dev Econ Sustain 1(2):13–27
Mattson DE (1986) Statistics-Difficult Concept of Understanding Explanations'. Bolchanzy Carducci Publishers Inc, p 281, 283.361,423
Ministry of Food and Agriculture, MoFA (2010) Agriculture in Ghana: facts and figures. Accra, Ghana, p 53
Mussei A, Mwanga J, Mwangi W, Verkuijl H, Mungi R, Elang A (2001) Adoption of improved wheat technologies by small-scale farmers in Mbeya District, Southern Highlands, Tanzania. International Maize and Wheat Improvement Centre (CIMMYT) and the United Republic of Tanzania, Mexico D.F
Nnadi FN, Akwiwu CD (2008) "Determinants of Youths' Participation in Rural Agriculture in Imo State, Nigeria". Journal of Applied Sciences 8(2): 328–333
Ofori IM (1973) Factors of agricultural growth in West Africa. ISSER, University of Ghana, Legon, Accra, Ghana
Olwande J, Mathenge M (2010) Market participation among the poor rural households in Kenya. Tegemeo Institute of Agricultural Policy and Development
Omamo SW (1998) Farm-to-market transaction costs and smallholder agriculture: Explorations with a non-separable household model. J Dev Stud 35:152–63
Pingali P, Meijer M, Khwaja Y (2005) Commercializing small farms: reducing transaction costs. www.ifpri.org/events/seminars/SmallFarms/ Accessed 10 October 2007
Rölling N, Leeuwis C, Pyburn R (2002) Beyond the aggregation of individual preferences: moving from multiple cognition in resource dilemmas. In: Leeuwis C, Pyburn R (eds) Wheelbarrows full of frogs: social learning in rural resource management. Koninklijke Van Gorcum, Assen, pp 25–47
Schultz TW (1945) Agriculture in an unstable economy. McGraw-Hill Book Company Inc, New York
Shepherd A (1997) Market information services: theory and practice. FAO (Food and Agriculture Organization), Rome
Southworth HM, Johnston BF (1967) Agricultural development and economic growth. Cornell University Press, UK
Stiglitz JE (1988) Economic organization, information and development. In: Chenery H, Srinivasan (eds) Handbook of development economics, vol. 1, chapter 5. Elsevier Science, Amsterdam
Thiele G, Devaux A, Reinoso I, Pico H, Montesdeoca F, Pumisacho M, Andrade-Piedra J, Velasco C, Flores P, Esprella R, Kurt Manrique T, Horton D (2011) Multi-stakeholder platforms for linking small farmers to value chains: evidence from the Andes". Int J Agr Sustain 423–433
Thiele G, Oros R, Velasco C, Ambrose K (2005) Por qué trabajar con plataformas? In: Thiele G, Bernet T (eds) Conceptos, Pautas y Herramientas. CIP, Lima, pp 66–70, 171
Vellema S, Ton G, Nakimbugwe D, Mwesige D, (2011) The Ugandan Oil Seed Sub-sector. Putting heads together: Agricultural innovation platforms in practice. ISBN 9789460221835. http://library.wur.nl/WebQuery/wurpubs/426831
World Bank (2006) Where is the wealth of nations? Measuring capital for the 21st century. World Bank, Washington, DC

10

Credit off-take from formal financial institutions in rural India: quantile regression results

Debdatta Pal[1*] and Arnab Kumar Laha[2]

* Correspondence: debdatta@iimraipur.ac.in
[1]Economics and Business Environment, Indian Institute of Management Raipur, GEC Campus, Raipur 352015, India
Full list of author information is available at the end of the article

Abstract: Extending financial services to unbanked population in India has remained a central part of the policy thrust of the Indian government for decades. To that effect, a widespread formal credit delivery mechanism has been established to meet the credit requirements of rural communities. However, the Indian government-backed formal financial sector has had limited success in providing resources to poor rural households, which has led to strong criticism of the policy and its implementation. In this study, we use data from 600 rural households spread across six Indian states to examine the changing distribution of credit off-take among borrowers of formal financial institutions. By using quantile regression, we find that even among rural households that could access loans from the formal banking sector, the distribution of credit off-take is skewed towards resource-rich households. We also find that even among borrowers in the upper quantiles of the conditional loan distribution, marginal farmers received substantially less loan amounts than those belonging to the category of medium and large farmers.

JEL classifications: C31, Q14

Keywords: Rural credit; Formal finance; Quantile regression; India

Background

The policy thrust to make formal credit inclusive has been a central focus of the Indian government. This may be attributed to two widely accepted facts. First, access to credit is a precursor for growth, and second, alternative credit sources are costly. Policy-makers in India have long pushed the formal financial sector to open more outlets and bankers to offer credit below the market clearing price (Burgess and Pande 2005). However, since credit is an inter-temporal contract, lenders face the dual problem of *adverse selection* (hidden information) and *moral hazard* (hidden action). If lenders, through screening, cannot identify borrowers based on their likelihood of default, willful default may rise. In response to this, informal lenders may add a *default premium* on top of the nominal interest rate, an option that is unavailable to formal financial sources, where interest rates are controlled by the government (Gonzalez-Vega 1984). Hence, formal lenders, namely banks and cooperatives, may ration credit, request additional documents from and frequent visits by the borrower (leading to higher borrower-side transaction costs), demand marketable collateral, and favor production credit over consumption loan. The void created out of the unmet credit demand from formal financial sources has led to the widespread dependence of Indian rural households on informal credit sources, namely moneylenders and traders (see Basu 2006 for empirical evidence).

Academic inquiries to identify the causes of such dismal performance on the part of the formal financial sector in agrarian economies can be classified under three broad groups: first, those that attempt to identify factors affecting the choice of credit sectors in rural economies (see, for example, Kochar 1997; Pal 2002; Sahu et al. 2004; Boucher and Guirkinger 2007; Datta and Ghosh, 2013; Pal and Laha 2013), second, those that explore the determinants of credit amounts received by rural households from the formal sector (see, for example, Datta 2003; Sahu et al. 2004) and third, those that estimate the effects of the lack of access to formal-sector credit on agricultural production (Carter 1989; Guirkinger and Boucher 2008). The first two groups of empirical investigations indicate that a large agricultural land size, good land quality, the presence of irrigation and regarding the loan applicant, membership in a higher caste and higher levels of educational attainment increase the applicant's chance of gaining access to formal credit sources as well as the quantum of credit disbursed by the formal sector. The third set consists of a growing body of empirical literature that suggest that credit constraints adversely affect farm production, farm profit, and farm investment.

This study contributes additional empirical evidence to support the findings of the second set of inquiries and attempts to answer the following question: among the rural households that have received loans from the formal sector, what is the changing distribution of credit off-take? Finding an answer to this question is pertinent from the following policy perspectives. First, it would help central bankers as well as policy makers understand variations in outreach of the formal banking sector within its own client group. Secondly, it would elucidate the extent of the gap between the credit off-take from the formal financial sector by the rich and poor households in the context of a developing country like India. Moreover, apart from including those variables mentioned in earlier studies, namely Datta (2003) and Sahu et al. (2004), this study also attempts to capture the effects of village-level attributes, such as the distance of households from commercial, financial, and educational institutions, on the loan amount.

These findings would complement the theoretical model advanced by Nakamura and Nakajima (2011) which shows that improvements in the credit market during the early stages of economic development is essential to escape poverty.

We used quantile regression, which was introduced by Koenker and Bassett (1978), to address this issue of improving the credit off-take from formal financial sources in rural India. Quantile regression is used when conditional quantile functions and even extremes are of interest, and it has been used in various areas of applied economics, such as health economics (see Winkelmann 2006, for example), financial economics (see Bassett and Chen 2001, for example), labor economics (see Buchinsky 1998, for a detailed survey), and education economics (see Eide and Showalter 1998). To the best of our knowledge, quantile regression has not been employed thus far to study changes in the distribution of credit off-take by rural households. Consistent with the theory of credit rationing, we observed that between resource-rich and resource-poor rural households, there is considerable variation in the quantum of loan off-take from formal financial institutions. By using quantile regression, we found that among borrowers in the 90th quantile of the conditional loan distribution, marginal farmers[a] received only approximately 80 percent of the loan amount of their counterparts belonging to the category of medium and large farmers.

Briefly, the structure of this paper is as follows. In Survey and data description, we describe the survey and dataset. Method presents the methodology used for analysis. Results and discussion contains the results and discussion, and in Conclusions, we indicate the policy implications.

Survey and data description

The survey was conducted between May and December 2010 as part of the research project of the Centre for Management in Agriculture, Indian Institute of Management Ahmedabad, titled "Assessing Policy Interventions in Agri-Business and Allied Sector Credit versus Credit Plus Approach for Livelihood Promotion," sponsored by the Ministry of Agriculture, Government of India. Six Indian states—Chhattisgarh, Maharashtra, West Bengal, Andhra Pradesh, Tamil Nadu, and Gujarat—were selected based on two state-level indicator criteria: first, the average population per branch of formal financial institution as of the end of June 2009, and second, incidences of indebtedness of rural households. Table 1 shows these two indicators against the national average.

Against the national average of 6554 people per rural branch of formal financial institution attempt has been made to choose states with even spread on this criterion. While a bank branch caters to 10410 people on average in Chhattisgarh, the number of people served by a bank branch in Maharashtra is as low as 2688. The average population per bank branch in West Bengal is the closest to national average. Therefore, variation on the level of branch penetration is well maintained in the sample. Similarly, the sample states are distributed evenly above, below, and at the average for the criterion of incidences of indebtedness of rural households. Through this process, we attempted to capture variations among the sample states. To ensure that various lending institutions are represented, we selected a village or cluster of villages within a single agro-climatic region of a state where all credit sources, namely, formal, semi-formal, and informal, are present (see Additional file 1: Appendix 1). Formal lending institutions are scheduled public or private sector commercial banks[c] (SCB), regional rural banks[d] (RRB), and cooperative banks,[e] namely, primary agricultural cooperative societies (PACS), multi-purpose PACS (MPACS), District Central Cooperative Bank (DCCB), and State Cooperative Banks. Semi-formal credit sources include self-help groups[f] (SHG) and microfinance institutions[g] (MFI). The informal sector includes moneylenders, input dealers, output traders, shopkeepers, and friends and relatives. A complete enumeration of all

Table 1 Indicators of branch penetration and rural indebtedness

State	Average population per rural financial institution branch as of March 2009[#]	Incidence of indebtedness of rural households (in %) as of June 2002[$]		
		Institutional	Non-institutional	All
Chhattisgarh	10410	14.4	6.7	19.8
Maharashtra	2688	22.8	7.2	27.5
West Bengal	6095	12.1	11.0	21.8
Andhra Pradesh	11784	14.9	32.9	42.3
Tamil Nadu	5976	13.9	21.3	31.3
Gujarat	3759	14.7	15.8	28.1
All of India	6554	13.4	15.5	26.5

[#]Authors' own calculation[b].
[$]Household indebtedness in India as of 06.30.02 (National Sample Survey Organisation 2005).

households[h] in a village was undertaken to form a sampling frame that was used to draw a representative sample of village households.

The survey was conducted in five villages in West Bengal, three villages each in Maharashtra and Chhattisgarh, and one village each in Andhra Pradesh, Gujarat, and Tamil Nadu. From each village in West Bengal, 30 households were chosen through random sampling, whereas in the other states, 50 households were randomly chosen from each village. Thus, the total sample size in this study consists of 600 households. Each sample household was administered a structured questionnaire to collect general information on household demographics, social characteristics, occupation (both primary and secondary), resource endowment, loan-related information (i.e., access to different credit sources between April 2009 and March 2010), and the terms and conditions of the loan facilities experienced by the borrowers.

Incidentally, besides respondents that borrowed from mutually exclusive sources (i.e., formal, semi-formal, or informal), this survey also included those that borrowed from multiple sources. Of the 600 sampled households, 231 met their full credit demand entirely from formal sources, 129 met part of their credit demand from formal sources, 91 respondents were excluded from formal sources but have used credit facilities from semi-formal and/or informal sources, 71 were dependent solely on informal credit sources, and 75 respondents did not avail any credit facilities during the corresponding period (see Table 2).

Of the 600 households, 525 availed themselves of at least one credit facility during April 2009 to March 2010. Out of these 525 borrower households, 293 households have used only one loan during this period. During the corresponding period, the numbers of households with two, three, and four loans are 140, 53, and 39, respectively. Therefore, the total number of loan cases gathered from this survey is 888 (see Table 3).

Of the 888 loan cases considered, 399 loans were offered by formal credit sources, 109 loans by semi-formal sources, and the remaining 370 loans by the informal sector. We followed the coding pattern suggested by Walter et al. (1987), which is given in Additional file 2: Appendix 2. To elucidate the characteristics of the villages and households covered in this study, we report summary statistics in Table 4.

Method

The analysis is based on 399 loan contracts extended by the formal financial sector among the 600 sample households. Earlier analyses (see, Datta 2003; Sahu et al. 2004)

Table 2 Categories of households based on loan sources

Source of loan			Category of household
Formal	Semi-formal	Informal	
Yes	No	No	Households met their full credit demand entirely from the formal sector
Yes	Yes	No	Households met part of their credit demand from the formal sector
Yes	No	Yes	
Yes	Yes	Yes	
No	Yes	No	Households excluded from the formal sector but which used credit facilities from semi-formal and/or informal sectors
No	Yes	Yes	
No	No	Yes	Households solely dependent on the informal sector
No	No	No	Households without any credit facilities

Table 3 Distribution of loans across households

Number of loans received during 2009-10	Number of households	Total loans
1	293	293
2	140	280
3	53	159
4	39	156
Total	525	888

of the determinants of the loan amount (*Y*) from the formal sector used ordinary least squares (OLS) regression methods. However, these OLS estimates of the various effects on the conditional mean of the loan amount, $E(Y \mid covariates)$, may not truly represent the nature and size of those effects on the lower tail of the loan amount distribution.

Here, we deviate from the earlier approach and use quantile regression (see, Koenker and Bassett 1978; Koenker 2005), which models the dependence of the conditional quantiles of *Y* given the covariates. This is helpful in models wherein extremes are vital. Here, quantile regression is expected to yield a comprehensive picture of the conditional distribution of loan off-take from formal creditors by rural customers given different village, household, and loan attributes for both the lower and upper quantiles.

Let $(y_i, \boldsymbol{x}_i)$, $i = 1,\ldots,n$, be a sample from a population, where $\boldsymbol{x}_i$ is a $(K \times 1)$ vector of regressors. Assuming that the τth quantile of the conditional distribution of y_i is linear in $\boldsymbol{x}_i$, let us denote $\xi(\boldsymbol{x}_i, \boldsymbol{\beta}_\tau)$ as the τth quantile of the distribution of $y_i \mid \boldsymbol{x}_i$. That is,

$$\xi(\boldsymbol{x}_i, \boldsymbol{\beta}_\tau) \equiv \inf[y : F_i(y|\boldsymbol{x}) \geq \tau]$$

Then, we assume that $\xi(\boldsymbol{x}_i, \boldsymbol{\beta}_\tau) = \boldsymbol{x}_i'\beta_\tau$, where, $\boldsymbol{\beta}_\tau$ is the unknown vector of parameters to be estimated for varying values of τ where $0 < \tau < 1$. By changing the value of τ, the entire distribution of y conditional on $\boldsymbol{x}$ can be traced. The estimator for $\boldsymbol{\beta}_\tau$ comes from:

$$min \sum_{i}^{n} \rho_\tau \left(y_i - \boldsymbol{x}_i'\boldsymbol{\beta}_\tau \right)$$

where $\rho_\tau(u)$ denotes the function defined as

$$\rho_\tau(u) = \begin{cases} \tau u & if\ u \geq 0 \\ (\tau - 1)u & if\ u < 0 \end{cases}$$

For further details, see Koenker (2005), pp. 28-67. In this study, values of τ are taken as $\frac{1}{10}$, $\frac{1}{2}$, and $\frac{9}{10}$, which results in the 10th quantile, median, and 90th quantile, respectively. Here, y_i is the amount of loan received by the sampled rural households from formal credit sources during 2009-10, and $\boldsymbol{x}_i$ is a vector of independent variables describing the village, household, and loan characteristics.

Results and discussion

As an outcome of the exchange mechanism, a credit transaction is governed not only by the demand for credit but also by the lender's willingness to extend credit. Furthermore, being an inter-temporal transaction, credit exposes the lender to post-contractual *default risks*. Therefore, the lender would take utmost care to ensure timely repayment of his or her dues. Hence, it is likely that the formal lender will consider a

Table 4 Summary statistics of exogenous variables used in the empirical analysis

A

Variable	Definition of variable	Mean	Minimum	Maximum	n
PIRRI	Percentage of agricultural land under assured irrigation in a particular village	58.60	0	100	14
PELECTRIC	Percentage of households having access to electricity in a particular village	72.73	3	98	14
AGRI	Average distance (in km) from agricultural infrastructures (e.g., Agriculture Produce Marketing Committee sub-yard, office of extension service providers, agricultural input retailers, and farm machinery dealers)	5.15	0.83	10.33	14
EDU	Average distance (in km) from educational institutions	2.23	0	6.5	14
FININS	Average distance (in km) from formal financial institutions (e.g., commercial banks and cooperatives)	3.76	0	8.33	14
FAMTOT	Number of key social leaders (e.g., *Panchayat Pradhan*[i], government extension officers, local school head masters, and financial institution officials) whom the household is familiar with	3.95	0	7	600
LOANAMT	Amount (in Indian Rupee - INR[j]) of loan made available	20656.19	200	600000	888
TCOST	Borrower's total transaction cost in INR	246.89	0	2960	888
RAINT	Annualized interest rate in reducing balance	18.47	0	61	888

B

Variable	Definition of variable	Frequency		Cumulative
		1	0	
CASTRDUM	Caste and religion:1 = Upper caste Hindu, 0 = Otherwise (Lower caste Hindu and minorities, such as Muslims)	186 (31.00)	414 (69.00)	600 (100)
LANDD1	Household is minimum from marginal farmer category (Yes = 1, No = 0)	536 (89.33)	64 (10.67)	600 (100)
LANDD2	Household is minimum from small farmer category (Yes = 1, No = 0)	199 (33.17)	401 (66.83)	600 (100)
LANDD3	Household is minimum from medium/large farmer category (Yes = 1, No = 0)	69 (11.50)	531 (88.50)	600 (100)
IRRISTATD1	At least a part of the agricultural land of the household has assured irrigation (Yes = 1, No = 0)	343 (57.17)	257 (42.83)	600 (100)
IRRISTATD2	Agricultural land of the household is covered under assured irrigation (Yes = 1, No = 0)	188 (31.33)	412 (68.67)	600 (100)
ALACDUM	Household engaged in allied activity as fisheries/orchard/farm forestry (Yes = 1, No = 0)	93 (15.50)	507 (84.50)	600 (100)
DWELLD1	Household possesses at least a mud built house (Yes = 1, No = 0)	580 (96.67)	20 (3.33)	600 (100)
DWELLD2	Household possesses at least a mud and brick built house (Yes = 1, No = 0)	285 (47.50)	315 (52.50)	600 (100)
DWELLD3	Household possesses a brick built house (Yes = 1, No = 0)	74 (12.33)	426 (87.67)	600 (100)
SANITD1	Household having at least an open toilet (Yes = 1, No = 0)	438 (73.00)	162 (27.00)	600 (100)

Table 4 Summary statistics of exogenous variables used in the empirical analysis *(Continued)*

SANITD2	Household with modern toilet facility (Yes = 1, No = 0)	271 (45.17)	329 (54.83)	600 (100)
PURPDUM	Loan is for production purpose (Yes = 1, No = 0)	525 (59.12)	363 (40.88)	888 (100)
SECDUM	Loan is secured by marketable collateral (Yes = 1, No = 0)	294 (33.11)	594 (66.89)	888 (100)

Note: Figures in parentheses are percentages.

variety of factors before deciding to approve a loan application. Further, for the loan applicant, the quantum of loan requirement would also be dependent on various factors. Below, we describe these factors and their relationship to the chance of loan approval by a formal lender as well as to the quantum of loan requirement.

Factors affecting the supply of credit

Since credit is extended to agricultural households who borrow on the basis of their land holding, such land holding acts as collateral from the lender's perspective. The lender is assured that borrowers with a larger acreage of land would produce sufficient output to generate income to repay loans without difficulty. An assured irrigation facility increases the reliability of higher crop production and is, in turn, associated with a higher profit potential of the farmer applicant. This would be perceived positively by the lender and would carry a higher probability that loan applications for higher amounts will be considered favorably by the formal lender. Lenders may also prefer households that engage in allied activities such as cattle rearing and fishery, which are alternative earning sources to farming, as such activities may lower the probability of default in the event of crop failure. Access to information from extension agents is expected to help farmers adopt improved farming technology, thereby increasing the probability of higher yields. Lenders may prefer farmers with a possibility of registering more income from higher yields.

The possession of landed house property as well as in-house toilet facilities would signal a resource-rich household in rural India, thereby increasing the household's chance of receiving a higher amount in loans from the formal lender. Familiarity with key social leaders is expected to improve the credentials of the loan applicant and, hence, may be considered favorably by the formal lender. Since large tracks of land are traditionally owned by resource-rich upper caste Hindu families in India, belonging to such a family may also create a favorable impression on the lender.

Indeed, banks may prefer households from villages with a higher coverage of assured irrigation and access to electrical power, as farmers from those villages are expected to undertake multiple-cropping and thus carry better profit potential as compared to households that do not have access to such public infrastructure support. Villages closer to agricultural infrastructure carry higher business potential and, thus, may be favored by formal lenders. Furthermore, proximity to formal financial institutions is expected to lead to a wider outreach of financial services as a result of minimum transaction costs incurred by the lender to service those clients. Banks may also prefer to lend in villages closer to educational institutions, as those villages would have a higher number of educated households, who may use loan funds more judiciously and carry better employability in non-farm sectors and, hence, have sufficient income to repay loans.

Lastly, formal creditors would prefer to lend for the purpose of production over that of consumption, since the former is associated with the possibility of generating positive surplus required to meet debt obligations.

Factors affecting the demand for credit

A large track of land with a good irrigation facility not only extends a positive signal to the formal lender but it also boosts the demand for credit to meet higher production expenses incurred from the purchase of improved varieties of seeds and other inputs, wage payments to laborers, the rental of agricultural machinery such as tractors, and electricity charges. Moreover, households with allied activities such as cattle rearing and fishery could assume a higher debt exposure, as income from alternative sources would act as a contingency for avoiding any *default risks* in the event of production failure. Households with access to information from extension agents are expected to adopt improved farming technology, thus, they may require a higher quantum of loan to meet higher production costs. Furthermore, upper caste Hindu families with large land holdings may demand higher loan amounts to meet higher costs of cultivation.

Peasant households belonging to villages with access to assured irrigation and electrical power are expected to undertake multiple-cropping and thus require a higher amount in loans compared to villagers farming under rain-fed mono-cropping systems. Villages closer to agricultural infrastructures are expected to carry a higher business potential and are therefore more likely to borrow money for expanding agricultural or other business activities. In addition, agricultural households from villages located near formal financial institutions are expected to perform more transactions and avail themselves of a higher quantum of loans due to the minimum transaction costs involved in dealing with formal lenders. Similarly, villages located near educational institutions are expected to have a higher number of educated households that are more informed about the lending process of formal financial institutions and thus have a better scope of obtaining higher loan amounts from the formal sector.

Households may apply for loans for agricultural production or other purposes, namely consumption purpose. Such loans may or may not be backed by marketable collateral. These, as well as the total *ex ante* monetary costs involved in negotiating the loan and annual interest rates, represent loan attributes that may affect both access to credit from the formal lender and the quantum of loan.

The variable definitions given in Table 4 include all of the variables mentioned above.

Initially, we outlined the effects of these variables on the loan amount, by using OLS. Next, three quantile regression models, for the 10th quantile, 50th quantile, and 90th quantile, were fitted with the loan amount as the dependent variable and the various village, household, and loan attributes as explanatory variables. The model parameters were estimated using the PROCREG and QUANTREG procedures of the SAS software.

State-level variations in both OLS and quantile regression were captured by state dummies. Although data were collected from 3-5 villages for Chhattisgarh, Maharashtra, and West Bengal, these villages are from a single cluster of an agro-climatic region.

Hence, we assume that the state-level dummy would be sufficient for capturing variability across villages.

Results of OLS regression analysis

The results of the OLS estimates are presented in Table 5. As expected, the presence of irrigation facilities and electrical power in a borrower's village positively boosted the credit off-take from formal financial institutions in that village. Close proximity of the borrower's village to formal financial institutions positively affected the loan amount, as expected. However, close proximity of the borrower's village to agricultural markets

Table 5 Ordinary Least Squares estimates of the linear regression model for loan amount

Explanatory variables	Parameter estimates	
	Coefficient	SE
Intercept	−209997***	44913.00
State dummy[1]		
Chhattisgarh (CGARH)	−32785*	16763.00
Maharashtra (MAHA)	−44746***	15840.00
West Bengal (WB)	−34541	23433.00
Andhra Pradesh (AP)	−81708***	20033.00
Tamil Nadu (TN)	−109030***	22474.00
Village attribute		
PIRRI	579.48*	328.02
PELECTRIC	1211.24***	312.82
AGRI	3075.69	4648.62
EDU	8582.52*	4590.04
FININS	−3606.29	4264.39
Household attribute		
LANDD1	1342.84	15353.00
LANDD2	7719.08	6557.80
LANDD3	28154***	8522.69
IRRISTATD1	2773.70	9178.17
ALACDUM	26352**	10331.00
INFORDUM	−25485*	7113.22
DWELLD2	−9211.10	7991.39
SANITD1	11445	8177.80
FAMTOT	5151.78**	2239.92
CASTRDUM	15235*	8132.51
Loan attribute		
PURPDUM	9801.21	11864.00
SECDUM	15093**	6910.41
TCOST	54.15***	7.91
RAINT	7745.93***	1234.58
R-square		0.5224
Number of observations		399

***Significant at a level of 1 percent; **significant at a level of 5 percent; *significant at a level of 10 percent significance.
[1]Gujarat (GUJ) state was used as the base state in this analysis.

and educational institutions in this OLS regression was found to negatively affect borrowing from the formal sector.

Unsurprisingly, farm households with a minimum of two hectares of land were found to be favored by formal lenders. Assured irrigation facilities do have a positive effect on the approval of loans from formal lenders. Moreover, households with additional income from allied activities were able to borrow greater amounts in loans from formal lenders. Familiarity with key social leaders, as expected, may have established a borrower's credentials on a strong footing for the formal lenders, therefore, these households could obtain higher amounts in loans from banks. Agricultural households belonging to the upper caste Hindu community, which traditionally controlled better production resources, attracted big ticket loans, as expected. In addition, formal lenders met the higher credit demands of households who asked for production loans and who offered marketable security to obtain such loans.

The above analysis is based on OLS regression analysis, which estimates the average loan amount from the financial sector, conditional on the covariates.

Results of quantile regression analysis

The next model, which concerns the estimates of loan amounts across quantiles, is presented in Table 6. As quantile regression analysis estimates the total loan amount in various conditional quantiles, we can develop more detailed and accurate information from different borrower groups, depending on the loan amount. We estimated three separate quantile levels, 10 percent, 50 percent, and 90 percent.

Here, we compared the results of the OLS regression with that of the quantile regression at three levels (i.e., 10th quantile, 50th quantile, and 90th quantile). We also checked whether the direction of the explanatory variables changed over quantiles and tried to interpret possible reasons for such variations across the borrower groups.

The extent of coverage of irrigation facilities in the sample villages was observed to have positively boosted the credit off-take from formal financial institutions across all quantiles, a result that is similar to the one in the OLS estimate. The extent of electrical power coverage, as expected, was found to have a positive and significant effect on loan amounts in both the 50th and 90th quantiles, which is in line with the OLS regression results. However, this independent variable was observed to have a negative effect on loan amounts in the 10th quantile. As one of the major uses of electricity in an agrarian economy is to supply power for mechanized water extraction, it seems plausible as households with loan amounts in the 10th quantile (i.e., with smaller loan amounts) may have used the groundwater-based irrigation facilities of affluent larger farmers on a payment basis, rather than possessing their own pumps.

Close proximity to agricultural infrastructures such as markets, contrary to expectation, was observed to affect the loan amount across the three quantiles negatively. The same variable was also shown to carry a negative effect on the loan amount in the OLS regression. This may be a result of a common practice in Indian villages, wherein traders pick up produce from farm gates. While the farmers save transaction costs in the form of transportation and wage loss, traders benefit from scale

Table 6 Estimated Quantile Regression models for different quantiles of loan amount

Explanatory variables	Parameter estimates					
	10th quantile		50th quantile		90th quantile	
	Coefficient	SE	Coefficient	SE	Coefficient	SE
Intercept	−3023.04	6145.14	−58343.10***	14224.51	−147202***	48695.88
State dummy[1]						
CGARH	−129.87	2022.37	−19829.60***	5339.15	−125685***	20686.72
MAHA	4978.60**	2350.93	−13951.00**	5607.84	−119181***	20987.51
WB	−3981.68	3444.37	−26902.60***	7848.67	−136140***	30152.03
AP	−934.78	4145.60	−41410.40***	6103.54	−146031**	66209.14
TN	8821.28*	4719.19	−29652.00***	7069.57	−185739***	29719.60
Village attribute						
PIRRI	79.70	52.57	277.56***	92.58	840.77**	357.02
PELECTRIC	−20.76	44.68	277.56***	97.64	939.31**	367.13
AGRI	804.46	761.64	2684.61**	1170.26	6420.22	4443.15
EDU	−264.58	664.07	2902.40***	987.40	8018.66	5027.19
FININS	−919.00	649.98	−2713.70***	999.26	−11031***	4025.82
Household attribute						
LANDD1	123.54	2446.33	−8176.26**	3462.95	−1756.14	14581.15
LANDD2	838.44	1159.54	5606.88***	1656.78	9743.56	6883.54
LANDD3	2439.05	1711.23	5507.91**	2679.73	47032.80***	16963.92
IRRISTATD1	−500.14	1332.19	5328.91**	2351.58	6506.94	8388.00
ALACDUM	52.82	1476.04	4681.41*	2605.45	29196.12**	12611.03
INFORDUM	−631.00	1218.70	−4227.71**	1983.89	−14808.40**	7058.17
DWELLD2	17.16	1134.54	−2243.86	1578.45	−13336.90*	8157.36
SANITD1	341.14	1112.72	5740.65***	1616.22	19080.75***	7270.81
FAMTOT	100.91	270.96	2048.82***	610.63	4138.44*	2414.25
CASTRDUM	911.02	1180.00	4869.94**	2062.23	18121.14**	9362.52
Loan attribute						
PURPDUM	−3536.50***	1348.02	7996.44***	2967.14	58381.59***	12673.53
SECDUM	380.90	955.75	5133.02***	1752.80	20225.97***	6662.52
TCOST	9.23	2.52	21.26***	4.22	64.07***	20.64
RAINT	656.33	217.79	2804.72***	450.08	9272.05***	1929.91
Wald test		13.4542		25.4063		9.6373
Likelihood ratio χ^2 (16)		32.6728		59.4395		72.8337
Number of observations		399		399		399

***Significant at a level of 1 percent; **significant at a level of 5 percent; *significant at a level of 10 percent significance.
[1]Gujarat (GUJ) state was used as the base state in analysis.

economics derived from the number of transactions from the same village as well as an assured supply of produce. Thus, even being far away from agricultural infrastructures such as markets, farmers would enjoy favor from formal lenders. Close proximity to formal financial institutions, as expected, boosted the loan off-take from the formal sector across quantiles and was significant at both the 50 percent and 90 percent levels. This regressor also had a similar effect on the regressand in the OLS regression.

Land holding, as expected, was also observed to have a positive and significant effect on loan amounts, which is in line with the OLS estimation results. While at the median level, households with a minimum of one hectare of land became significant, at the 90th quantile level, land holdings of a minimum of two hectares were shown to be significant. This clearly indicates that borrower households with larger operational land holdings are given larger loans from formal creditors. Availability of assured irrigation had a negative though insignificant effect on the loan amount in the lower quantile but a positive and significant effect at the median level. This may be explained by the possibility that households with lesser loan amounts are also marginal or tenant farmers who engage in cultivation during rainy seasons only. Because of the costly nature of crop cultivation during the winter season, such farmers may not be involved in winter crop production, which is dependent primarily on irrigation, and they may be working as wage laborers in the fields of large farmers. Unsurprisingly, alternative income from allied activities boosted the loan off-take from formal financial institutions, a result that is observed both in the OLS and across quantiles. Familiarity with key social leaders was also found to have a positive effect on the loan amount in all three quantiles, as expected. Belonging to the upper-caste Hindu community had a positive effect on the loan amount in all three quantiles but gained statistical significance in the 50th and 90th quantiles only. Both variables also showed a positive and significant effect on formal sector loans in the OLS framework.

Borrowing for production purposes was found to have a positive and statistically significant effect on loan amounts in the OLS and under quantile regression at the 50 percent and 90 percent levels. Interestingly, this variable showed a negative and significant effect on the loan amount in the 10th quantile. This may be attributed to the provision of a small quantum of loans from the formal sector for consumption purposes, which is non-productive in nature. It is worth noting that formal credit sources resort primarily to project-based lending, where there is visible cash flow and adequate surplus to meet debt obligations. The formal sector's appetite for collateral-backed finance was reflected in both the OLS and quantile regression, as availability of collateral was shown to have a positive and significant effect on the loan amount.

We also found that larger loan sizes were positively and significantly associated with higher transaction costs, which means that borrowers may need to incur considerable expenses in frequent visits to bank branches to obtain credit facilities. Furthermore, higher interest rates were found to have a positive and significant effect on loan size. This may be due to prevailing interest rate regulations where, within an interest cap, banks charge more on higher credit limits. These trends were observed in both the OLS and quantile regression (for the 50th and 90th quantiles). Both variables have a positive (though statistically insignificant) effect on loan amounts at the 10 percent level under quantile regression.

The coefficients of the explanatory variables across the 10th, 50th, and 90th quantiles in the quantile regression are listed in Table 6, which indicates the variations in loan off-take from formal credit sources across borrower categories (i.e., marginal farmer, small farmer, and medium and large farmers. Because of the ordinal nature of this variable, the category of landless farmer was included as the reference category. We discuss the findings of this study by illustrating two situations, one favorable and the other unfavorable.

Suppose agricultural household type *A*, which has more than two hectares of land, hails from a relatively prosperous village in Chhattisgarh state. In this village, 90 percent of the agricultural land has access to assured irrigation and the same percentage of households have access to electrical power. The village is located near critical infrastructures, such as agricultural input and output markets, educational institutions, and financial institutions, with the average distance between the village and these facilities being 4 km, 2 km, and 2 km, respectively. Household type *A* belongs to the upper caste Hindu community, enjoys assured irrigation for its agricultural land, is engaged in allied activity, possesses a brick house, and has in-house toilet facilities. The household is familiar with four key social leaders and receives information on agricultural activity from extension agents. It can receive a loan for agricultural production at an annual interest rate of 4 percent. For negotiating this loan contract, this household must incur an *ex ante* transaction cost of INR 300, and because of the household's higher profit potential, it could receive this loan facility without offering any marketable collateral.

In contrast, suppose agricultural household type *B*, which has less than one hectare of land, hails from a backward village in Chhattisgarh state. In this village, only 50 percent of the agricultural land has assured irrigation, and an equal percentage of households have electrical power. The village is located far from critical infrastructures such as agricultural input and output markets, educational institutions, and formal financial institutions, with the average distance between the village and these facilities being 10 km, 6 km, and 8 km, respectively. Household type *B* belongs to either a lower caste Hindu or minority community, has only a part of its land under irrigation, is not involved in allied activities, and lives in a mud house without in-house toilet facilities. The household is familiar with only two key social leaders and does not receive any information about production activity from extension agents. It can receive a loan for production facility at an annual interest rate of 10 percent. To access this loan facility, household type *B* has to bear an *ex ante* transaction cost of INR 500, and because of its poor profit potential, the household must arrange marketable collateral to receive the loan from the formal sector.

Table 7 shows six cases in each state, along with their corresponding 10th quantile, 50th quantile, and 90th quantile of the conditional distribution of the loan amount. These six cases comprise both the favorable and unfavorable conditions described above across the categories of marginal, small, and medium and large farmers. Thus, there are a total of 30 cases across the five states.

In Table 7, we see from Case 1 that the middle 80 percent of type *B* households have a loan off-take between INR 7089 and INR 75653, with a median off-take of INR 30934. From Case 6, we observe that in contrast to type *B* households, 80 percent of type *A* households have a loan off-take between INR 7894 and INR 130012, with a median off-take of INR 36351. We see that there is considerable variation in the amount of loan off-take among both types of households, with the variation being larger for type *A* households. Being resource rich, type *A* households are able to negotiate a higher amount in loans. This stands to be 17.5 percent if we compare the median off-take for both types of households. This is troubling from a state-policy

Table 7 Loan amount across quantiles under varying situations

Case no.	State	PIRRI	AGRI	EDU	FININS	PELECTRIC	LANDD1	LANDD2	LANDD3	IRRISTATD1	ALACDUM	INFORDUM
1	CGARH	50	10	6	8	50	Yes			Yes	No	No
2	CGARH	50	10	6	8	50		Yes		Yes	No	No
3	CGARH	50	10	6	8	50			Yes	Yes	No	No
4	CGARH	90	4	2	2	90	Yes			Yes	Yes	Yes
5	CGARH	90	4	2	2	90		Yes		Yes	Yes	Yes
6	CGARH	90	4	2	2	90			Yes	Yes	Yes	Yes
7	MAHA	50	10	6	8	50	Yes			Yes	No	No
8	MAHA	50	10	6	8	50		Yes		Yes	No	No
9	MAHA	50	10	6	8	50			Yes	Yes	No	No
10	MAHA	90	4	2	2	90	Yes			Yes	Yes	Yes
11	MAHA	90	4	2	2	90		Yes		Yes	Yes	Yes
12	MAHA	90	4	2	2	90			Yes	Yes	Yes	Yes
13	WB	50	10	6	8	50	Yes			Yes	No	No
14	WB	50	10	6	8	50		Yes		Yes	No	No
15	WB	50	10	6	8	50			Yes	Yes	No	No
16	WB	90	4	2	2	90	Yes			Yes	Yes	Yes
17	WB	90	4	2	2	90		Yes		Yes	Yes	Yes
18	WB	90	4	2	2	90			Yes	Yes	Yes	Yes
19	AP	50	10	6	8	50	Yes			Yes	No	No
20	AP	50	10	6	8	50		Yes		Yes	No	No
21	AP	50	10	6	8	50			Yes	Yes	No	No
22	AP	90	4	2	2	90	Yes			Yes	Yes	Yes
23	AP	90	4	2	2	90		Yes		Yes	Yes	Yes

Table 7 Loan amount across quantiles under varying situations *(Continued)*

24	AP	90	4	2	2	90			Yes	Yes	Yes	Yes
25	GUJ	50	10	6	8	50	Yes			Yes	No	No
26	GUJ	50	10	6	8	50		Yes		Yes	No	No
27	GUJ	50	10	6	8	50			Yes	Yes	No	No
28	GUJ	90	4	2	2	90	Yes			Yes	Yes	Yes
29	GUJ	90	4	2	2	90		Yes		Yes	Yes	Yes
30	GUJ	90	4	2	2	90			Yes	Yes	Yes	Yes

Table 7 Loan amount across quantiles under varying situations *(Continued)*

Case no.	DWELLD2	SANITD1	FAMTOT	CASTRDUM	PURPDUM	SECDUM	TCOST	RAINT	LOAN AMT		
									10%	50%	90%
1	No	Yes	2	No	Yes	Yes	500	10	7089	30934	75653
2	No	Yes	2	No	Yes	Yes	500	10	7804	44717	87152
3	No	Yes	2	No	Yes	Yes	500	10	9405	44618	124442
4	Yes	Yes	4	Yes	Yes	No	300	4	5579	22667	81223
5	Yes	Yes	4	Yes	Yes	No	300	4	6294	36450	92723
6	Yes	Yes	4	Yes	Yes	No	300	4	7894	36351	130012
7	No	Yes	2	No	Yes	Yes	500	10	12198	36813	82157
8	No	Yes	2	No	Yes	Yes	500	10	12913	50596	93656
9	No	Yes	2	No	Yes	Yes	500	10	14513	50497	130946
10	Yes	Yes	4	Yes	Yes	No	300	4	10687	28546	87727
11	Yes	Yes	4	Yes	Yes	No	300	4	11402	42329	99227
12	Yes	Yes	4	Yes	Yes	No	300	4	13003	42230	136516
13	No	Yes	2	No	Yes	Yes	500	10	3238	23861	65198
14	No	Yes	2	No	Yes	Yes	500	10	3952	37644	76697
15	No	Yes	2	No	Yes	Yes	500	10	5553	37545	113987
16	Yes	Yes	4	Yes	Yes	No	300	4	1727	15594	70768
17	Yes	Yes	4	Yes	Yes	No	300	4	2442	29377	82268
18	Yes	Yes	4	Yes	Yes	No	300	4	4042	29278	119557
19	No	Yes	2	No	Yes	Yes	500	10	6284	9353	55307
20	No	Yes	2	No	Yes	Yes	500	10	6999	23136	66806
21	No	Yes	2	No	Yes	Yes	500	10	8600	23037	104096
22	Yes	Yes	4	Yes	Yes	No	300	4	4774	1086	60877
23	Yes	Yes	4	Yes	Yes	No	300	4	5489	14869	72377

Table 7 Loan amount across quantiles under varying situations *(Continued)*

24	Yes	Yes	4	Yes	Yes	No	300	4	7089	14770	109666
25	No	Yes	2	No	Yes	Yes	500	10	7272	55445	230534
26	No	Yes	2	No	Yes	Yes	500	10	7987	69228	242034
27	No	Yes	2	No	Yes	Yes	500	10	9588	69129	279323
28	Yes	Yes	4	Yes	Yes	No	300	4	5709	42497	206908
29	Yes	Yes	4	Yes	Yes	No	300	4	6423	56280	218408
30	Yes	Yes	4	Yes	Yes	No	300	4	8024	56181	255697

perspective, since the financial inclusion policy set by the government of India aims to ensure adequate and affordable credit flow even to resource-poor households. We see from this comparison that the loan off-take is skewed towards resource-rich households, whereas it should ideally be in the other direction. Table 7 also shows that even among borrowers in the 90th quantile of the conditional loan distribution, marginal farmers received only approximately 80 percent of the loan amount of their counterparts who belong to the category of medium and large farmers (compare case 25 with case 27).

Conclusions

This study was motivated by the growing debate concerning the large-scale exclusion of rural households from the ambit of formal credit delivery channels, despite the continuous thrust of the state towards inclusive growth. Unlike studies that cover only limited geographical areas, this study includes six Indian states with varying institutional settings.

Several conclusions related to policy can be drawn from this analysis. First, the formal credit market in rural areas is sensitive to the characteristics of the village and the applicant's household and production. Second, there is considerable variation in the quantum of loans received from formal creditors. Third, even among borrowers in the 90th quantile of the conditional loan distribution, marginal farmers received only approximately 80 percent of the loan amount of their counterparts belonging to the category of medium and large farmers. In other words, even where poor clients managed to access formal-sector creditors, the quantum of credit off-take is lower than that of their resource-rich counterparts. Therefore, despite government pushes for financial inclusion, the distribution of credit from the formal financial sector remains skewed towards resource-rich rural households.

Endnotes

[a]Households with operational land holdings of up to 1 hectare, above 1 hectare and up to 2 hectare, above 2 hectare and up to 4 hectare, and more than 4 hectare are termed marginal, small, medium, and large farmer, respectively. Though medium farmer and large farmer are separate categories in government documents, we have placed them together in this study as medium and large farmers (i.e., households with operational landholdings of more than 2 hectares).

[b]Mid-year rural population (source: Centre for Monitoring Indian Economy) divided by the total number of a commercial bank's rural branches (source: Quarterly Statistics on Deposits and Credit of Scheduled Commercial Banks, Reserve Bank of India) and cooperative societies (source: Trend and Progress of Banking in India, Reserve Bank of India).

[c]Scheduled commercial banks (SCB) comprise those banks that are registered under the second schedule of India's central bank (i.e., Reserve Bank of India's Act, 1934). They include both public and privately owned banks. Their operation is mostly spread over multiple districts and even across states.

[d]Regional Rural Banks (RRBs) came into existence in 1975, with their primary goal being to cater to rural clients. The operation of RRBs is limited to few

districts. RRBs are jointly owned by the Government of India, the respective State Government, and the sponsor bank, which is either an SCB or State Cooperative Bank.

[e]Cooperative banks are financial institutions wherein the customers are the owners. Primary agricultural cooperative societies (PACS) are localized units owned by local people who share a common interest and who are involved in deposit mobilization and lending activities. In cases where cooperatives are also engaged in non-financial activities such as the selling of agricultural input, the running of consumer stores, and the trading of agricultural output, these banks are known as Multi-purpose Cooperative Societies (MPACS).

[f]Self Help Group (SHG) is a group of approximately 20 financially challenged people who belong to a common socioeconomic strata that have come together initially to save money and to use the corpus for lending among themselves. After a certain time period of savings, SHGs may approach the bank for a loan. This process obviates the need for marketable collateral, which banks usually ask to secure repayment, by instituting peer pressure and a group guarantee mechanism.

[g]Microfinance Institutions (MFI) form groups of poor people and act as lenders to the groups after sourcing loan from other financial institutions such as commercial banks.

[h]If a village is too large in terms of the number of households, some representative hamlets of that village (not exceeding 300 households) were used for complete enumeration. In such cases, only those households covered under complete enumeration constitute the population from which the sample was drawn.

[i]The elected head of a local village-level governance structure.

[j]During 2009-10: US$1 = Indian Rupee (INR) 47.

Additional files

Additional file 1: Appendix 1. Distribution of lenders across sample villages in certain states in India.
Additional file 2: Appendix 2. Coding of ordered exogenous variables.

Competing interests
The authors declare that they have no competing interests.

Acknowledgements
The authors would like to thank the editor, two anonymous referees, Jeffrey B. Nugent and Vasant P. Gandhi for their valuable comments, which have led to substantial improvements to this paper. Authors have benefitted from conversations on this topic with Samar K. Datta and M.S. Sriram, as well as from the comments of the participants of the 88th Annual Conference of the Western Economic Association held at Seattle. Authors are grateful to Samar K. Datta for making available the dataset used in this study. We also acknowledge doctoral fellowship and dissertation support grant extended to the first author by Indian Institute of Management Ahmedabad.

Author details
[1]Economics and Business Environment, Indian Institute of Management Raipur, GEC Campus, Raipur 352015, India. [2]Production and Quantitative Methods, Indian Institute of Management Ahmedabad, Vastrapur, Ahmedabad 380015, India.

References
Bassett G, Chen H (2001) Quantile style: return-based attribution using regression quantiles. Empir Econ 26(1):293–305
Basu P (2006) Improving Access to Finance for India's Rural Poor. Manuscript No. 36448 The World Bank, Washington, D.C

Boucher SR, Guirkinger C (2007) Risk, wealth, and sectoral choice in rural credit markets. Am J Agric Econ 89(4):991–1004
Buchinsky M (1998) Recent advances in quantile regression models: a practical guideline for empirical research. J Hum Resour 33(1):88–126
Burgess R, Pande R (2005) Do rural banks matter? Evidence from the Indian social banking experiment. Am Econ Rev 95(3):780–795
Carter MR (1989) The impact of credit on peasant productivity and differentiation in Nicaragua. J Dev Econ 31(1):13–36
Datta SK (2003) An institutional economics approach to the problems of small farmer credit in India. Working Paper 2003-07-01, Indian Institute of Management, Ahmedabad
Datta S, Ghosh A (2013) Explaining access to credit by rural households: results based on a study of several states in India. Working Paper 2013-02-03. Indian Institute of Management, Ahmedabad
Eide E, Showalter MH (1998) The effect of school quality on student performance: a quantile regression approach. Econ Lett 58(3):345–350
Gonzalez-Vega C (1984) Credit-rationing behavior of agricultural lenders: the iron law of interest-rate restrictions. In: Dale W, Adams D, Graham H, Von-Pischke JD (ed) Undermining Rural Development with Cheap Credit. Westview Special Studies in Social, Political and Economic Development series (pp.78-95). Westview Press, Boulder, Colo, London
Guirkinger C, Boucher SR (2008) Credit constraints and productivity in Peruvian agriculture. Agric Econ 39(2):295–308
Kochar A (1997) An empirical investigation of rationing constraints in rural credit markets in India. J Dev Econ 53(2):339–371
Koenker R, Bassett G (1978) Regression Quantiles. Econometrica 46(1):33–50
Koenker R (2005) Quantile Regression, 1st edition. Cambridge University Press, Cambridge
Nakamura H, Nakajima T (2011) A credit market in early stages of economic development. Econ Lett 112(1):42–44
National Sample Survey Organisation (2005) Household Indebtedness in India as on 30-06-2002, All-India Debt and Investment Survey, NSS 59th Round (January-December 2003), Report No.501 (59/18.2/2). Ministry of Statistics and Programme Implementation, Government of India. http://mospi.nic.in/Mospi_New/site/inner.aspx?status=3&menu_id=31
Pal D, Laha AK (2013) Sectoral Choice of Credit in Rural India. IIMA Working Paper 2013-03-01 Indian Institute of Management, Ahmedabad
Pal S (2002) Household sectoral choice and effective demand for rural credit in India. Appl Econ 34(14):1743–1755
Reserve Bank of India Quarterly Statistics on Deposits and Credit of Scheduled Commercial Banks, various issues. http://www.rbi.org.in/scripts/QuarterlyPublications.aspx?head=Quarterly%20Statistics%20on%20Deposits%20and%20Credit%20of%20Scheduled%20Commercial%20Banks
Reserve Bank of India Trend and Progress of Banking, various issues. http://www.rbi.org.in/scripts/AnnualPublications.aspx?head=Trend%20and%20Progress%20of%20Banking%20in%20India
Sahu GB, Madheswaran S, Rajasekhar D (2004) Credit constraints and distress sales in rural India: evidence from Kalahandi District, Orissa. J Peasant Stud 31(2):210–241
Walter SD, Feinstein AR, Wells CK (1987) Coding ordinal independent variables in multiple regression analyses. Am J Epidemiol 125(2):319–323
Winkelmann R (2006) Reforming health care: evidence from quantile regressions for counts. J Health Econ 25(1):131–145

11

Family farm business and access to rural development polices: a demographic perspective

Luca Bartoli* and Marcello De Rosa

* Correspondence: bartoli@eco.unicas.it
Department of economics and Law, University of Cassino and Southern Lazio, Via Folcara Cassino (Fr) 03043, Italy

Abstract

This article aims to show the ability of family farms in obtaining funds from rural development policies. Under the hypothesis that family farms' characteristics influence their access to policies, the perspective of analysis is demographic. In the paper two demographic viewpoints are investigated, to emphasize the role of the family composition in fostering strategic farms' decisions: the localization in the life cycle and the presence of assistants. The empirical test is provided for an Italian region: the funds of rural policies gained by family farms are examined, divided up into the three main strategic axes of the rural development plan. The results of the analysis highlight some interesting differences between the two demographic perspectives; the presence of assistants influences a farm's strategic process and increases the access to rural policies. This induces normative consequences in terms of the possible articulation of policies aimed at sustaining family farms.

Keywords: Rural development policies; Family farm business

Background

The aim of this paper is to analyze funding strategies adopted by family farms with reference to rural development plans. More precisely, it looks into the influence of demographic variables on obtaining funds from the rural development policy: we define as "consumption of policy" the farm's ability to obtain funds from rural development policies.

A general characteristic of the agricultural sector in many European countries is the presence of an overlap between the productive and domestic spheres, that is farms' strategies depend on the family situations (Jervell 1999). According to Gasson *et al.* (1988), family farm businesses are related to situations where:

a) *the principals are related by kinship or marriage,*
b) *business ownership is usually combined with managerial control, and*
c) *control is passed from one generation to another within the same family.*

In this context, *any boundary between productive and reproductive work in the farm household is artificial* (Errington and Gasson 1993) and conditions farms' strategies and aptitude to invest: the number of family members and localization in the life cycle could be relevant variables of influence. In a lot of cases, demographic transition, either vertical or oblique (Bergstrom 1996), is stimulated by policies aiming at favoring intergenerational transmission. Besides, other investments are necessary to maintain a farm's profitability and its persistence over time. In order to cope with an even more competitive scenario and to secure a family farm's resilience, a mix of strategies has to be carried out (Darnhofer 2010).

Recent rural development polices make a set of instruments available in order to sustain farm competitiveness and to diversify agricultural income. The possibility to obtain funds is influenced by a set of constraints which, in many cases, reduce the consumption of policies. In this situation, an interesting field of analysis is the influence of demographic variables in accessing rural policies.

After presenting a brief theoretical overview, the paper proposes an empirical test within an Italian region (Lazio), where the odds of gaining access to rural development policies will be examined through a logistic regression model. The paper sustains the hypothesis of differentiation in the access to policies, due to demographic factors in the composition of family farms.

As Offutt (2002) points out, *since farm households are demonstrably diverse, analysts would seem obliged to investigate hypotheses about differential response and impact.* One of these differences concerns demographic variables: family contexts are particularly favorable to set up a farm venture (Jervell 2011): an abundant economic literature has emphasised the strict connection between farm household strategies and style of farming (among others, Whatmore, 1994), by demonstrating the persistency of family farms (Sabbatini 2011). Their ability to survive over time witnesses the relevance of F-connection[1] in fostering lower levels of transaction costs and a higher aptitude to adapt (Ben-Porath 1980; Pollack 1985; Corsi 2009).

Social approaches to this theme underline the relevance of non-profit decisions in maintaining farming activity (Kuehne *et al.* 2010). For example, some researchers emphasize the relevance of family identity in strengthening the links between the future and past family history (Burton, 2004). Furthermore, a sentimental attachment to the place (place dependence and place identity) reinforces the links between farmer and farming and delays the eventual decision to retire (Kuehne 2012). However, in many other cases financial constraints are determining factors, thus inducing the abandonment of activity. To avoid this, agricultural policy offers financial support to promote farm adjustments.

Some authors have recently stressed the role of the Common Agricultural Policy in preserving small farms (Koutsou et al. 2011), through the process of the functional repositioning of agriculture, as a result of deepening, broadening and regrounding, and through the creation of rural webs (van der Ploeg and Marsden 2008).

As a matter of fact, family farms are expected to play an important role even in the new scenario of "modern rurality" designed within the rural development policies (Abdelmalek, 2000). Family units are requested to face competitive challenges through the definition of farm strategies, which are generally overlapping with family strategies (Caillavet *et al.* 2005; Abdel 1997). In this context a relevant subject of analysis is related to the capabilities of gaining access to rural development policies, which contain a set of measures to sustain either farm competitiveness or diversification, along sectorial or territorial trajectories of development. This ability could be decreased both on the supply and the demand side. From the supply standpoint, a large literature taking origin from Downs (1957) work has demonstrated that deviations emerge in a context of public choice. As far as demand is concerned, collective action theories evidence the relevance of pressure groups in conditioning policy decisions (Olson 1965; Becker 1983)[2].

The impacts of agricultural policies on farms' structural adjustments are analysed also within agent-based models (Balmann 1997; Albisser and Lehmann 2007; Balmann et al. 2003), which endogenise the importance of psychological and social factors which interfere in the farmer's strategy definition. They are considered to be a useful tool to evaluate agricultural policy impact (Berger 2001; Happe 2004). More recent studies have emphasized the relevance of territory, social capital and socio-economic farm characteristics in obtaining access to rural policies and fostering local rural development (Sabbatini, 2008; Meert *et al.*, 2005; Marotta, 2007; Casieri *et al.*, 2010). The inclination to adopt rural policy, that is to receive subsidies within the rural development framework, is variable and could depend on socio-demographic characteristics of the farm: family size, presence of young farmers and localization in the life cycle are influencing factors that should be examined in depth. As a matter of fact, consuming policy is a costly activity in terms of time spending and opportunity costs; in addition, informational asymmetries represent an important barrier to access. To overcome these obstacles, farmers have to improve their attention: neo-Austrian perspectives permit to get a better comprehension of the decision processes through the concept of entrepreneurial alertness (Kirzner 1973).

In the Kirznerian perspective, the entrepreneur is characterised by the aim to increase profits: to this end, he or she must pay attention to opportunities. Kirzner (1997) makes reference to the concept of *alertness* to summarize the entrepreneur's aptitude to take advantage of opportunities. Access to rural development policies becomes a way to support farm adjustments.

The decision about consuming or not and about what measures are better to adopt could be influenced by demographic variables. As thoroughly demonstrated in literature, demographic variables are commonly recognized as explanatory factors influencing farm strategies: family composition and localization in the life cycle are determinants of different paths of development. What does the term *demographic* really mean? Should one refer to all the members of the family, or to the members effectively employed on the farm? Are there any differences if considering one or the other typology of family farms? Our paper sets in this context and aims at analysing the influence of demographic variables on the farm's decision to adopt rural policies. By proposing a two-way classification of family farms, the hypothesis under analysis is that possible differences in the access to rural development policies could emerge. The research is conducted through interdisciplinary approaches which privilege a demographic perspective of analysis, where the fundamental elements are related to farm family traits.

To this end, after a brief methodological note our analysis proposes an empirical test of the access to rural policies, within an Italian region (Lazio). The hypothesis under study is that family composition influences the consumption of policies.

Methods

Family farms are the object of our analysis, marked out by three methodological steps. The first one concerns the demographic classification of family farms from a double perspective:

a. one is essentially based on demographic variables: it considers just age and the size of family as relevant aspects to be analysed;
b. the other classification takes into account farm activity and the composition of family work.

The first classification makes reference to previous publications (Bartoli and De Rosa 2007), where family composition and its localization in the life cycle are the key variables. The second one emphasizes the role of farm's assistants[3] exclusively employed within the farm or not. Due to the different perspective, the family cycle is a bit different: in the demographic typology, the young phase threshold is 35 years. Otherwise, in the second typology the age of reference is 40 years, because it is the threshold to gain access to rural development policies for generational change[4].

Of course, this difference could raise possible dissimilarities in the total amount of young farms[5].

I typology

- < 35 years
- 35–64 years
- > 65 years

II typology

- < 40 years
- 40–64 years
- > 65 years

Accordingly, the following typologies of farms will be analyzed (Table 1):

The differences between the two types of family farms concern a possible different degree of involvement of family members: in the first type, family members may not be involved in the farm business and may have interests in working outside the farm; in the second case the decision process stems from people who may not share similar objectives but who are involved in farming activity; they participate and cooperate to define the farm's strategy.

The second methodological step tries to link family types and access to the rural development policy, by focusing attention on the consumption of rural development policies in the region of Lazio (Italy): the measures under observation belongs to the three main axes of regional rural program:

1. competitiveness of the agricultural sector;
2. environment and landscape;
3. quality of life and diversification in rural areas.

Information and data are taken from the questionnaire of the last census of Italian agriculture, at Section 1, point 7[6]. As a consequence, our analysis is necessarily partial: indeed, census data take as reference year 2010, so this covers a maximum of two/three years compared to the duration of the rural development programme (7 years).

Table 1 The two considered typologies

I typology	*II typology*
Only farmer	Young farmer and a not young assistant
- young	- p/e
- mature	- np
- older	
Childless couples	Young farmers without assistant
- young	
- mature	
- older	
Couples with children	Mature farmer and a young assistant
- young	- p/e
- mature	- np
- older	
Extended families and other	Mature farmers without assistant
	Older farmer and a young assistant
	- p/e
	- np
	Older farmers without assistant

p = prevailingly employed in the farm.
np = not prevailingly employed in the farm.
e = exclusively employed in the farm.

The third methodological step aims at testing the (eventual) higher probability to gain access to the rural development policy on the basis of type of family farm. To this end, we apply a binomial logistic regression with categorical explanatory variables: more precisely the explanatory variables were divided into *n-1* dichotomous variables (Knoke and Burke 1980). This methodology is based on the assumption that the probability for the dependent variable Y to be equal to 1 is a function of the considered explanatory variables[7]. Since these variables are polichotomous (belonging to one of the types of that particular issue), they are treated as if each mode is in turn an explanatory dichotomous variable that helps to increase or decrease the success probability of the dependent variable. The function studied by the model is the following:

$$Y = \log\ it(x) = \ln[odds\ \ (x)] = \beta_0 + \beta_1 x_1 + \ldots \beta_k x_k$$

where Y is the dichotomous dependent variable "adoption of policies" that equals 1 for farms that have gained funds and 0 for those that have not obtained any fund. The estimation function measures the success probability related to each modality of the independent variable. For each modality, the probability to get access to funds is estimated with respect to a defined modality.

The model is based on the concept that the logit is a linear function of each regressor's parameters. In order to distinguish the two different demographic perspectives, two different models will be estimated for each type of family farms.

Results and discussion

The recent publication of the Italian census of agriculture reveals some structural adjustments but confirms two key factors characterising the agriculture of Lazio region.

First of all, the presence of a fragmented structure of family farms can be observed: even though the number of family farms has reduced by 41%, with respect to the previous census in 2000, in 2010 96% of all farms are directly managed by farmers, i.e., an increase of two percentage points. The second aspect of interest is the low presence of young farmers: Table 2 illustrates the prevalence of mature and older farms with respect to the younger ones. Less than 9% of farms is managed by younger farmers (<40 years); the share sensibly reduces in farms managed by farmers under 30. In the perspective of our analysis this contributes to fix the agricultural sector and impede processes of structural adjustment.

In what follows, we investigate the role of policies in fostering farm transformation and the eventual differences on the basis of demographic variables.

Consumption on the basis of family characteristics

An important result of our analysis confirms previous studies on limited access to rural polices on behalf of Italian farms (Bartoli and De Rosa, 2011). Very small percentages of farms gain access to funds; in this context some differences on the basis of demographic variables emerge. Most of the first type of family farms does not use rural policy measures to sustain farm growth (Table 3). On average, 95% of farms have not used any measure of the rural development plan. However, the presence of young farmers raises the percentage of policy consumption, but within the group of young farmers different aptitudes stand out. Single young farmers display the highest percentage of access to policies (14.4%). On the other hand, moving to other young families (childless and couple with children), the percentages decrease, even if they remain above the average (respectively 9% and 10%). In the remaining family farms, statistically significant results are found in extended families and partially in single mature farms. In the other typologies a systematic lower level of access is evident, above all in the childless

Table 2 Age of farm entrepreneur (%)

	2010	2000
< 20 years	0.04	0.15
20 – 24 years	0.52	0.36
25 – 29 years	1.35	1.10
30 – 34 years	2.50	2.65
35 – 39 years	4.46	4.70
40 – 44 years	6.83	7.16
45 – 49 years	9.22	9.61
50 – 54 years	11.06	13.00
55 – 59 years	12.60	11.08
60 – 64 years	14.35	13.24
65 – 69 years	10.27	12.58
70 – 74 years	10.82	11.12
75 – 79 years	8.10	8.01
> 80 years	7.89	5.24
Total	100.00	100.00

Source: Region Lazio, 6th Italian census of agriculture.

Table 3 Access to rural policies in the first type of family farms

I family type	Total consumption		Total
	NO	YES	
Only farmer Y	1,466	246	1,712
Only farmer M	14,603	952	15,555
Only farmer O	12,154	589	12,743
Childless couple Y	780	77	857
Childless couple M	18,694	800	19,494
Childless couple O	11,115	367	11,482
Couple with children Y	753	89	842
Couple with children M	19,830	1,135	20,965
Couple with children O	3,638	146	3,784
Extended families and other	13,516	1,178	14,694
Total	96,549	5,579	102,128
%	**Total consumption**		**Total**
	NO	**YES**	
Only farmer Y	85.6	14.4	100.0
Only farmer M	93.9	6.1	100.0
Only farmer O	95.4	4.6	100.0
Childless couple Y	91.0	9.0	100.0
Childless couple M	95.9	4.1	100.0
Childless couple O	96.8	3.2	100.0
Couple with children Y	89.4	10.6	100.0
Couple with children M	94.6	5.4	100.0
Couple with children O	96.1	3.9	100.0
Extended families and other	92.0	8.0	100.0
Total	94.5	5.5	100.0

Source: data processed from the database of Region Lazio.

older families: in this case the reduced propensity to consume policies is evidently due to the limited perspective of vertical transmission.

The second typology highlights a similar composition in the access to polices, which points out a higher percentage in the case of young families: but, in this case, a significant contribution to the phenomenon is given by the presence/absence of an exclusive assistant (Table 4). In young families with an exclusively employed (even not young) assistant, the percentage of access is relevant: 18.3% of farms have obtained funds. In the other two cases of young family farms the shares of access are certainly statistically significant, but lower (about 11%). The transition toward older stages of the life cycle reduces the access to rural policies, above all in the presence of assistants non-exclusively employed in the farm. The role of assistants seems relevant in obtaining access to funds, even when this role is decreasing in older phases of the life cycle.

By comparing the two types of family farms we can add other considerations: in each phase of life cycle, the presence of an assistant (either prevalent or exclusive) enhances the access for funds and gives a strong contribution to decision process. In order to validate previous results, an econometric model is proposed in the next paragraph.

Table 4 Access to rural policies in the second type of family farms

II family type	Total consumption		Total
	NO	YES	
Young farmer and a not young assistant p/e	1,669	374	2,043
Young farmer and a not young assistant np	641	84	725
Young farmer without assistant	6,483	840	7,323
Mature farmer and a young assistant p/e	4,754	382	5,136
Mature farmer and a young assistant np	2,018	109	2,127
Mature farmer without assistant	44,461	2,264	46,725
Older farmer and a young assistant p/e	1,240	86	1,326
Older farmer and a young assistant np	894	60	954
Older farmer without assistant	34,389	1,380	35,769
Total	96,549	5,579	102,128
%	**Total consumption**		**Total**
	NO	**YES**	
Young farmer and a not young assistant p/e	81.7	18.3	100.0
Young farmer and a not young assistant np	88.4	11.6	100.0
Young farmer without assistant	88.5	11.5	100.0
Mature farmer and a young assistant p/e	92.6	7.4	100.0
Mature farmer and a young assistant np	94.9	5.1	100.0
Mature farmer without assistant	95.2	4.8	100.0
Older farmer and a young assistant p/e	93.5	6.5	100.0
Older farmer and a young assistant np	93.7	6.3	100.0
Older farmer without assistant	96.1	3.9	100.0
Total	94.5	5.5	100.0

Source: data processed from the database of Region Lazio.

The logistic model

As highlighted in Table 5, the results of the logistic regression model are acceptable and statistically significant, even if the pseudo R-squared does not have a high value. From the table we can reject the zero value hypothesis. Thereby, family farms typology influences the access to rural development policies.

Tables 6 and 7 shows the results obtained from the model, comparing the likelihood of access among the two typologies of family farms. Demographic variables introduce further elements of reflection with respect to previous analysis: the logistic regression highlights a strong correlation between localization in the life cycle and access to policies. Family farms in which the entrepreneur is a young farmer are more probable to gain access to policies; in mature and elderly families an inverse correlation with the access to rural policies is evident. The more a family farm is in the advanced stage of the life cycle the less will be the probability to consume policy. In the Kirznerian perspective, single young farmers are more alert and determined to invest and to exploit the opportunities, with respect to young couples with or without children. In relation to extended families the probability for a single young farmer is 91% higher (Table 7), while, in young couples the same probability is 13% higher and in couples with children it is 35% higher: the presence of children increases the access to policies, presumably to increase the ability of the next generation to take over the farm. On the other hand,

Table 5 Zero value global hypothesis test

	I typology		
Test	*χ2*	*DF*	*Pr > χ2*
Likelihood ratio	6,614.769	9	<.0001
Score	7,450.192	9	<.0001
Wald	6,976.323	9	<.0001
	Association of predicted probabilities and observed answers		
Concordant percentage	52.1	D Sommers	0.189
Discordant percentage	33.2	Gamma	0.222
Linked percentage	14.8	Tau-a	0.020
Couples	538,646,871	c	0.594
Pseudo R-squared	0.07		
	II typology		
Test	*χ2*	*DF*	*Pr > χ2*
Likelihood ratio	11,253.678	8	<.0001
Score	14,730.895	8	<.0001
Wald	13,103.435	8	<.0001
	Association of forecasted probabilities and observed answers		
Concordant percentage	45.8	D Sommers	0.208
Discordant percentage	25.0	Gamma	0.294
Linked percentage	29.2	Tau-a	0.021
Couples	538,646,871	c	0.604
Pseudo R-squared	0.07		

families in the mature and older phase of the life cycle have a lower probability to exploit policy opportunities.

The logistic model applied to the second type of family farms gives other interesting insights, due to the presence of an assistant (Tables 8 and 9). If, on the one hand, young farmers confirm their aptitude to obtain funds for rural development, on the other side, the presence of an assistant partially modifies the picture. An assistant prevailingly or exclusively employed to support farming activity gives the farm higher opportunities of being funded. Moreover, in the case of negative estimates, the presence of prevalent or

Table 6 Logistic regression for the first demographic typology

	Estimation	Standard error	χ2 Wald	Pr > χ2
Intercept	−2.7072	0.0218	15404.318	<.0001
Only farmer Y	0.9222	0.0654	199.0056	<.0001
Only farmer M	−0.0233	0.0370	0.3949	0.5298
Only farmer O	−0.3198	0.0436	53.8447	<.0001
Childless couple Y	0.3917	0.1090	12.9005	0.0003
Childless couple M	−0.4442	0.0390	129.9321	<.0001
Childless couple O	−0.7035	0.0522	181.4673	<.0001
Couple with children Y	0.5717	0.1026	31.0505	<.0001
Couple with children M	−0.1534	0.0349	19.2762	<.0001
Couple with children A	−0.5084	0.0786	41.8620	<.0001

Table 7 Odd ratio (vs extended families)

	Estimation
Only farmer Y	1.925
Only farmer M	0.748
Only farmer O	0.556
Childless couple Y	1.133
Childless couple M	0.491
Childless couple O	0.379
Couple with children Y	1.356
Couple with children M	0.657
Couple with children A	0.460

exclusive assistants reduces the negative effect on the access to rural development policies. Table 9 makes this well-rendered. Firstly, the transition towards older steps of the life cycle induces a reduction in the probability to obtain funds. Secondly, it confirms the importance of a prevalent or exclusive presence of assistants in farming activity.

The comparison between the two methodologies of classification of family farms has revealed its utility: in fact, discriminating the presence of assistants is a relevant factor in stimulating the ability of consuming policy. The next paragraph details the main measures used by farms.

The consumed measures

In this paragraph we will detail the consumption of rural development policies by articulating them on the basis of the three strategic axes. The synthetic results are presented in Table 10, while in the following tables the same results are distinguished by family farm typology.

The lowest share of funded farms concerns the incentives for farm diversification (third axis): here the percentage of farms is less than 1%. In the first axis the level of adoption is very low and involves 2.4% of total farms (Table 10), while the rate of access to the measure included in the second axis is higher (3.4%).

The high costs of using European policies, in terms of bureaucracy, transaction costs, etc. could be a first motivation to explain low rates of access. On the other side, the

Table 8 Logistic regression for the second demographic typology

	Estimation	Standard error	$\chi 2$ Wald	Pr > $\chi 2$
Intercept	−25.083	0.0277	8216.9520	<.0001
Young farmer and a not young assistant p/e	1.0126	0.0575	309.6402	<.0001
Young farmer and a not young assistant np	0.4760	0.1060	20.1646	<.0001
Young farmer without assistant	0.4647	0.0426	119.2140	<.0001
Mature farmer and a young assistant p/e	−0.0131	0.0545	0.0576	0.8104
Mature farmer and a young assistant np	−0.4103	0.0910	20.3114	<.0001
Mature farmer without assistant	−0.4692	0.0336	195.4155	<.0001
Older farmer and a young assistant p/e	−0.1603	0.1022	2.4610	0.1167
Older farmer and a young assistant np	−0.1931	0.1208	2.5544	0.1100

Table 9 Odd ratio (vs extended families)

Odd ratio (vs older farmer without assistant)	*Estimation*
Young farmer and a not young assistant p/e	5.584
Young farmer and a not young assistant np	3.266
Young farmer without assistant	3.229
Mature farmer and a young assistant p/e	2.002
Mature farmer and a young assistant np	1.346
Mature farmer without assistant	1.269
Older farmer and a young assistant p/e	1.728
Older farmer and a young assistant np	1.672

low share of access is particularly relevant in the case of 3rd axis, which contains new tools for sustaining quality of life and economic diversification in rural areas[8].

A detailed consumption of policies is illustrated in Table 11: the table clearly discriminates young versus older families: policies falling in the first axis are generally consumed by young farms; besides, when young farmers have an assistant the aptitude to invest is systematically higher (6 percentage points);

Elderly families prevailingly consume measures under the second axis, which are exclusively related to environmental payments[9]. Finally, the table points out a very low percentage of farms having used measures of quality of life and diversification of rural economy (3rd axis). However, even in this case, the second type of family farms located in younger phases display a higher percentage of access, thanks to the presence of an active assistant.

Conclusions

The paper aimed at showing the relevance of demographic factors in influencing the way family farms gain access to funds for rural development. The demographic perspective has been introduced through two typologies of classification of family farms. Empirical analysis has shown important insights:

- first of all, a reduced share of consumption of rural policy emerges, above all in terms of new tools foreseen within the third axis of the rural development programme;
- a second element of reflection stems from the consideration of demographic variables: family farms managed by a young entrepreneur demonstrate a higher probability to obtain funds, thanks to a longer life expectancy, which raises farmers' alertness and his propensity to invest;
- finally, the two adopted demographic perspectives confirm the importance of the assistants, above all exclusive or prevalent, in taking investment decisions and

Table 10 Rate of access to rural development policies

	n. farms	*% on total farms*
axis 1	2,463	2.4
axis 2	3,479	3.4
axis 3	101	0.1

Table 11 Consumption of rural development policies (%)

	I axis	*II axis*	*III axis*
Only farmer Y	9.5	6.5	0.3
Only farmer M	23.3	3.6	0.1
Only farmer O	2.1	2.7	0.0
Childless couple Y	6.0	3.5	0.0
Childless couple M	1.6	2.8	0.1
Childless couple O	1.1	2.3	0.0
Couple with children Y	5.6	6.5	0.1
Couple with children M	2.1	3.7	0.1
Couple with children O	1.1	3.0	0.1
Extended families and other	4.0	4.7	0.2
	I axis	*II axis*	*III axis*
Young farmer and a not young assistant p/e	12.6	7.7	0.7
Young farmer and a not young assistant np	7.3	5.7	0.6
Young farmer without assistant	7.0	5.7	0.4
Mature farmer and a young assistant p/e	3.0	4.9	0.2
Mature farmer and a young assistant np	1.8	3.6	0.0
Mature farmer without assistant	1.9	3.3	0.1
Older farmer and a young assistant p/e	3.1	4.1	0.2
Older farmer and a young assistant np	2.1	4.4	0.0
Older farmer without assistant	1.5	2.6	0.0

obtaining access to rural policies to sustain farm adjustment. The lower level of access registered in the first typology of farms emphasises the relevance of the present/absent assistant farmer, recently underlined in literature on the subject (Koutsou et al. 2011).

According to the neo-Austrian perspective adopted by this paper, it is possible to sustain the hypothesis that the family farm's alertness is significantly higher in the case of either young farms or young farms where assistants support the farmer's activity, above all in the early stages of life cycle. However, the explanation for the ability of young farmers of obtaining funds from rural development policies cannot overlook the fact that the subjective condition of a "young farmer" is a priority in the allocation of resources for rural development. There is also a specific measure for the settlement of young farmers and an implicit resource reservation. On the other side, that means that "targetisation" works well in improving generational renewal. In this perspective, the introduction of a demographic viewpoint has given important insights, in order to fully understand the decision making mechanism to consume rural policies. The differences due to the two adopted demographic perspectives support the necessity to discriminate the supply of policy by taking these factors into account. More precisely, the discovered impact of the presence of family members working as assistants should encourage policies to foster their permanence in the farm and to avoid the search of off-farm jobs. Surely, rural development policies aimed at encouraging a farm's boundary shift (Banks et al. 2002) are in the agenda, but they are not exploited enough, owing to a set of motivations well described in other studies on the subject. To this end, more efforts are necessary to increase the percentage of access and, through this, raise the probability of higher levels of persistency of family farms.

Endnotes

[1]Families, friends and firms.

[2]See, among others, Rausser 1982; de Gorter and Swinnen 2002, Peltzman 1976.

[3]The assistant is a family member who helps the farmer.

[4]The average age is related to the woman, when she is present in the farm.

[5]Dissimilarities are eventually attributable to the presence of young members even within extended families.

[6]See www.istat.it for more details.

[7]The value is equal to 1 for farms that get access to funds and 0 for farms that do not.

[8]For the evaluation of policy consumption under Axis 3 it is necessary to precise that a part (often considerable) of the resources allocated to this axis is not devoted to farmers, but to other entities, both public and private.

[9] See the data concerning the progress of the spending on www.reterurale.it.

Competing interests
Both authors declare that they have no competing interests.

Authors' contributions
LB and MDR prepared the data, performed the statistical analysis, drafted, read and approved the final manuscript together.

Acknowledgements
This research was conducted with support from the Region Lazio, Department of Agriculture and Rural Development. We would like to thank Roberto Aleandri, Stefano Risa, Laura Baldini, Manuela Corleto (Region Lazio) and the anonymous reviewers for helpful comments and suggestions on an earlier version of the manuscript.

References
Abdel HT (1997) Accumulation productive dans l'agriculture familiale : origines et modalités. Options Méditerranéennes 12:151–165

Abdelmalek AA (2000) L'exploitation familiale agricole: entre permanence et évolution. Ècomomie Rurale 255–256:40–52

Albisser G, Lehmann B (2007) Agent-based modelling of structural change in agriculture: experiences from the implementation of cooperation and collaboration. 103rd EAAE seminar. Adding value to the agrofood supply chain in the future Euromediterranean space, Barcelona, Spain, April 23rd - 25rd

Balmann A (1997) Farm-based modelling of regional structural change. European Review of Agricultural Economics 25 (1):85–108

Balmann A, Happe K, Kellermann K, Kleingarn A (2003) Adjustment costs of agri-environmental policy switchings: a multi-agent approach. In: Janssen MA (ed) Complexity and ecosystem management: the theory and practice of multi-agent approaches. Mass., Edward Elgar Publishers, Northampton

Banks A, Long J, van der Ploeg JD (eds) (2002) Living Countrysides: Rural Development Processes in Europe - The State of the Art. Elsevier, Doetinchem

Bartoli L, De Rosa M (2011) L'adoption des politiques pour le développement agricole et rural dans les systèmes territoriaux italiens. Revue d'économie régionale et urbaine 3:601–624

Bartoli L, De Rosa M (2007) Différenciation typologique entre familles agricole set fermes dans l'agriculture italienne. New Medit 1:36–45

Becker G (1983) A theory of competition among pressure groups for political influence. Q J Econ 3:371–400

Ben-Porath Y (1980) The F-connection: family, friend and firm, and the organization of exchange. Population and Development review 6(1):1–30

Berger T (2001) Agent-based spatial models applied to agriculture: a simulation tool for technology diffusion, resource use changes and policy analysis. Agricultural Economics 25(2):245–260

Bergstrom T (1996) Economics in a family way. Journal of Economic Literature 34(4):1903–1934

Burton R (2004) Seeing through the 'good farmer's' eyes: towards developing an understanding of the social symbolic value of 'productivist' behavior. Sociologia Ruralis 44(2):195–215

Caillavet F, Facchini F, Moreddu C (2005) Introduction. Économie Rurale 289–290:7–9

Casieri A, Nazzaro C, Roselli L (2010) Trust building and social capital as development policy tools in rural areas. An empirical analysis: the case of the LAG CDNISAT. New Medit. Mediterranean Journal of Economics, Agriculture and Environment, IX(1):24–30

Corsi A (2009) Family farm succession and specific knowledge in Italy. Rivista di Economia Agraria LXIV(1–2):13–29

Darnhofer I (2010) Strategies of family farms to strengthen their resilience. Environmental Policy and Governance 20 (4):212–222

de Gorter H, Swinnen J (2002) Political Economy of Agricultural Policy. In: Gardner B, Rausser G (eds) Handbook of Agricultural Economics vol. 2. Elsevier, Amsterdam

Downs A (1957) An economic theory of democracy. Harper and Row, New York

Errington A, Gasson R (1993) The farm family business. Cab International, Wallingford

Gasson R, Crow G, Errington A, Hutson J, Marsden T, Winter DM (1988) The farm as a family business: a review. Journal of Agricultural Economics 39(1):1–41

Happe K (2004) Agricultural policies and farm structures - Agent-based modelling and application to EU-policy reform. In: Studies on the Agricultural and Food Sector in Central and Eastern Europe vol. 30, Halle (Saale). , , http://www.iamo.de/dok/sr_vol30.pdf accessed on: 05.04.2011

Jervell AM (2011) The family farm as a premise for entrepreneurship. In: Alsos GA, Carter S, Ljunggreen E, Welter F (eds) Handbook of research on entrepreneurship in agriculture and rural development. Elgar, Cheltenham

Jervell AM (1999) Changing Patterns of Family Farming and Pluriactivity. Sociologia Ruralis 39(1):110–116

Kirzner IM (1973) Competition and entrepreneurship. University of Chicago Press, Chicago

Kirzner IM (1997) Entrepreneurial discovery and the competitive market process: an Austrian approach. Journal of Economic Literature XXXV:60–85

Knoke D, Burke P (1980) Log Linear Models. Sage, Beverly Hills, CA

Koutsou S, Partalidou M, Petrou M (2011) Present or absent farm heads ? A contemporary reading of family farming in Greece. Sociologia Ruralis 51(4):404–419

Kuehne G (2012) My decision to sell the family farm. Agriculture and Human Values September::1–11. doi:10.1007/s10460-012-9393-7

Kuehne G, Bjornlund H, Loch A (2010) Why do farmers make non-profit decision? RIRDC Publication, Australia

Marotta G (2007) Consumo di politiche e prospettive delle aree rurali in Campania. In: Falessi A, Galietta M, Marotta G (eds) Consumo di politiche e prospettive delle aree rurali nel Mezzogiono d'Italia. Milano, F.Angeli

Meert H, van Huylenbroeck VT, Bourgeois M, van Hecke E (2005) Farm household survival strategies and diversification on marginal farms. Journal of Rural Studies 21(1):81–97

Offutt S (2002) The future of farm policy analysis: a household perspective. American Journal of Agricultural Economics 84(5):1189–1200

Olson M (1965) The logic of collective action: public goods and the theory of groups. Harvard University Press, Cambridge

Peltzman S (1976) Toward a more general theory of regulation. Journal of Law and Economics 19(2):211–240

Pollack RA (1985) A Transaction Cost Approach to Families and Households. Journal of Economic Literature 23(2):581–608

Rausser GC (1982) Political economic markets: PERTs and PESTs in food and agriculture. American Journal of Agricultural Economics 64(5):821–833

Sabbatini M (ed) (2011) Pressione socioeconomica e strategie emergenti delle aziende agricole. F.Angeli, Milano

Sabbatini M (2008) Competitività e strategie emergenti delle imprese agricole. In: Boggia A, Martino G (eds) Agricoltura e mercati in transizione. Proceedings of the XLIII congress of SIDEA, Assisi, 7–9 September 2006

Van der Ploeg JD, Marsden T (eds) (2008) Unfolding webs. Assen, Van Gorcum

Whatmore SJ (1994) Farm Household Strategies and Styles of Farming: Assessing the Utility of Farm Typologies. In: Ploeg JDVD, Long A (eds) Born from Within: Practice and Perspectives of Endogenous Rural Development Assen, Royal van Gorcum

12

Socioeconomic determinants of organic cotton adoption in Benin, West Africa

Epiphane Sodjinou[1*], Laurent C Glin[2], Gian Nicolay[3], Silvère Tovignan[4] and Jonas Hinvi[5]

* Correspondence: sodjinoue@gmail.com
[1]Ecole Nationale des Sciences et Techniques Agronomiques de Djougou, Université de Parakou, 03 BP 412, Porto-Novo, République du Benin
Full list of author information is available at the end of the article

Abstract

Organic cotton relies on ecological processes and the use of natural resources to sustain the production system, unlike conventional cotton, mainly characterized by massive utilization of synthesis chemicals. In West Africa, where rural livelihoods are particularly vulnerable, organic cotton is expected to contribute not only to poverty reduction but also to strengthen households' resilience. The objective of this study was to assess institutional and socioeconomic factors determining farmers' decisions to adopt organic cotton. For this purpose, we applied a probit model on empirical data collected from producers of the Centre and the Northern parts of Benin. Overall, we found that organic cotton adoption is mainly determined by farmers' socioeconomic characteristics, the physical distance between farm and house, and contact with extension and advisory services. Organic farming is more attractive to women compared to conventional farming. This because such type of cotton farming enables women to hold a separate cotton farm and thus increase their economic independence, whereas with the conventional system they depend mainly on the farm of the (male) head of the household. Older, less educated and low-income farmers who express environmental concern are more likely to adopt organic cotton. Subsequently, organic cotton should be considered as a prospective policy option to reach the poor and strengthen their livelihoods conditions while contributing to preserve the environment and natural resources. Furthermore, farmers who have their farm near home are more likely to adopt organic farming than those who have the farm far from their home. It also came out that organic farmers have more contacts with advisory and extension services. Finally, the study noted that there is still a need to enhance the extension system by: (1) exploring, designing, and upgrading innovative pedagogic tools such as videos and mobile phone technology to foster learning; and (2) strengthening organic farmer's organizations and the linkage with agricultural research organizations for technology development.

Keywords: Organic farming adoption; Low-income farmer; Gender; Probit; Benin

Background

Cotton plays a significant economic role for many countries of West Africa. Indeed, in addition to its contribution to the GDP and to total merchandise exports (from 25% to 45%), this sector represents more than half the income for roughly two millions of households in West Africa (Baffes 2007; Moseley and Gray 2008). Despite this economic importance, cotton is a controversial crop, which is at the centre of debates on globalization, structural adjustments, food security, biotechnology, agricultural subsidies, poverty alleviation, and sustainable development (Moseley and Gray 2008). In particular, Benin is among the countries most dependent on cotton (Gergely 2009). In fact, cotton accounts for 40% of foreign exchange earnings, employs 45% of rural households, and provides income to more than one third of the population (MAEP 2011). Accordingly, the Government of Benin gave a significant place to this sector in its development strategy as a tool for poverty alleviation.

For long, the cotton sector was governed by a parastatal company called SONAPRA (*Société Nationale pour la Promotion Agricole*) which had the monopoly over most of the commercial roles relating to cotton production, such as purchase and distribution of seeds and fertilizers as well as pesticides, direct purchasing of cotton from the farmer, and determination of price (Kütting 2004). But, since the beginning of the 1990s the sector has undergone some reforms including the privatization of input supply, introduction of private ginners and creation of interprofessional bodies to take over the sector management, through a highly regulated system precluding competition among ginners (Gergely 2009). However, during the last three years the new institutional arrangement collapsed due to a number of crises that resulted in distrust and mistrust between the government and the private stakeholders (Glin 2014). Despite all the upheavals and changes experienced by the sector, the method of production of conventional cotton has remained virtually unchanged with a strong use of synthetic chemical inputs.

Indeed, in order to obtain high quality and competitive cotton fiber on international market, producers are encouraged to use extensively mineral fertilizers and synthesis chemical pesticides. This was particularly encouraged by structural adjustments and neoliberal policies that stressed agricultural intensification as a strategy to achieve food security and poverty alleviation (Glin et al. 2012). In addition to official chemical inputs, other pesticides banned in Benin, including lindane and dieldrin, are smuggled from Nigeria and illegally used in cotton production (Pazou et al. 2006a). Many of these official or illegally used chemical products have adverse effects on environment and both the human and animal health. For instance in 2000, 147 persons were poisoned by cotton pesticides of which 10 fatalities (OBEPAB 2002). In certain localities, the cotton farms are located near rivers and pesticides could "distribute among the river ecosystem components, such as water and sediment, and accumulate in biota" (Pazou et al. 2006b).

This context came together with international developments (e.g., the 1992 United Nations Conference on Environment and Development) targeting environmental issues and created a favorable situation for alternatives to the conventional farming system (Glin et al. 2012). Taking advantage of this context, OBEPAB (*Organisation Béninoise pour la Promotion de l'Agriculture Biologique*) with the support of international and national networks introduced organic cotton production in Benin in the mid-1990s as an alternative to address the above mentioned problems. Currently, other organizations including Helvetas Swiss Intercooperation-Benin are active in the organic cotton sector in Benin.

The organic cotton relies only on ecological processes and the use of natural resources. It could increase the income of poor farmers through higher sales prices. One important requirement of the organic production is the certification, which entails a regular and effective control and auditing of the production and processing practices to verify and ensure the traceability of the organic products and the compliance with the international organic standard. The cost of certifying smallholdings of organic farmers, particularly in developing countries, favors certification at the level of farmer groups or cooperatives, instead of individual farmers (Glin et al. 2013: 549, see also Lockie et al. 2006; Ton et al. 2007).

In West Africa, where rural livelihoods are particularly vulnerable, organic cotton is expected to contribute not only to poverty reduction but also to strengthen households' resilience (Glin et al. 2013). Furthermore, given its ability to improve soil fertility (soil structure, biological activity, etc.), the risk of loss of production is relatively low in organic farming (Hulsebusch et al. 2007). Besides, several studies proved that organic cotton is more profitable than conventional cotton (e.g., Gbesso 2000; Pacini et al. 2002; Matthess et al. 2005; Eyhorn et al. 2007), provides benefits regarding biodiversity (Hole et al. 2005) and climate protection (Skinner et al. 2014).

However, the development of organic cotton has some drawbacks, including the structural difficulties (both actual and perceived) of doing business in sub-Saharan Africa; lack of transport or difficulties to transport organic fertilizer, a shortage of export oriented knowledge (Ferrigno et al. 2005). Furthermore, the development of organic cotton in most sub-Saharan African countries is enhanced by projects funded by Western donors (Ferrigno et al. 2005; Justus 2009).

Despite these shortcomings, the production of organic cotton has experienced significant growth over the past decade, moving from 185.2 tons during the 2002–2003 agricultural campaign to 824.3 tons in 2011–2012, representing an annual increase of about 34.5% (Figure 1)[a]. Currently around 2,500 farmers are engaged in organic cotton production, gaining approximately USD 600,000 as export value for Benin economy. Moreover, during the last five years the organic cotton chain is undergoing a progressive transformation from an experimental, small-scale and donor-dependent initiative towards a market-oriented, large-scale and self-financing transnational commodity network (Glin et al. 2012). The growth of the organic cotton sector eventually convinced the Government to target organic cotton in Benin's Agriculture Sector Development Plan (MAEP 2011). Over the past two agricultural campaigns, the Government through SONAPRA (a state owned company) was also engaged in the marketing of the organic cotton.

Unfortunately, like in most developing countries, very few data are available on organic farming in Benin and West Africa (Kristiansen et al. 2006). In particular, there

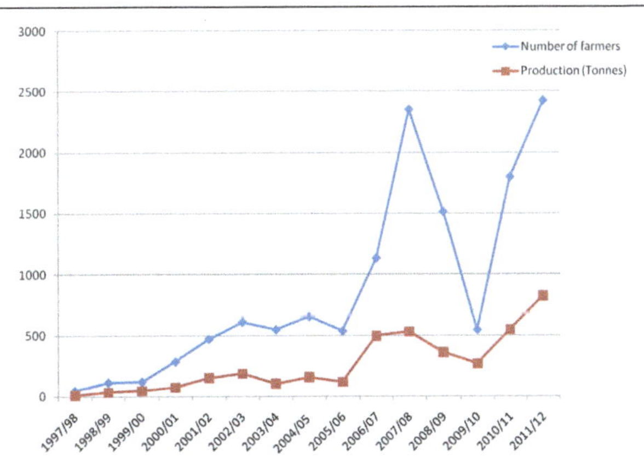

Figure 1 Trends of organic cotton production in Benin (OBEPAB, 1996–2012).

are very few studies that investigated farmers' decisions to adopt organic cotton in West Africa. Thus, this study aims to contribute to the understanding of factors underlying farmers' decision to adopt organic cotton, based on empirical data collected from producers of the Centre and the Northern parts of Benin. In other words, the objective of this study is to assess institutional and socioeconomic factors influencing the adoption of organic cotton in Benin. The identification of factors promoting or hinder the adoption of organic cotton can help to better define policy options and strategies for further dissemination of this type of cotton farming.

The remainder of the paper is organized as follows. The second section, related to data and methods, begins with the description of the sampling and data collection techniques, and then outlines the analytical framework. The results and discussion section (section 3) starts with a descriptive analysis of data collected on conventional and organic cotton producers, and ends with the results of the probit model, interpreted, and discussed in three sub-sections. The fourth section concludes and provides the policy implications of our findings.

Methods

Sampling and data collection

The study was carried out in the Republic of Benin, a tropical West African country situated between the 1st and 4th Meridian of longitude and between the 6th and 12th parallels of north latitude. Data were collected, during the agricultural campaign 2012–2013, from three research sites of the Project SYPROBIO (SYstèmes de PROduction BIOlogique diversifiés), a EU funded regional and multi-stakeholder research project involving national research organizations, farmers' organizations, and Helvetas Swiss Intercooperation. The project promotes farmer-driven innovations to address food security, poverty alleviation and resilience to climate change. The project is led by FiBL (Research Institute of Organic Agriculture) and is implemented in Benin, Burkina Faso and Mali.

The first of our research sites is located in the Centre of Benin, one of the largest cotton production areas with a low pressure on land[b]. The second site, Northeast of Benin, belongs to the oldest and largest cotton zone of Benin and is characterized by a high pressure on land. The third site is in Northwest-Benin, a marginal zone of cotton production with also a high pressure on land. In each of the three categories of areas, a village were selected, taking into account criteria such as the coexistence of organic farming (certified) and conventional cotton farming, the year of introduction of organic cotton (at least 3 years) and physical accessibility in all seasons of the year. Thus, Aklampa in Glazoué District, was selected for the Centre of Benin, Bonhanrou (District of Banikoara) for the Northeast-Benin, and Batia (in Tanguieta District) for the Northwest-Benin (cf. Figure 2).

At each site we made a census of all households and then classified these households into four classes (rich, medium, poor, very poor) according to the Wealth Ranking approach (Ranjani et al. 2008). The Wealth Ranking was carried out using, for each site, five key-informants who have a thorough understanding of the socio-cultural and economic context of the site. It is worth noting that our wealth ranking approach primarily accounts of the living conditions of the household as a whole and does

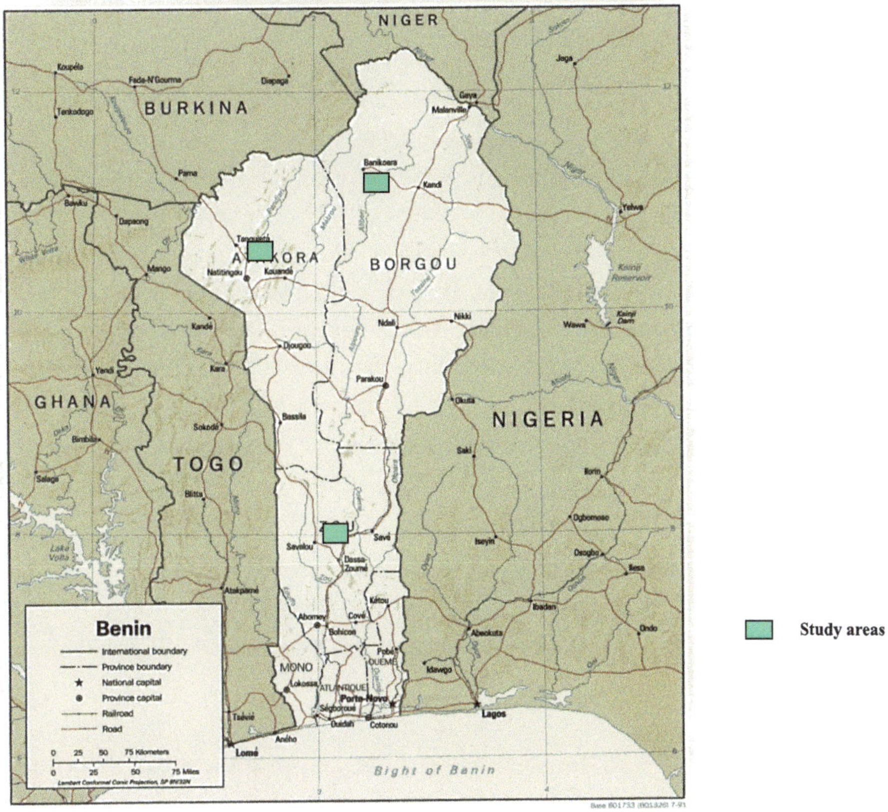

Figure 2 Study areas.

not discriminate individual members (be it household head or not, man or woman) within household. We assume that all individual members within a household share the same livelihood conditions. Thus, they are attributed accordingly the same wealth ranking score.

About 81 households were selected at each site. The number of households selected in each (wealth) class is based on a quota defined by the formula: $q = n/N$, with N the number of households identified in the village, q the sampling coefficient and n (with $n \approx 81$) the total number of households to be selected in the village (Table 1). For each (wealth) class with n_i households, the number of members selected was obtained as followed: $(q \times n_i)$. These households were randomly selected using random numbers generated with Microsoft Excel software.

Table 1 Distribution of the sampled farmers per research site

Class	Research site			Total
	Batia	Bonhanrou	Aklankpa	
Rich	12	4	14	30
Medium	13	7	27	47
Poor	28	50	26	104
Very poor	28	22	11	61
Total	81	83	78	242

Finally, in each sampled households, all member producing cotton were interviewed separately.

Analytical framework

The model used

We assume that the decision to adopt or not to adopt organic cotton is voluntary and that the differences in responses at the household level are often due to the fact that producers have, e.g., different cultures, different levels of education, different endowments in resources, different constraints, different farming objectives and preferences, and experience different socio-economic conditions. As consequence, some farmers will adopt and others will not, depending upon the utility they expect from the innovation (the organic farming). In other words, let U_i^a be the expected utility that a given farmer i would receive from adopting the organic cotton and U_i^b the expected utility gained from using the alternative, i.e. conventional cotton. "The observed choice between the two reveals which one provides the greater utility" (Greene 2008: 777). Hence, farmers adopt organic cotton if the utility they expect to obtain (U_i^a) by adopting this activity is higher than what they would have if they adopt conventional cotton (U_i^b), i.e. if ($U_i^a > U_i^b$). Following Greene (2008: 777), a common formulation is the linear random utility model,

$$U_i^a = x_i^{'}\beta_a + \varepsilon_{ia} \text{ and } U_i^b = x_i^{'}\beta_b + \varepsilon_{ib} \quad (1)$$

Following Verbeek (2004: 192), for each farmer i, we can write the utility difference between adoption and non-adoption as a function of observed characteristics (x_i) and unobserved characteristics (ε_i), i.e.:

$$y_i^* = U_i^a - U_i^b = x_i^{'}\beta + \varepsilon_i \quad (2)$$

with y_i^* a latent variable which is not observable, and $\beta = \beta_a - \beta_b$. Models, often derived from equation 1, determine the probability that farmers adopt organic cotton ($y_i = 1$). In practice, the probit or logit models are used to determine the probability that $y_i = 1$. The probit is based on standard normal distribution while the logit is based on standard logistic distribution. According to Greene (2008), these two models often lead to the same conclusion and it is difficult to make a choice between the probit and the logit on theoretical bases.

In this study, the probit model was used and is given by the following formula (Greene 2008: 773):

$$\text{Prob}\left(y_i = 1 \middle| x_i\right) = \int_{-\infty}^{x'\beta} \phi(t)dt = \Phi\left(x_i^{'}\beta\right), \quad (3)$$

where, $\phi(\cdot)$ and $\Phi(\cdot)$ correspond respectively to the density function and the cumulative distribution function of the normal distribution. The dependent variable is the adoption status of the producer, with 1 for farmers who adapt organic cotton, and 0 for farmers who do not adopt it.

Hypothesized factors influencing organic cotton adoption

There are several models or paradigms employed in the literature to explain the decision of smallholder farmers to adopt new technologies (Negatu and Parikh 1999;

Moumouni et al. 2013). We identified here four models: (1) the technology diffusion model that considers the characteristics of technology as determinants of adoption (e.g., Sarker and Itohara 2008); (2) the economic constraints model, which emphasizes the importance of economic and institutional factors in the adoption of technology (e.g., Bolwig et al. 2009); (3) the technology user's context model that posits that farmers' characteristics and subjective perceptions of technology influence adoption (e.g., Sarker and Itohara 2008) and (4) the integrated model that combines the features of the three above models. For this integrated model five categories of factors may influence farmers' decision to adopt a given innovation or technology (e.g., Chilonda and van Huylenbroeck 2001; Sodjinou 2011). These include characteristics specific to farmers and their households (e.g. age, knowledge, motivation/objective), characteristics related to their farm (land available, labor, water, etc.), biophysical factors (diseases/parasites, climatic factors, etc.), institutional factors (contact with extension services, access to credit, information sources, government agricultural policy, etc.) and economic factors (e.g. markets and prices of outputs and inputs). We adopted this model to capture the complexity of organic farming and related adoption factors. Accordingly, we used the following explanatory variables in our probit model (Table 2):

- *the age of the farmer* (AGE). The relationship between farmers' age and the decision to adopt an innovation or technology is not clear in the literature. Indeed, some authors found that older farmers are more likely to adopt an innovation, while others found that young people are open to adopt new technologies (Rogers 2003). Thus, in this study, we expect that the sign of the variable AGE can be positive or

Table 2 Variables used in the probit model with expected sign on organic cotton adoption

Variable	Label	Expected signs
AGE	Age of the producer (year)	+/−
AGE2	Square of AGE (years^2)	+/−
SEX	Sex of the producer (1 = male, 0 = female)	+
EDUC	Formal education of the producer (year)	+
EXPCOT	Experience in cotton production in general (year)	+/−
CLASSP	Class of poverty as perceived by producers (1 = very poor, 2 = poor, 3 = medium, 4 = rich)	-
CENTRE	Regional dummy (1 = Farmers located in the Centre of Benin, 0 = otherwise)	+/−
NORTHW	Regional dummy (1 = Farmers located in the northwest-Benin, 0 = otherwise)	+/−
DISTEX	Distance between the producer's house and his farm (km)	+/−
DISTEX2	Square of DISTEX (km^2)	+/−
HHSIZE	Household size (number of members)	+
SUPTOT	Amount of land available in ha	+/−
OXEN	Ownership of oxen (1 = yes)	+
DISTM	Distance between the village and the nearest market place where the producer sells its products (km)	-
ACRED	Obtained credit (1 = yes, 0 = no)	-
COOP	Membership of producers' organization (1 = yes, 0 = no)	+
NVISIT	Contact with extension agents (number of visits between January and October 2012)	+

negative. Besides, Sodjinou (2011: 36) argued that producers might be opened to new technologies until a certain age after which they become less open until they reach old age. The opposite may also occur in some cases. Accordingly, in order to allow this nonlinear relationship, we include the square of the farmers' age in the adoption model;

- *the gender of the producer* (SEX), with 1 for male and 0 for female. We used the gender of the producer rather than the gender of the household head, the conventional practice in most adoption studies. Indeed, in cotton production regions of Benin, it is not common to find adult men in female-headed households, while male-headed households may include one or many women. However, adopting the gender of the household head could be dismissive of the reality at individual level. In fact, decisions made within households depend upon the characteristics of each producer rather those of the household head (Doss et al. 2003), particularly as far as the adoption of organic cotton is concerned. In the Sub-Saharan African context, sometimes the attitude and position of the household head regarding organic is different from that of the other members. Thus, the household head can adopt organic while some individual members can remain conventional and vice-versa. Therefore, the utilization of farmer's gender instead of the gender of the household head allows us to assess the behavior of female farmers regarding organic cotton adoption in female-headed as well as male-headed households. For Doss and Morris (2001), controlling for everything else, men and women farmers made the same adoption decisions. However, in rural areas of Benin, women often face problems in the adoption of new technologies due to the lack of time or funds and poor control over productive resources. Thus, the variable SEX is supposed to have a positive influence on the organic cotton adoption, meaning that male farmers will be more likely to adopt organic cotton than female producers;
- *year of schooling* (EDUC). Educated producers are able to read manuals and other extension materials. They also have easily access to information, particularly through the media and can communicate easily with extension services. Education may also enhance farmers' ability to efficiently allocate inputs across competing uses and to gain more knowledge about adverse effects of conventional cotton farming (see Polson and Spencer 1991). For this reason, we expect that education will have a positive effect on the probability to adopt organic cotton;
- *producers' experience in cotton farming in general* (EXPCOT). Farmers' experience in cotton production can allow them to appreciate the advantages and disadvantages associated with this activity. Put differently, knowledge gained over time by farmers, from working in an uncertain and uncontrolled cotton production environment, may help in evaluating information, thereby influencing their adoption decisions (Sall et al. 2000). In addition, some producers who have spent several years in conventional cotton production can become accustomed or create/integrate networks that oblige them to continue conventional farming. Due to this relatively ambiguous situation, we expect that the effect of this variable on the adoption of organic cotton can be positive or negative.
- *the poverty level of farmers' household* (CLASSP), as identified by key informants through the wealth ranking, with values 1 (very poor), 2 (poor), 3 (medium) and 4 (rich). Rich farmers have as well more land as well as financial resources and can

hardly manage the weeding without herbicides. Moreover, the issue of pesticides and mineral fertilizers is also financial because they represent a relatively high cost often difficult to be borne by the poor peasants (Guilmo et al. 2006). These poor farmers will be more likely to practice organic cotton farming than to rich farmers. Thus, we expect this variable to have a negative sign meaning that poor producers will be more likely to adopt organic cotton;

- *regional dummies (variables CENTRE and NORTHW).* As noted above, data were collected in three regions: Centre, Northeast and Northwest of Benin. These three regions are located in three different agro-ecological zones and have different socio-cultural realities. To account for this and to control for agro-climatic differences that could affect the household attitudes; we include two regional dummies in the probit model, namely the *CENTRE,* with 1 for farmers from the Centre of Benin and 0 otherwise, and the *NORTHW,* with 1 if the farmer is located in the Northwest-Benin and 0 otherwise. The Centre of Benin (notably Glazoué) is a region with low human pressure on agricultural land while the two other regions (notably Tanguiéta and Banikoara) experience high land pressure. Thus, the variable *CENTRE* also controls for pressure on land. The pressure on land in addition with socio-cultural variability and agro-climatic conditions inherent to each research site can affect farmers' attitude toward organic cotton adoption. We expect that the variables *CENTRE* and *NORTHW* could have positive or negative signs;
- *the distance (km) between producers' house and their farms* (DISTEX). Producers who have their farm close to their house will have less difficulty to transport organic fertilizers and other inputs used in the organic cotton production. This means that peasants who have their farms close to their house will be more likely to adopt organic farming. However, farms closed to producers' houses are generally smaller in size, reducing thus the possibility to practice some cropping systems strongly encouraged in organic agriculture such as crop rotation and inter-cropping. In such a condition, producers who have their farms far from their houses would be more likely to practice organic cotton. In short, we hypothesized that the sign of the variable DISTEX will be positive or negative. Furthermore, we introduced the square of DISTEX in our probit model in order to check whether there is nonlinearity in the relationship between organic cotton adoption and the distance between producers' house and their farm. In other words, we tested whether the relationship between the distance and the adoption of organic cotton is U-shaped or hill-shaped;
- *household size* (HHSIZE), i.e. the number of members in the household. The production of organic cotton is more labor-intensive than conventional cotton farming especially for transportation, application of organic inputs (pesticides, fertilizers, etc.) and weed control. This implies that the availability of labor will positively affect the adoption of organic cotton; this implies that this variable may have a positive sign;
- *amount of land available* (SUPTOT) in ha. According to the literature, smallholder farmers are more likely to adopt organic farming because they can easily mobilize the necessary organic inputs (particularly organic manure) and labor force for their small size of operation. The bigger the farm size the more difficult it is to appropriately handle organic farming. However, we may also predict that

households with large land area could easily adopt certain practices of soil fertility improvement, such as inter-cropping, crop rotation, agro-forestry and improved fallow. Indeed, these technologies promoted in organic production require the mobilization of land for a certain period of time, which producers who have small land area cannot afford. In this regard, farmers who have large areas of land will be more likely to adopt these organic technologies. In light of the above, we expect that SUPTOT will have positive or negative signs;

- *possession of oxen* (OXEN), with 1 for producers who own oxen and 0 otherwise. Oxen are important means of tillage as they allow producers to sow large areas. The oxen also produce cow dung widely use as fertilizer in the production of organic cotton. Oxen can also help producers in the transport of organic fertilizers and other products necessary for the production of organic cotton. Therefore, possession of oxen may have a positive influence on the adoption of organic cotton;
- *the distance (in km) between producers' house and the nearest rural market* (DISTM) where they use to sell their produce. At market places, farmers often interact with other producers and exchange on agricultural activities and technologies. These places can therefore foster discussion about the benefits of organic cotton and lead certain farmers to adopt this technology (Sodjinou and Henningsen 2012). Accordingly, we expect this variable to have negative sign, i.e. peasants living close to local market will be more likely to adopt organic cotton;
- *access to credit* (ACRED), with 1 for the producer who has obtained credit and 0 otherwise. In the literature, access to credit generally has a positive effect on innovations adoption. Indeed, having access to financial means allows producers to acquire the required inputs for the implementation of the new technology. However, the situation is different in the particular case of organic farming given the reliance on locally available resources instead of market-based inputs. Overall, organic farming is less dependent on credit than conventional farming. Besides, the measure of 'access to credit' often causes problems, i.e. this variable might be endogenous. To overcome this problem, Doss (2006) suggests including a measure of whether the farmer had ever received credit. However, this measure of access to credit is still not perfect, but it is a better measure of access to credit than the simpler question of whether the farmer used credit in the current period (Doss 2006). Accordingly, we use this approach in this study, and we expect that access to credit will negatively influence the adoption of organic cotton;
- *membership of cotton producers' organization* (COOP), with 1 = yes and 0 = no. Cotton producers' organization allows farmers to be in touch with their colleagues. This allows them not only to exchange with colleagues on new technologies but also to have access to agricultural inputs. These producer organizations and the meetings they organize are channels for the dissemination of innovations. So belonging to a producer organization is expected to have a positive influence on the adoption of organic cotton;
- *contact with extension agents* (NVISIT). Extension agents are sources of information on new agricultural technologies. Thus, we expect that farmers who have contact with extension agents will be more likely to adopt organic cotton.

Results and discussion

Farmers' involvement in cotton production

Before analyzing the results of the probit model used in this study, it is important to mention that the proportion of households who continued with organic cotton since they converted is almost the same as for households that still grow conventional cotton. Stated differently, since the introduction of organic cotton in the mid-1990s, 38% of households have finally adopted it, while 38% of households interviewed remained in conventional cotton (Table 3). Nearly 5% of households have abandoned organic cotton of which about 3% returned to conventional cotton and 2% decided to no longer produce any type of cotton. In the same vein, among the households interviewed, 7% have never produced cotton while 9% produce both conventional and organic cotton (i.e. the husband produces the conventional cotton while the wife produces organic cotton, or vice versa). The latter corroborates the finding of Bassett (2010) for whom certain households commonly grow conventional cotton and organic cotton simultaneously, a practice forbidden by organic certifiers.

For all the interviewed producers, the major reasons for converting into organic cotton farming are mainly related to indebtednesses and health problems encountered in the conventional cotton system. Other reasons motivating farmers' decision to adopt organic cotton (Table 4) include: the desire to diversify their sources of income (85% of interviewees), the willingness to make money (84%) and to be financially independent. Thus, it clearly appears that financial reasons next to health considerations are the major determinant of farmers' decision to convert to organic cotton. This is in line with several other studies that pointed out financial reason as a major motivational factor to converting into organic in Sub-Saharan Africa context (see Ton 2002; Tulip and Ton 2002; Dowd 2008: Glin et al. 2012). However, Niemeyer and Lombard (2006) argue that financial reasons such as higher selling prices or reduced input costs are minor in the decision to adopt organic farming. In fact, producers are also concerned with their health, their environment and the fertility of their soil. For Niemeyer and Lombard (2006), in Germany, producers reported having abandoned conventional agriculture to organic farming for reasons such as: decline in farm income, soil exhaustion and soil erosion, increase in pesticide costs and aversion to the use of pesticides. For this reason Niemeyer and Lombard (2006) argue that there is a strong need for a greater national involvement in organic farming.

Table 3 Distribution of surveyed households according to their involvement in cotton production (%)

	Rich	Medium	Poor	Very poor	All
Overall wealth distribution	12.40	19.41	42.98	25.21	100
Conventional cotton	36.70	29.80	41.30	41.00	38.40
Organic cotton	46.70	48.90	33.70	32.80	38.00
Has left organic cotton for conventional cotton	3.30	0.50	1.00	0.00	2.50
Has left conventional cotton and is not at all in cotton production	0.00	0.00	2.90	6.60	2.90
Has left the organic cotton and is not at all in cotton production	0.00	0.00	1.90	4.90	2.10
Has never produced cotton	3.30	4.30	9.60	8.20	7.40
Production of the two type of cotton	10.00	8.50	9.60	6.60	8.70

Table 4 Other reasons for adopting organic cotton

Reasons	Rich	Medium	Poor	Very poor	All
Advice from neighbor/friend	76.0	68.0	66.0	61.0	67.0
To make money	82.0	89.0	80.0	87.0	84.0
Love for this activity	29.0	26.0	30.0	30.0	29.0
Learned that organic cotton is profitable	94.0	75.0	66.0	78.0	75.0
Want to do something new	24.0	29.0	18.0	17.0	21.0
In order to valorize a non-used land	29.0	32.0	43.0	43.0	38.0
Lower yields in conventional cotton	0.0	18.0	9.0	9.0	10.0
To diversify sources of income	76.0	96.0	84.0	78.0	85.0
To be financially independent	59.0	78.0	82.0	87.0	78.0
Other (to avoid debt, to protect the environment, etc.)	47.0	48.0	34.0	39.0	40.0
Achievement of objectives that motive organic cotton adoption	100.0	96.0	88.0	83.0	91.0

We focused our analysis mainly on the households involved in conventional (38% of households interviewed) and in organic (38% of households) cotton.

Characteristics of organic cotton producers versus conventional cotton producers

The average age of organic cotton producers is relatively higher than that of conventional cotton producers, 44 years against 41 years (Table 5). More than half of organic cotton producers are females, against only 8% of women in the conventional cotton. This is in line with Bassett's (2010: 53) argument that women are typically excluded from conventional cotton growing because of its high costs and discrimination by extension agents and men. The gender analysis of producers

Table 5 Socioeconomic characteristics of organic and conventional farmers: descriptive statistics[(a)]

	Conventional (n = 98)	Organic (n = 93)	Total (n = 191)	Statistical test[(b)]
Age of the producer (year)	40.5 (11.5)	43.5 (12.5)	42.0 (12.0)	−1.75*
Sex of the producer (% of male)	91.8	49.5	71.2	41.79***
Formal education of the producer (year)	3.0 (3.7)	1.1 (2.4)	2.1 (3.3)	4.23***
Experience in cotton production in general (year)	13.7 (10.4)	10.8 (9.6)	12.3 (10.1)	1.97*
Distance between the producer's house and his farm (km)	5.0 (7.4)	2.5 (2.0)	3.8 (5.6)	3.20***
Household size (number of members)	10.0 (7.2)	8.8 (5.2)	9.4 (6.3)	1.31
Land available for the household (ha)	17.4 (20.0)	16.1 (25.8)	16.8 (23.0)	0.39
Distance between the village and the nearest market place where producer sell their products (km)	6.3 (8.7)	8.0 (11.1)	7.1 (9.9)	−1.18
Obtained credit (% of yes)	28.6	25.8	27.2	0.18
Membership of producers' organization (% of yes)	57.1	90.3	73.3	−26.84***
Contact with extension agents (% of yes)	60.2	90.3	74.9	−23.01***
Contact with extension agents (number of visits between January and October 2012)	1.9 (2.2)	7.0 (8.8)	4.4 (6.8)	−5.44***
Number of oxen owned by the producer for animal traction	1.1 (2.0)	0.5 (1.1)	0.8 (1.7)	2.78***

[(a)]Figures in parenthesis are standard deviations;
[(b)]Statistical test: Pearson chi-square for dummies and t test others.
***Significant at 1%, *Significant at 10%.

growing organic cotton within each category of households according to the welfare status (Figure 3) also shows the dominance of females. Indeed, in 60% of the surveyed households, organic cotton is produced only by females. Intra-household analysis shows that in the class of poor households, organic cotton is produced solely by females in 29% of cases, while in 47% of cases organic cotton is practiced by both male and female. Females are also the majority among producers of organic cotton in the class of rich and medium households. In wealthy households, females and males are almost equally represented in the production of organic cotton. Overall, the predominance of females among adopters of organic cotton is interesting given that they are often the most marginalized stratum of society, with less access to land.

The level of education of organic cotton farmers is relatively low (1 year) compared to that of conventional cotton producers (3 years). Similarly, the number of years of experience in cotton production is relatively higher for producers of conventional cotton than for organic cotton producers. Conventional cotton farmers also have their farm twice far from their houses than organic cotton producers. In other words, it stands out that, compared to conventional farms, organic farms are closer to the farmer's housing (Table 5). This allows a more intense labor and eases as well the transportation of the necessary resources, particularly cow manure and organic matters which are of key importance in organic farming.

Table 5 also indicates that household size of conventional cotton producers is higher than that of households of organic cotton farmers. This also applies to the total area available. However, the difference between the two groups is not significant at 5% level.

Almost 27% of respondents have access to credit, with about 29% for conventional cotton producers against 26% for organic cotton producers. 90% of organic cotton producers are members of a cooperative against 57% for producers of conventional cotton. This indicates that organic farmers are better organized and have more contacts with advisory and extension services. The organic cotton producers also have more access to extension services than conventional cotton farmers.

Overall, we can retain that conventional cotton producers are mainly men with relatively more productive resources (available land, labor, oxen) unlike organic cotton households. However, they are relatively younger and have their farms two times far from their houses compared to organic cotton producers.

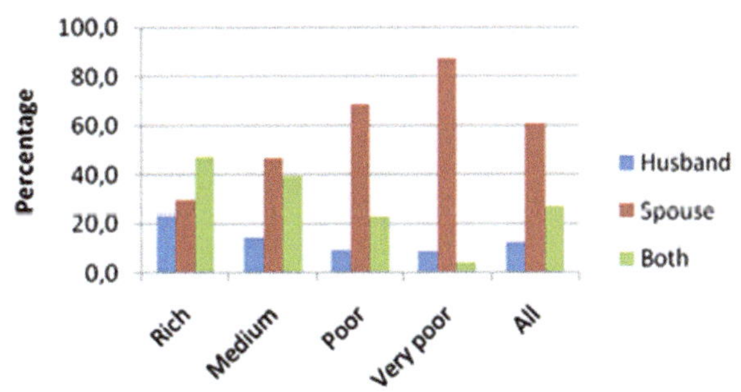

Figure 3 Gender of the member producing organic cotton in the household of organic cotton farms.

Farmers' characteristics and organic cotton adoption

The results of the probit model are presented in Table 6. It is worth noting that in our initial probit model, the squared of the variable AGE (AGE^2) and the squared of the distance between producers' house and their farms ($DISTEX^2$) had no significant effects on the adoption of organic cotton farming. Indeed, the Likelihood Ratio (LR) tests indicated that these two variables had no significant influence of the adoption of organic cotton.[c] Thus, we re-estimated the probit model without these two variables.

Table 6 indicates that producers' age positively influences the adoption of organic cotton. In other words, older farmers are more likely to adopt organic cotton than younger producers. The probability that a producer adopts organic cotton increases, *ceteris paribus*, by 1.0 percentage point when farmer's age increases by one year. This result is contrary to the findings of several authors (Duram 1997; Marenya and Barrett 2007; Läpple and Van Rensburg 2011; see also Niemeyer and Lombard 2006) who argued that younger farmers are more open to the adoption of organic farming than older farmers.

Gender significantly influences the adoption of organic cotton. The probability of organic cotton adoption tends to be, *ceteris paribus*, 49 percentages points higher for female farmers. This indicates that organic farming is more attractive to women compared to conventional farming. In fact, organic cotton farming enables women to hold a separate cotton farm and thus increase their economic independence, whereas with the conventional system they depend mainly on the farm of the (male) head of the household (Glin et al. 2012: 344; see also Tovignan 2005). The use of inputs locally available is also key factor that foster women decision to practice organic farming (Assogba et al., 2014). In most cases, females also perform more effectively in organic farming than males because their farms are smaller in size and often closer to the compounds. This eases the mobilization of organic inputs as well as the provision of the necessary cares and the overall management of the organic farm. This corroborates the findings of Bjørkhaug (2006), for whom male-headed farms tend to be large with a strong commercial focus, while women-headed farms tend to be smaller in size, with lower output volume and higher contribution of off-farm income to total farm income. Moreover, women are particularly concerned with the health effects of the highly toxic pyrethroid insecticides used in conventional cotton farms (Bassett 2010). There is a belief that toxic chemical products affect woman fertility and fetal development in case of pregnancy. All this in addition to the high inputs cost and discrimination in conventional (cotton) farming system makes organic the best and preferred option for women.

In addition, contrary to our expectation, the probability that a farmer adopts organic cotton tends to be, *ceteris paribus*, 5 percentage points higher when the level of education drops a unit (Table 6). This means that less educated farmers are more likely to adopt organic cotton farming than highly educated farmers. This result is contrary to those generally available in earlier studies of the adoption of organic farming (e.g., Demiryurek 2010; Mzoughi 2011) where education often has a significant positive effect on organic farming adoption. For Läpple and Van Rensburg (2011) the level of education has no significant effect on the adoption of organic farming. Our result is however in line with Burton et al. (2003). According to this author the negative effect of education on organic farming adoption may be due to differences in country and commodity coverage. In Benin, the level of education is relatively low in rural

Table 6 Determinants of organic cotton adoption: results of the probit model

Variable	Label	Coefficient[a]	Marginal effect[a][b]
AGE	Age of the producer (year)	0.025* (0.014)	0.010* (0.005)
SEX	Sex of the producer (1 = male, 0 = female)	−1.542*** (0.441)	−0.490*** (0.107)
EDUC	Formal education of the producer (year)	−0.118** (0.053)	−0.046** (0.021)
EXPCOT	Experience in cotton production in general (year)	−0.034* (0.021)	−0.013* (0.008)
CLASSP	Class of poverty as perceived by producers (1 = very poor, 2 = poor, 3 = medium, 4 = rich)	−0.102 (0.206)	−0.039 (0.080)
CENTRE	Regional dummy (1 = Farmers located in the Centre of Benin, 0 = otherwise)	1.346** (0.641)	0.451*** (0.173)
NORTHW	Regional dummy (1 = Farmers located in the northwest-Benin, 0 = otherwise)	−0.318 (0.473)	−0.124 (0.185)
LNDISTEX	Logarithm of the distance between the producer's house and his farm (km)	−1.112*** (0.326)	−0.431*** (0.126)
HHSIZE	Household size (number of members)	0.052* (0.031)	0.020* (0.012)
LNSUPTOT	Logarithm of the amount of land available (ha)	−0.235 (0.237)	−0.091 (0.092)
OXEN	Ownership of oxen (1 = yes)	0.452 (0.572)	0.168 (0.202)
LNDISTM	Logarithm of the distance between the village and the nearest market place where producer sell their products (km)	0.075 (0.117)	0.029 (0.045)
ACRED	Obtained credit (1 = yes, 0 = no)	−0.340 (0.341)	−0.133 (0.134)
COOP	Membership of producers' organization (1 = yes)	0.554 (0.371)	0.217 (0.144)
NVISIT	Contact with extension agents (number of visits between January and October 2012)	0.355*** (0.073)	0.138*** (0.026)
CONS	Constant	0.192 (0.738)	

Number of observation = 191; LR chi2(14) = 144.43***, Log likelihood = −60.110, Pseudo R2 = 0.5457.
[a]Figures in parenthesis are standard errors; [b]The partial derivative is given by $\frac{\partial E[y|x]}{\partial x} = \phi(x'\beta)\beta$ continue variables and $\text{Prob}[y = 1|\bar{x}_{(d)}, d = 1] - \text{Prob}[y = 1|\bar{x}_{(d)}, d = 0]$, for dummy variables (Greene 2008: 775), with $\bar{x}_{(d)}$ the means of all the other variables in the model.
***Significant at 1%, **Significant at 5%, *Significant at 10%.

areas; and the educated people living in cotton production areas are generally involved in the management of conventional cotton producers' cooperative (Alidou, 2014). They are relatively rich and can hardly accept losing the privileges and the social prestige associated with the management of conventional cotton cooperatives (see also Alidou, 2014).

The negative effect of the level of education on the decision to produce organic cotton could probably be linked to the fact that more than half of organic producers in the sample are women. To test this, we re-estimated the probit model separately for females and for males. The results obtained and presented in Table 7 indicate that the level of education has a negative influence on the adoption of organic cotton for males and for females. However, the influence of education on the adoption of organic cotton is not statistically significant for females ($p > 0.10$) while it is significant for males. In short, the negative effect of the level of education on the adoption of organic cotton is not due to the fact that more than half of organic producers in the sample are women.

In the same vein, there is a negative (but insignificant) relationship between the adoption of organic cotton and the level of household wealth (Table 6). This means that poor farmers and wealthier peasants have almost the same probability to adopt organic cotton farming. Thus, beyond the poverty level, there are other factors that lead farmers to become organic cotton producer. For example, Moumouni et al. (2013) argued that beyond profitability, attitudinal aspects and social dynamics (in the household and the society) also influence producers' decisions to adopt organic farming methods. Although poor and rich farmers have statistically the same likelihood to adopt the

Table 7 Determinants of organic cotton adoption: results of the probit model for women and men

Variables	Label	Women Coefficient[(a)]	Men Coefficient[(a)]
AGE	Age of the producer (year)	0.044 (0.034)	0.025 (0.017)
EDUC	Formal education of the producer (year)	−0.144 (0.401)	−0.118** (0.057)
EXPCOT	Experience in cotton production in general (year)	−0.010 (0.055)	−0.043* (0.026)
CLASSP	Class of poverty as perceived by producers (1 = very poor, 2 = poor, 3 = medium, 4 = rich)	−0.343 (0.554)	−0.059 (0.233)
CENTRE	Regional dummy (1 = Farmers located in the Centre of Benin, 0 = otherwise)		1.327* (0.741)
NORTHW	Regional dummy (1 = Farmers located in the northwest-Benin, 0 = otherwise)	−0.693 (0.852)	−0.895 (0.901)
LNDISTEX	Logarithm of the distance between the producer's house and his farm (km)	−1.487 (1.208)	−1.077*** (0.355)
HHSIZE	Household size (number of members)	0.162 (0.138)	0.044 (0.035)
LNSUPTOT	Logarithm of the amount of land available (ha)	0.158 (0.463)	−0.257 (0.290)
OXEN	Ownership of oxen (1 = yes)		0.930 (0.899)
LNDISTM	Logarithm of the distance between the village and the nearest market place where producers sell their products (km)	−0.087 (0.248)	0.137 (0.141)
ACRED	Obtained credit (1 = yes, 0 = no)	−0.407 (0.965)	−0.369 (0.399)
COOP	Membership of producers' organization (1 = yes)	0.998 (0.661)	0.440 (0.528)
NVISIT	Contact with extension agents (number of visits between January and October 2012)	0.447* (0.236)	0.370*** (0.083)
CONS	Constant	−0.805 (1.730)	−1.232 (0.880)

[(a)]Figures in parenthesis are standard errors;
***Significant at 1%, **Significant at 5%, *Significant at 10%.

organic cotton, Lapple (2010) argued that, when these low-income farmers convert to organic farming they are more likely to continue in the future. Giovannucci et al. (2005) explained low-income farmers' attitudes by the fact that organic cotton production is almost an effective risk management tool that reduces their input costs, and provides them improved income.

Finally, it stands out that the adoption of organic cotton significantly depends on the number of years that the producer spent in (conventional) cotton farming (Table 6). In other words, farmers with few years of experience in cotton in general tend to adopt organic cotton. Thus, the probability that a farmer adopts organic cotton tends to be, *ceteris paribus*, about 1 percentage point higher when the number of years of experience in cotton in general drops a unit.

Effect of farm and biophysical factors on organic cotton adoption

The variable NORTHW has no significant effect on the adoption of organic cotton, whereas variable CENTRE significantly influences the adoption of organic farming. The producers of Center-Benin are much more likely to adopt organic cotton than producers of the two other regions (Northwest-Benin and Northeast-Benin). Thus, as the variable CENTRE also control for land pressure, one can say that farmers who live in areas with low human pressure on land are more likely to adopt the organic farming than their counterpart farmers who live in areas with high human land pressure. In other words, the probability that producers of Centre-Benin (area with low human pressure on land) adopt organic cotton is, *ceteris paribus*, 45 percentage points higher than the producers of two other regions.

Producers who have their farm near home are more likely to adopt organic farming than those who have the farm far from their homes. This is due to the fact that producers who have their farm close to their house have less difficulty to transport organic fertilizers and other inputs used in the organic cotton production. The probability that farmers adopt the organic cotton is 4.3 percentage points higher when they live a kilometer close to farm.

It also came out that household size has a positive and significant effect ($p > 0.10$) on the adoption of organic cotton. This can be explained by the fact that organic cotton production is more labor-intensive than conventional cotton. Put differently, this reflects the importance of family labor in the adoption of organic farming given resource-poor farmers have financial constraints for hiring wage laborers (Marenya and Barrett 2007). As a result, farmers with relatively large amount of labor are more likely to engage in organic cotton farming.

Finally, the area of land available for the household has a negative (but insignificant) influence on the adoption of organic cotton. The analysis of the land area under cotton production indicates (Table 8) that the area under conventional cotton is almost the double of that of organic cotton. Likewise, within the different classes of prosperity, the area cultivated in conventional cotton is higher than in organic cotton. For instance, in the class of rich households, the average size in conventional cotton is almost the triple of organic cotton. Whatever the level of prosperity, organic cotton farmers grow less than 10 ha (Table 8). Most of the organic farmers (78.6%) grow less or not more than 2 ha of cotton. This is particularly true for farmers of very poor conditions. In the class

Table 8 Amount of land under organic and conventional cotton

Type of cotton farming	Rich	Medium	Poor	Very poor	All
Conventional cotton (ha)[(a)]	8.63 (10.64)	3.23 (2.28)	2.75 (2.09)	1.64 (1.21)	3.30 (4.59)
Distribution (%)					
[0, 2] ha	20.0	42.9	57.4	86.2	57.1
]2, 5] ha	33.3	42.9	33.3	10.3	29.4
]5, 10] ha	26.7	14.3	9.3	3.4	10.9
>10 ha	20.0	0.0	0.0	0.0	2.5
Organic cotton (ha)[(a)]	3.21 (2.17)	2.60 (1.98)	0.79 (0.72)	0.56 (0.37)	1.56 (1.73)
Distribution (%)					
[0, 2] ha	41.2	53.6	97.7	100.0	78.6
]2, 5] ha	47.1	35.7	2.3	0.0	17.0
]5, 10] ha	11.8	10.7	0.0	0.0	4.5
>10 ha	0.0	0.0	0.0	0.0	0.0

[(a)]Figures in parenthesis are standard deviations.

of medium households, more than half of organic cotton producers grow less than 2 ha of cotton. Overall, it stands out that (i) the area under organic cotton is positively associated with the level of wealth of households, with the area under cotton for the rich being almost six times that of the poorest; (ii) the average size produced in conventional cotton is almost the double of that produced in organic cotton. The reason of this is twofold. First, organic cotton farming is more labor-intensive than conventional cotton and can only be adequately handled on relatively small scale, particularly due to the weeding process. Secondly, organic farming relies mainly on natural resources and materials (cow manure, neem seeds, etc.) for soil fertility and pest management and it is difficult to mobilize enough for a large size of operation. The same situation is observed for organic agriculture in countries like Canada, New Zealand and the USA (see Niemeyer and Lombard 2006).

Finally, it is worth noting that the ownership of oxen and "the distance between the village and the nearest market place where producers sell their produce" have no significant effect ($p > 0.10$) on the organic cotton adoption.

Effect of institutional and economic factors on organic cotton adoption

The institutional factors considered include: access to credit, support from extension services, and membership in farmer's association.

It came out that access to credit has no significant influence on the adoption of organic cotton, meaning that credit is not a determinant factor in the dissemination of organic cotton (Table 6). In fact, organic farming is less dependent on marketed inputs compared to conventional cotton farming. Moreover, there is no credit incentive in the institutional support provided by the several organic cotton programs in Benin. Furthermore, because of the indebtedness they witnessed in the conventional cotton system, most of the organic farmers are more suspicious about credit.

With respect to support from extension services, the number of visits of extension agents is positively and significantly associated with the adoption of organic cotton farming. Organic cotton farmers benefit from intensive extension services from NGOs and development organizations, which are the major drivers of organic value chains in Benin and in West Africa in general. This confirms our expectations regarding the effect of extension on the adoption of organic cotton. Indeed, organic farming is knowledge intensive and requires regular interactions between farmers and extension services.

Membership in a farmers' cooperative has positive but insignificant effect on the adoption of organic cotton (Table 6). In other words, the decision to engage in the production of organic cotton is not dependent on whether the producer belongs to a cooperative or not. However, although membership in cooperatives has no significant influence on the adoption of organic cotton, various studies have highlighted the importance of cooperatives and other farmer associations in the adoption of innovations. Thus, Sodjinou and Henningsen (2012) showed that cooperatives are not only channels for the dissemination of innovations but also institutions where producers share the innovations and the various problems encountered. Indeed, "joint problem solving, and farmer-to-farmer exchange can improve community relations and lead to greater involvement and commitment of producers" (Giovannucci et al. 2005). Accordingly, as noted by Giovannucci et al. (2005), farmers' organizations could be considered as a central aspect of a strategy aimed at using organic agriculture as a tool for poverty alleviation in rural areas.

Basically, organic farming is 'knowledge intensive' and requires the mobilization and building of effective socio-technical networks allowing interactive and social learning among peer organic farmers, extension and advisory services, and even research organizations (van Elzakker 1999; Glin et al. 2012; Moumouni et al. 2013; Nicolay and Baker 2012). Besides, the social control and internal control systems that are at the heart of the certification mechanism of organic farming require collective action to control the traceability of materials and resources used (sprays equipments, organic inputs, etc.) and ensure the compliance with the organic standard.

Conclusion

The objective of this study was to assess socioeconomic and institutional factors determining farmers' decisions to adopt organic cotton. In fact, the organic cotton initiative was developed as an alternative to the environmental, health, and socioeconomic concerns of conventional cotton farming with its high reliance on chemicals. Thus, the organic agenda may well fit in the current policy debate on green economy, climate smart agriculture and resilience by reconciling economic development with environmental and social principles. In Benin, the government is strongly concerned with the fulfillment of the Millennium Development Goals, particularly halving poverty while sustaining environment and natural resources. From the results of this study, it clearly appears that organic cotton farming is more attractive to women, older, less educated, and poorer producers, all those being the most vulnerable categories in society. Subsequently, organic cotton should be considered as a prospective policy option to reach the poor and strengthen their livelihoods conditions while contributing to preserve the environment and natural resources. According to FAO (2010), increased investments in agricultural development targeting the smallholder farmers in developing countries is an important element in improving food security in rural areas. Moreover, the increasing trends of chemical inputs (pesticides and mineral fertilizers) costs in the context of uncertainty and a fragile global economy make the organic cotton a refuge to protect vulnerable farmers against external shocks. Furthermore, organic farming enables women to hold a separate cotton farm and thus increase their economic independence contrary to the conventional system where they depend mainly on the farm of the (male) head of the household.

Lastly, the institutional factors that determine the adoption of organic cotton include mainly the support from extension services and to some extent farmers' membership in cooperatives. Despite the relative strength of the organic cotton system in these areas, there is still a need to enhance the extension system by: (1) exploring, designing, and upgrading innovative pedagogic tools such as videos and mobile phone technology to foster learning; and (2) strengthening organic farmer's organizations and the linkage with agricultural research organizations for technology development.

Endnotes

[a]The decline in organic cotton production in 2010 is due to the unsuccessful marketing over the years 2008 and 2009 saw organic farmers being paid later than conventional ones, which created distrust among many organic farmers and quite a few subsequently withdrew from organic cotton production during the year 2009/2010 (Glin et al. 2012).

[b]Pressure on land means overuse of agricultural land, and also high/increased demand of land for agriculture. The land available is not often sufficient to satisfy the increased demand.

[c]For the test of the variable AGE2, LR chi2 = 0.00 (with P-val = 0.987), and For the test of the variable DISTEX2, LR chi2 = 2.75 (with P-val = 0.097).

Competing interests

The authors declare that they have no competing interests.

Authors' contributions

All authors: conceived the study, involved in its design and participated in field data collection. ES: prepared the data, performed the statistical analysis. ES and LCG drafted the manuscript together. NG: read and commented the first draft. All authors read and approved the final manuscript.

Author details

[1]Ecole Nationale des Sciences et Techniques Agronomiques de Djougou, Université de Parakou, 03 BP 412, Porto-Novo, République du Benin. [2]FiBL West Africa, BP 16 IER/CRRA Sikasso, République du Mali. [3]FiBL (Research Institute of Organic Agriculture), 5070 Frick, Switzerland. [4]Faculté d'Agronomie, Université de Parakou, Parakou, République du Benin. [5]Institut National des Recherches Agricoles du Benin, Natitingou, République du Benin.

References

Alidou GM (2014) Networking, Social Capital and Gender Roles in the Cotton System in Benin. Wageningen University, Dissertation

Assogba SC-G, Tossou RC, Lebailly P, Magnon Y (2014) Sustainable intensification of agriculture in Benin: myth or reality? Lessons from organic and cotton made in Africa production systems. Int J Agric Innovation Res 2(5):2319–1473

Baffes J (2007) The "cotton problem" in West and Central Africa: the case for domestic reforms. CATO Institute. *Econ Dev Bull No. 11*. Available via http://www.cato.org/pubs/edb/edb11.html. Accessed 30 Dec 2013

Bassett T (2010) Slim pickings: fairtrade cotton in West Africa. Geoforum 41(1):44–55

OBEPAB (Organisation Béninoise pour la Promotion de l'Agriculture Biologique, 2002) Le Coton au Bénin: rapport de consultation sur le coton conventionnel et le coton biologique au Bénin. PAN UK's Pesticides, Poverty and Livelihoods project. Available via http://www.pan-uk.org/Projects/Cotton/pdfs/Benin.pdf. Accessed 29 Dec 2013.

Bjørkhaug H (2006) Is there a female principle in organic farming? An interpretation of data for Norway. In: Holt G, Reed M (eds) Sociological Perspectives of Organic Agriculture: From Pioneer to Policy. CAB International, London, pp 195–209

Bolwig S, Gibbon P, Jone S (2009) The economics of smallholder organic contract farming in tropical Africa. World Dev 37(6):1094–1104

Burton M, Rigby D, Young T (2003) Modelling the adoption of organic horticultural technology in the UK using duration analysis. Australian J Agr Resource Econ 47(1):29–54

Chilonda P, van Huylenbroeck G (2001) A conceptual framework for the economic analysis of factors influencing decision-making of small-scale farmers in animal health management. Rev sci tech Off int Epiz 20(3):687–700

Demiryurek K (2010) Analysis of information systems and communication networks for organic and conventional hazelnut producers in the Samsun province of Turkey. Agr Syst 103:444–452

Doss CR (2006) Analyzing technology adoption using microstudies: limitations, challenges, and opportunities for improvement. Agric Econ 34:207–219

Doss CR, Morris ML (2001) How does gender affect the adoption of agricultural innovations? The case of improved maize technology in Ghana. Agric Econ 25:27–39

Doss CR, Mwangi W, Verkuijl H, de Groote H (2003) Adoption of Maize and Wheat Technologies in Eastern Africa: A Synthesis of the Findings of 22 Case Studies. CIMMYT Economics Working Paper 03-01. Mexico: CIMMYT

Dowd B (2008) Organic Cotton in Sub-Saharan Africa. A new Development Paradigm? In: Moseley W, Gray L (eds) Hanging by a Thread: Cotton, Globalization and Poverty in Africa. Ohio University Press, Athens, OH, pp 251–271

Duram A (1997) A pragmatic study of conventional and alternative farmers in Colorado. Prof Geogr 49(2):202–213

Eyhorn F, Ramakrishnan M, Mader P (2007) The viability of cotton-based organic farming systems in India. Int J Agric Sustain 5(1):25–38

FAO (2010) The State of Food Insecurity in the World 2010, Addressing Food Insecurity in Protracted Crises. Food and Agriculture Organization of the United Nations, Rome

Ferrigno S, Ratter SG, Ton P, Vodouhê DS, Williamson S, Wilson J (2005) Organic cotton: a new development path for African smallholders? The Gatekeeper Series, 23p.

Gbesso CA (2000) Etude comparée de la rentabilité socioéconomique du coton biologique et du coton conventionnel. Université d'Abomey-Calavi, Mémoire de Maitrise

Gergely N (2009) The Cotton Sector of Benin. World Bank, Washington

Giovannucci D, Eyhorn F, Han Z, Joensen L, John M, Mehta S, Meng F, Ramakrishnappa K, Reddy SST, Thimmaiah A, Xi Y, Zong H (2005) Organic agriculture and poverty reduction in Asia: China and India Focus. IFAD Office of Evaluation, Rome. Available via http://www.ifad.org/evaluation/public_html/eksyst/doc/thematic/organic/asia.pdf. Accessed 31 Dec 2013.

Glin L C (2014) Governance of global organic agro-food networks from Africa. PhD thesis, Wageningen University, the Netherlands.

Glin LC, Mol APJ, Oosterveer P, Vodouhê DS (2012) Governing the transnational organic cotton network from Benin. Global Netw 12(3):333–354

Glin LC, Mol APJ, Oosterveer P (2013) Conventionalization of the organic sesame network from Burkina Faso: shrinking into mainstream. Agric Hum Values 30(4):539–554

Greene WH (2008) Econometric Analysis, 6th edn. Pearson, Upper Saddle River NJ

Guilmo E Navarro E, Ribet C, Stanicka L, Vaugeois M (2006). Coton Bio ou coton Bt : Une solution pour les paysans pauvres ? 11p., Available: http://www.museum.agropolis.fr/pages/savoirs/cotontransgenique/rapport_controverse.pdf. Accessed 16 Nov 2014.

Hole DG, Perkins AJ, Wilson JD, Alexander IH, Grice PV, Evans AD (2005) Does organic farming benefit biodiversity? Biol Conserv 122:113–130

Hulsebusch C, Wichern F, Hemann H, Wolff P (eds) (2007) Organic agriculture in the tropics and subtropics current status and perspectives. Supplement No 89, J Agr Rural Dev Trop. Kassel University Press GmbH. Available via http://www.uni-kassel.de/upress/online/frei/978-3-89958-263-5.volltext.frei.pdf. Accessed 31 Dec 2013.

Justus O (2009) Value Chain Analysis of Organic Cotton Sub sector in Tanzania. Department of Agricultural Economics and Agribusiness Management, Egerton University, Kenya. Avaible via http://www.iiste.org/Journals/index.php/DCS/article/viewFile/14424/14733. Accessed 27 Aug 2014

Kristiansen P, Taji A, Reganold J (2006) Organic Agriculture: A Global Perspective. CABI Publishing, Wallingford

Kutting G (2004) Globalization and the Environment: Greening Global Political Economy. State University of New York Press, Albany

Lapple D (2010) Adoption and abandonment of organic farming: an empirical investigation of the Irish Drystock Sector. J Agr Econ 61(3):697–714

Läpple D, Van Rensburg T (2011) Adoption of organic farming: are there differences between early and late adoption? Ecolog Econ 70:1406–1414

Lockie S, Lyons K, Lawrence G, Halpin D (eds) (2006) Going Organic: Mobilizing Networks for Environmentally Responsible Food Production. CABI, Oxfordshire and Cambridge

MAEP (Ministère de l'Agriculture de l'Elevage et de la Pêche (2011) Plan stratégique de relance du secteur agricole (PSRSA) au Bénin. MAEP, Cotonou

Marenya PP, Barrett CB (2007) Household-level determinants of adoption of improved natural resources management practices among smallholder farmers in western Kenya. Food Pol 32:515–536

Matthess A, van den Akker E, Chougourou D, Midingoyi S (2005) Compétitivité et durabilité de cinq systèmes culturaux cotonniers dans le cadre de la filière. GTZ-MAEP, Cotonou

Moseley WG, Gray LC (2008) Hanging by a Thread: Cotton, Globalization, and Poverty in Africa. Ohio University Press, Athens, OH

Moumouni MI, Baco MN, Tovignan S, Gbèdo F, Nouatin GS, Vodouhê SD, Liebe U (2013) What happens between technico-institutional support and adoption of organic farming? A case study from Benin. *Organic Agriculture*. DOI 10.1007/s13165-013-0039-x.

Mzoughi N (2011) Farmers adoption of integrated crop protection and organic farming: Do moral and social concerns matter? Ecolog Econ 70:1536–1545

Negatu W, Parikh A (1999) The impact of perception and other factors on the adoption of agricultural technology in the Moret and Jiru Woreda (district) of Ethiopia. Agric Econ 21:205–216

Nicolay G, Baker B (2012) Challenges and opportunities for organic research and extension. *Rural 21*-03/2012. Available via http://orgprints.org/23505/1/rural2012_03-S32-34.pdf. Accessed 29 Dec 2013.

Niemeyer KB, Lombard JP (2006) Emerging Scared: An Analysis of Socioeconomic Data on Conversion in South Africa. In: Holt G, Reed M (eds) Sociological Perspectives of Organic Agriculture: From Pioneer to Policy. CAB International, London

Pacini C, Wossink A, Giesen G, Vazzana C, Huirne R (2002) Evaluation of sustainability of organic, integrated and conventional farming systems: a farm and field-scale analysis. Agric Ecosyst Environ 95:273–288
Pazou EYA, Boko M, van Gestel CAM, Ahissou H, Lalèyè P, Akpona S, van Hattum B, Swart K, van Straalen NM (2006a) Organochlorine and organophosphorous pesticide residues in the Ouémé River catchment in the Republic of Bénin. Environ Int 32:616–623
Pazou EYA, Lalèyè P, Boko M, van Gestel CAM, Ahissou H, Akpona S, van Hattum B, Swart K, van Straalen NM (2006b) Contamination of fish by organochlorine pesticide residues in the Ouémé River catchment in the Republic of Bénin. Environ Int 32:594–599
Polson RA, Spencer DSC (1991) The technology adoption process in subsistence agriculture: the case of cassava in southwestern Nigeria. Agr Syst 36:65–78
Ranjani KM, Sagayam J, Sudha N, Rengalakshmi R (2008) Gender, efficiency, poverty reduction, and empowerment: reflections from an agriculture and credit programme in Tamil Nadu, India. Gene Dev 16(1):101–116
Rogers EM (2003) Diffusion of Innovations, 5th edn. The Free Press, New York
Sall S, Norman D, Featherstone AM (2000) Quantitative assessment of improved rice variety adoption: the farmer's perspective. Agr Syst 66:129–144
Sarker A, Itohara Y (2008) Factors influencing the extent of practice of organic farming technologies: a case study of Tangail district in Bangladesh. Am J Agr Biol Sci 3(3):584–590
Skinner C, Gattinger A, Muller A, Mäder P, Fliessbach A, Stolze M, Niggli U (2014) Greenhouse gas fluxes from agricultural soils under organic and non-organic management: A global meta-analysis. Sci Total Environ 468–469: 553–563
Sodjinou E (2011). Poultry-based intervention as tool for poverty reduction and gender empowerment: empirical evidence from Benin. Dissertation, University of Copenhagen. Available via http://curis.ku.dk/ws/files/33324451/Thesis_Sodjinou_February_2011_VF.pdf. Accessed 01 June 2013.
Sodjinou E, Henningsen A (2012) Community-based management and interrelations between different technology adoption decisions: innovations in village poultry farming in Western Africa. FOI Working Paper 2012/11. Institute of Food and Resource Economics, Copenhagen. Available via http://okonomi.foi.dk/workingpapers/WPpdf/WP2012/WP_2012_11_village_poultry_farming.pdf. Accessed 01 June 2013.
Ton P (2002) Organic Cotton Production in Sub-Saharan Africa: The Need of Scaling up. Pesticide Action Network, London
Ton G, Bijman J, Oorthuizen J (2007) Introduction. In: Ton G, Bijman J, Oorthuizen J (eds) Producer Organizations and Market Chains: Facilitating Trajectories of Change in Developing Countries. Wageningen Academic Publishers, Arnhem, pp 11–19
Tovignan DS (2005) Gender Perspectives in the Adoption of Organic Cotton in Benin: A Farm Household Modeling Approach. University Of Giessen, Dissertation
Tulip A, Ton P (2002) Organic Cotton Study: Uganda Case Study. Pesticide Action Network, London
van Elzakker B (1999) Organic Cotton Production. In: Myers D, Stolton S (eds) Organic Cotton: From Field to Final Product. Intermediate Technology Publications, London
Verbeek M (2004) A Guide to Modern Econometrics, 2nd edn. John Wiley & Sons Ltd, West Sussex

13

Agricultural districts in the Italian regions: looking toward 2020

Daniela Toccaceli

Correspondence:
daniela.toccaceli@unifi.it
Department of Science for Economics and Business, University of Florence Via delle Pandette, 9 I-50127Florence

Abstract

Agricultural, quality agro-food, supply-chain and rural districts were introduced by Italian law in 2001. With reference to European structural policies, especially rural development policy, and to the districts' originality and relationship to the LEADER method and programmes, we wonder whether they can still form a useful part of the new programming period 2014-2020. The question is analysed in the wider context of international literature on socio-economic and economic policy, so as to stimulate broader interest, with a view to the future. A survey on national scale is presented, limited to the districts legally recognised by the Regions. A detailed historical reconstruction is given, offering a 'reconstructive reading' through which we clarify the actual nature of these districts, including their connections to clusters and industrial districts. We also discuss how to better link the theoretical and the legal frameworks. And we examine the connection between the basic economic mechanisms of the district theory and how the Regions have implemented them in reality. To do this, we summarised the overall conceptual framework into a few basic components for our desk research. This has led to development of a methodological definition of districts in agriculture.

Finally we give a concise analysis of the main outlines of the structural policies relating to this essential notion of 'district', in order to reach some conclusions of general validity.

Keywords: Agricultural; Quality agro-food; Supply-chain and rural districts; Industrial districts; Cluster; Business network; LEADER method; Rural governance and multi-level governance; Rural development policy; European structural and investment funds; Community-led local development; Italian Regions

Background

The new approach for European structural policy for the period 2014–2020 gives the local scale a strategic role in implementing interventions in support of participatory local development. This is to be based upon multi-level governance (Committee of the Regions 2009) and expansion of the LEADER method, especially in order to implement *community-led local development* (CLLD) and multi-fund strategies (European Commission's Directorates General AGRI, EMPL, MARE and REGIO 2013, European Union 2013:357–359).

In Italy the Regions[a] are responsible for implementing these policies through their own programming and for integrating them using other policy tools. Many Regions had already recognised agricultural districts under the Italian law of 2001[b.] that empowered them to identify Districts in Agriculture[c.] (DAs) in the form of rural or quality agro-food districts.

At the time of its introduction, that law aimed to boost agricultural modernisation, following the new deal promoted by Agenda 2000. Today, with the new programming cycle of European structural policies and common agricultural policy, the important question is, how useful are the DAs and how can they be inserted into the new regional programming framework in order to implement rural development programmes?

Despite many studies into regional experiences of DAs[d.], organic knowledge is still lacking about implementation of the law on a national level, which is still ongoing. Indeed, researching on a national scale has proved very complex for three reasons: i) DAs have become very different "things" in each regional interpretation, so it is difficult to make comparisons among them; ii) DAs are still an open chapter of economic and agricultural research, because of the particular theoretical framework that has been adopted in Italy; iii) the theoretical and legal frameworks are still not fully integrated (Musotti 2001b).

We have addressed these difficulties by reconstructing the legislative situation stratified over a twenty-year period – the 'reconstructive reading' according to Albisinni (2010) - and by using a fresh approach. Our research aims to create the conditions for comparing experiences which have arisen in concurrence with very different regulatory frameworks and political contexts. Our article also tries to recreate an updated geographical mapping of the phenomenon at national level and seeks to outline the main economic attributes assumed by DAs. On the basis of our findings we offer our initial thoughts on the potential and weaknesses of future use of DAs in the coming programming period 2014–2020.

Although this article focuses on Italian agricultural, quality agro-food, supply-chain and rural districts, the issue is placed in the context of international socio-economic literature on the subject. The research may thereby be of more general interest, firstly because clusters and industrial districts are widespread at global level in the agricultural sector and also because of the original approach used for analysing regional policies.

Some different instruments, similar to DAs even if not derived from a 'district' concept, exist in other European countries, as mentioned below.

A brief literature review of clusters and Marshallian Industrial Districts

Agglomerations of companies closely interconnected, specialised and co-located in a specific place have been the subject of empirical observations and researches over many years and in many countries. The earlier Marshallian Industrial District (MIDs) notion originated what is nowadays considered the core of the theoretical framework for explaining their persistence and diffusion: external economies and "industrial atmosphere" (Becattini et al. 2009). Though they have changed over time, they have been noted because of their success as well as, sometimes, for the excellent reputation of some product-place associations and specialisations, for example the "Prosciutto di Parma".

According to Becattini et al. (2009) and Porter and Ketels (2009), clusters and MIDs are neither synonyms nor concepts conceived in same scientific context or with reference to identical economic and societal environments, although they have "common roots".

Coming from a business strategy approach (Porter 1998) to understand the determinants of competition, Porter and Ketels (2009) define clusters as "geographic agglomerations of companies, suppliers and service providers, and associated institutions in a particular field, linked by externalities and complementarities of various type" (2009:172). These authors consider them "by design, a broader concept and a more general one" (2009:182) than MIDs, which are instead considered as "one type of a cluster".

Conversely, Becattini's MIDs are defined as "socio-territorial entities characterised by the active presence of both a community of people and a population of firms in one naturally and historically bounded area with a dominant industrial activity" (Pyke et al. 1990, cit. in Porter and Ketels 2009:172). Becattini adds, however, that "the population of firms and the related market institutions are only one, though central, part of it" (Becattini et al. 2009: xviii) and consequently clusters are considered only the industrial core of a more complex district process.

For the purpose of this paper, it is worth underlining one significant difference and some essential, common key points.

While the concept of MIDs is strictly related to industrial manufacturing sectors as well as to a system of small-medium enterprises, the notion of cluster is "by design" recognised in any production or services sector (Porter 1998). So agricultural clusters are identified in rural regions, are studied with the same methodological approach and are involved in identical considerations about their role in the value chain and about public policies (Porter 2003; Porter et al. 2004; Feser and Isserman A. 2009). Instead, justifying the translation of the MID concept to agriculture has required great efforts on the part of Italian researchers, as discussed below.

In any case, both converge on these issues: i) the core of the theoretical framework rests on a few basic conceptual nodes: the proximity (spatial dimension) of companies and institutions, interrelated in vertical, horizontal and diagonal ways, allows for better organisation of the value chain (business dimension); more added value is provided by a special local mix of externalities resulting from (in MID terminology) industrial atmosphere or (in cluster terminology) business environment (socio-economic dimension); ii) both put at the core of their conceptual construction the strategic function of the relationships between companies, local institutions in a wide sense, and the governance mechanisms for better management of those relationships; at the same time both recognise the twofold presence of an internal (local) network as well as of an external network, each playing an important role in the district process (network and organising dimension); iii) both are positively correlated with local development processes and not only to the growth of the enterprises involved; iv) for both, researchers have addressed policy recommendations aimed at supporting their growth and the development of the place/region where they are located (policy dimension).

So, clusters and MIDs are recognised in many countries which have introduced specific policies.

In the discussion of the results we analyse how these basic components work together in the Italian regional patterns.

Cluster policy approaches

Porter (2003) survey on economic performance of U.S. regions claims that regional development policies have to be particularly attuned to traded clusters because of their capability to support higher wages but also because they drive local employment and affect regional economic performances. On the other hand, Feser and Isserman A. (2009) found that rural clusters play an important role in the national value chain. They maintained that public policies should support these clusters, not only to sustain rural development but also to achieve an overall goal of regional and national economic development. Still focusing on rural regions in the U.S., Porter et al. (2004) state the importance of "cluster thinking" in rural economic development, remarking that policy has to be set at local and regional level because of the heterogeneity of rural areas and that community-based planning is needed. A "cluster-based approach" is thought essential for involving the private sector, and an "overarching organisational structure" for economic development is deemed helpful to coordinate and drive the process. However, they note, the role of institutions in a wider sense in the process of economic development in rural areas requires further research to be better understood.

In many European countries cluster policies - including industrial district policies - have existed since the 90s. They have been addressed to systems of small and medium enterprises in various production sectors, as well as to networks in the cluster, either for developing them or sometimes for using a "cluster lens" to boost other instruments e.g. for innovation processes or regional development. An Oxford Research (2008) found it difficult to identify a clear pattern among the 27 countries investigated. The complex nature of a cluster entails that cluster policies are complex too, given their multiple dimensions and the multiple actors involved in the process of programming and implementing the policies. An important reason for this is that the structural policies of the European Union have come to be delivered through multi-level governance processes involving both local actors and, vertically, more distant subjects and administration levels. Meanwhile, principles of subsidiarity and proportionality[e.] were entering the political and legal systems and processes of decentralisation towards the most peripheral levels of administration were occurring in many countries, Italy included.

Borràs and Tsagdis (2008) recognised five types of cluster policy, ranging from three top-down types that aim to create clusters from scratch ('creationist') or to support them without encompassing a bottom-up process ('narrow' and 'top-down'), to the two more complex types: the 'evolutionary' one that considers clusters tend to evolve in a cumulative and gradual way and policy is embedded in the network which shapes the direction of these changing dynamics, and the 'network' policy. The latter is based on multi-level governance because it recognises that the role of public authorities is inscribed into a relationship between private and public actors in the territory where they try to solve problems collectively. 'Network' policy is confident not only regarding success by public authorities but also the strength of the network.

And other related policies for rural areas in Europe

In European rural areas some other policies have been implemented that involve the concepts of cluster and districts, in their approaches based on the concept of network, territorial value and governance, and their focus on innovation in the rural environment. In this paper we consider only policies for qualifying agriculture and agro-food products and on LEADER[f.] programmes and methods. Although not cluster-specific, these policies have played a specific role in the integrated mix of main policies implemented in those areas, especially with reference to the rural districts and the quality agro-food districts which have developed in Italy, as further discussed below.

Stemming from the 1988 reform of structural funds, several LEADER programmes have developed continuously, undergoing various adjustments and growing in importance in the scope of rural policies, until they became integrated in rural development programmes in the past and in the current period. This was done without changing the basic methodological principle to which the specific 'LEADER added value' is attributed: bottom-up approach, territorial and integrated approach to rural development, public-private partnership in local action groups (LAGs), territorial networking and transnational cooperation to stimulate rural innovation processes. In other words, LEADER is a "network policy" in the sense that it looks at different types of networks with a multi-purpose intent. Firstly, at the 'social' network, which addresses the intangible factor of development through reinforcing the sense of local identity of the rural population, as well as the external image of the territory. This boosts marketing of local products (Ray 1998), because in LEADER a thread connects the sociological entity 'community' and the geographical entity 'territory' "that can be linked with an identifiable cultural heritage and identity" (Lee et al. 2005:275), as can the socio-economic entity 'social capital'. Secondly, the 'horizontal' network, as Murdoch (2000) put it, "linked to more general and also non-agricultural processes", which aims at integrated development of rural economy in a defined area. Thirdly, the 'vertical' network, which links actors of a specific area because of their involvement in the agro-food sector (Murdoch 2000).

Governance is therefore a fundamental process for the success of this policy, which has often attracted criticism about its capability to concretely and effectively achieve the prefigured added value (European Court of Auditors 2010).

For several years voluntary certification schemes have been in use in Europe[g.] offering protected status to foodstuffs with a certified local identity. This is a mechanism for adding value to such foods, thereby aiding local economies and supporting to some extent rural development (Tregear et al. 2007). In these policies too, it is the role of network and governance that is interesting for our purposes. Regarding the designation of origin supply chains (*filière*) Perrier-Cornet and Sylvander (2000) focus, firstly, on the development of collective strategies through coordination by *filière* actors in various ways. The aim of these strategies is to gain a collective, competitive advantage linked to a specific place and shared ways of working and traditions. This advantage depends on quality territorial revenue (Pecqueur 2001; Mollard 2001). Secondly, they discuss the more complex territorial strategies developed using various mixes of sectorial and territorial governance related to different degrees of territorial anchorage of the supply chains.

The overlap seems self-evident between the two types of European policies directed at farms and agro-food business supply chains, which sometimes may be spatially concentrated in local systems, if not coinciding with a recognised district. Sforzi and Mancini (2012), however, suggest it is worth bearing in mind the difference between a 'simple' localised agro-food system (Courlet and Pecqueur 1992) and a district, which implies a socioeconomic process (local community involvement) that is not present in the former.

Finally, worth mentioning is comparative research (Pacciani and Toccaceli 2014; Toccaceli 2014) into some case studies in several European member states, seeking to identify methodological analogies between DAs and other instruments developed to manage rural governance. This found considerable similarities with the French *contrat de Pays* (Toccaceli 2012:28) and the functioning of the *collectivités territoriales* (Rouger 2014), as well as with other systems in Belgium (Collignon 2014) and LEADER-based experiences in Spain (Gallego Moreno 2014).

DAs in the debate by Italian agricultural economists

In Italy the MID notion began to be applied to the agricultural sphere from the end of the '80s, as a result of contamination between industrial and agricultural economists (Angeli 2000), in a context both of visible change in agricultural structure and of progressive differentiation of production in the countryside. Meanwhile, European rural development policy started to be formulated with the European Commission's communication "The Future of Rural Society" (Commission of the European Communities 1988). In Italy, Iacoponi (1990) began[h.] by putting the question as simply an extension to agriculture of the Marshallian district concept re-elaborated by Becattini (1987, 1989) for Italian industrial districts.

However, the debate soon became an occasion for rethinking (and disputing) the comparison between "two opposite economic paradigms*[i.]" (Iacoponi 1990; Cecchi 1992; Becattini 2000 and Becattini 2001; Iacoponi 2001): the neoclassical and the districtual. As a result, the "systemic" stream of economic agricultural thought was revisited, deriving from the "territorial" approach (De Benedictis 2000; De Filippis 2000, Fabiani 2000; Angeli 2000; Musotti 2001a) observing that "the reality was beginning to escape the theory*" (Favia 2000). But only some agricultural economists were willing to welcome the 'district' paradigm, so limited but essential contributions referring to the notion of agricultural district were produced.

Iacoponi (2001) considered the paradigm of the industrial district proposed by Becattini to be composed of two core concepts, interacting dynamically in the local district system: the production-organisational (typically Marshallian) and the socio-economic (typically Becattinian). Such an artificial separation is useful in neatly clarifying the different notions of districts in agriculture: the agro-industrial district and the agricultural district, both related to the production-organisational approach (Iacoponi 1990, Cecchi 1992); and the rural district that better reflects the socio-economic approach (Romano 2000; Iacoponi 2002).

The notions of agricultural district and agro-industrial district

Iacoponi based his assessment of the compatibility between industrial and agro-industrial districts on a theoretical elaboration aimed at reconnecting the company-oriented economic approach typical of economic agrarian tradition, with the "contribution of the modern theory of the firm and developing it in such a direction as to include production systems organised at the local level*" (1990:711). According to an integrated vision of agriculture (Casati 1988) he took into account the overall production structures of agriculture and observed that only few farms are autonomous from an organisational point of view (1990). All the others are integrated either into the general economic system of the territory, or into the agro-industrial system. In the latter case, depending on the agribusiness structure, integration can be either at the level of the local system of enterprises (agricultural district), or at the level of product market (supply-chain integration), or at the level of the local system of firms and product markets (agro-industrial district). As an example of agro-industrial district Iacoponi cited the case of Parmigiano Reggiano.

Cecchi accepted Iacoponi's idea and elaborated it with more attention to the farms' territorial value. He therefore argued that: "The designation of agricultural district has to be found in the centrality of agricultural production: that is, it must be able to identify local agriculture as the "engine" of a district; so even in the presence of agro-industrial activities, the latter exists and thrives because it represents a processing of local agricultural products*" (1992:96). As an example of agricultural district he cited the Chianti wine district. He then goes on to provide a definition of agro-industrial district, which is when "the main distinguishing element is represented by the fact that the processing industry will use mainly agricultural products from outside the district itself*" (1992:97). The example referred to is the Po milk chain, which (evolving from an agricultural district) transforms milk mainly from external sources.

The notion of rural district

The rural district is analysed in relation to the various forms of socio-economic organisation that can characterise local systems in the countryside (Montresor 2000; Romano 2000; Forleo 2000). This covers not only areas where agricultural production is prevalent, but also local systems of rural development or peri-urban areas where production differentiation has not occurred (Cecchi 2000) or agro-districts and, lastly, rural districts. Cecchi described patterns of farm integration into the overall system of the territory, where the development of economic organisation solutions has enhanced the contextual knowledge and the capabilities of cooperation and social integration of local actors. He therefore distinguished "the districts of the areas of agricultural specialisation from those of rural local systems*" (2000:125).

Consequently, rather than connect the rural to the Marshallian district, according to the basic principles set out by Becattini (2000), Cecchi uses the "district approach" to separate social processes from economic ones. He thus highlights their specific contribution to and support of production differentiation of the countryside and sees the consolidation of the rural economy as having a plurality of economic activities rather than a predominance of agriculture. On this basis, he asserts that the possibility of identifying rural districts rather than agro-industrial districts is "more likely", although it is not easy to "perceive" a rural district because of the greater "physical distance between people and businesses" (2000:128). He thereby underlines how the spatial concept of proximity is changeable in the rural context and very different from the industrial one, especially in relation to MIDs.

Iacoponi, on the other hand, states that "the hypothesis that rural development is a local, sustainable, integrated and endogenous development leads us to the concept of the rural district, which is derived from the concept of industrial district (Becattini 1987; Bellandi 1987) and agro-industrial district (Iacoponi 1990; Fanfani and Montresor 1998;

Cecchi 1992; Favia 1992).*" (2002:74). Following Becattini's approach, he distinguishes firmly the notion of rural district from the weaker notion of a geographically significant network, because the first "is a strengthening and intensification of relations historically settled in the territory thanks to agriculture and activities related to them*" (2000:76). In fact, he notes that the main character of a rural district is the ability to provide districtual synergies for the economic activities of small and medium enterprises which have a sporadic presence in the territory, where "the community market and the atmosphere of the rural district can perform the miracle of integrating the 'missing branches' of the local economy.*" (2002:75–76), So Iacoponi highlights the basic socio-economic engine of the rural district.

Both scholars wrote taking as reference the case of the rural district of Maremma (Pacciani 1997; Pacciani 2002a and Pacciani 2003; Belletti and Marescotti 2004 and Belletti and Marescotti 2010; Pacciani and Toccaceli 2010; Toccaceli 2012) promoted by Pacciani in 1996. Pacciani bypassed the scientific debate to enter the experimental and operational dimension oriented towards planning and implementation of the rural district, in its first prototypal form. His approach was independent of the idea of "transposing" the notion of industrial district to agriculture. Rather, he reworked and combined certain elements in the political and cultural debate of the '90s, including the gradual strengthening of European rural development policy and local development support (De Rita and Bonomi 1998) to be pursued through specific instruments of negotiated planning (CNEL 1999; De Meo and Nardone 2002). Pacciani's contribution (Pacciani 2002b) is particularly significant because he suggested a conceptual link between the rural district and LEADER, just as strong as (if not stronger than) the one with Marshallian industrial districts.

For the economists mentioned above, it is self-evident that the overlap of the concepts of local and rural development in the notion of rural district is strictly linked to all the new functions of agriculture (economic, environmental and socio-cultural). Later both Quality Agro-Food Districts (DAQ) and Rural Districts (DIRU) came to be considered a positive response to the rapid development of globalisation (Tinacci Mossello 2002) which is pushing for the relocation of many industrial districts (Corò and Grandinetti 1999). DAs, by contrast, as a "nemesis of history" (especially quality agro-food districts, see De Rosa and De Vincenzo 2002:14), react by putting in place a strategy deeper rooted in the territory, conducted by enhancing typical products and multi-functionality of agriculture.

The political approach in Italy

Pacciani defines the rural district of Maremma "a complete implementation of the Cork Declaration*", introducing it into the agrarian-economic debate as a "development model and method of governance*" in rural areas (Pacciani 2003:48). In this sense, the rural district was also taken into account by the government.

With his formulation, Pacciani offers a response to the policy issues raised by Iacoponi and Cecchi. Iacoponi (1990) believes that the Common Agricultural Policy (CAP) should strengthen existing districts and encourage the growth of those new ones. Cecchi considers that regulatory measures will be perceived differently, as if through "filters", in different places and this implies the existence of a local demand for policies that should find a "strong political mediation, implemented on the basis of territorial articulation" (1992:88).

The issue of how such mediation may legitimately be conducted, although noted, was left open while in practice the rural district as a "method of governance" was already providing some local responses.

In 2001 the economic debate was bypassed by government intervention placing the DAs in the context of legislation regarding modernisation of agriculture[j]. This regarded DAs as new tools for increasing competitiveness of the primary sector according to European guidelines for reviewing the CAP and rural development policy.

Because of its pragmatic aim, the government only partially followed indications arising from the scientific debate, rather considering pilot experiences autonomously occurring in Maremma and Piedmont (Toccaceli 2012).

The government was particularly driven by the progressive intertwining of legislation on industrial districts and negotiated planning, whose applicability was extended to all sectors, including agriculture, thus opening the door to district legislation in that sector (Albisinni 2010, 2011a and Albisinni 2011b). In the legal rationale, in fact, the DAs – and particularly the rural districts - were also taking on a second, but not secondary, meaning as tools for supporting local development of rural areas.

The news was not welcomed unreservedly by a part of the scientific world. Although extensive preparatory work had been carried out, the final text does not appear to have fully taken it into account and Musotti (2001b) observed that this law would scarcely help in identifying and properly "recognising" the district situation.

The controversy highlighted a misconception that was implicit in national law and that was to become increasingly more evident in the regional laws. Economic theory had studied those districts that had formed spontaneously, therefore the law should enable their recognition. Instead, in a 'creationist' interpretation – well evident in several regional laws – the DAs were to become subjects to be set up for implementing policies at local level.

In this blurry legislative context, the Regions have made and are still making very different choices about timing, as well as about the definition and aims of DAs and the rules of implementation to be adopted.

The DAs in National and Regional legislation

Because DAs refer to a composite regulatory framework that was created over a twenty-year period, a "reconstructive reading" (Albisinni 2010) was needed to evaluate the complexity of the phenomenon and the political intentions behind it. Bearing in mind that interest in DAs in Italy ran parallel to the evolution of European rural development policy, the succession of national regulatory interventions might be seen as falling into three periods, corresponding to different political and economic approaches. (See Table 1).

The first period corresponds to the years between the first law on industrial districts in 1991 and the Law of Orientation of 2001. At this stage, the action of the government was inspired by the constant intertwining of the initial scientific debate on industrial districts and the possibility of their extension to agriculture, the progress of the

Table 1 Evolutionary path of the European and Italian regulations referring to the DAs and related definitions

EU political references	Italian National laws		Legal notions	Definitions
C.I LEADER (1988–2006)	1991-2001	First law on industrial districts	**Industrial Districts** Law 317/91 art.36 para.1	*Local territorial areas with high concentration of small enterprises, with particular reference to relationship between presence of enterprises and resident population, as well as to production specialisation of enterprises as a whole.*
		Regulation of negotiated planning	Negotiated planning L. 104/1995	*Regulations agreed between public entities, or with competent public entity, and party or parties, public or private, for implementation of different interventions related to a unique development purpose that requires comprehensive assessment of activities pertaining to it.*
EU Rural development policy			1998 Del. CIPE	*Applicability of negotiated planning to rural, agricultural and fisheries districts*
		Second law on industrial districts	**Local productive systems (LPS)** Law 140/99 art.6 para.8	*Homogeneous productive contexts with high concentration of industrial enterprises and specialisation of business systems*
		The Orientation Law	**Rural districts** Legislative decree 228/2001 art.13 para.1	*Local production systems referred to in Article 36, paragraph 1, of Law of 5 October 1991 n. 317 [see first row above] and subsequent amendments with homogeneous identity from a historical and territorial point of view arising from integration between agricultural activities and other local activities, as well as production of goods or services of particular specificity, consistent with traditions and natural and territorial vocations.*
			Quality agro-food districts Legislative decree 228/2001 art.13 para.2	*Local production systems, including interregional, with significant economic presence and productive interrelationship, and interdependence of farms and agri-food enterprises, and one or more certified or protected products in accordance with applicable Community or national regulations, or from traditional or typical products.*
LEADER AXIS (2007–2013)	2002-2008	The Finance Law 2006	**Productive Districts** Law 266/2005 art.1 para.366	*Free combinations of enterprises articulated territorially and functionally, aiming at enhancing development of areas and sectors, improving effective organisation and production, according to principles of vertical and horizontal subsidiarity, also identifying ways of working with entrepreneurial associations.*
				Freely accessible by industrial companies, services, tourism and agriculture and fisheries.
	2008-2013	Business networks and network contracts	**Business network** L.133/2008 art.6bis	*Free aggregations of single production centres cohesive in unitary development of industrial policies, also in order to improve their presence in international markets*
Extension of LEADER method to ESI FUNDS	2014-2020	**What role for the DAs?**		

negotiated planning experience (Adornato 2005; Pacciani 2003; Albisinni 2011a) and the acknowledgment of some pilot experiments.

The early definition of industrial district (Table 1) well represents the basic components of the scientific notion in the spatial, business and social dimensions. It is limited to the industrial sector and only thanks to the formulation of the local production system notion did it pave the way for the extension of the district into different areas, such as rural, agro-food and fisheries. Similarly, in that period, negotiated planning instruments were being extended to cover agriculture and fisheries, services and tourism. The concept of negotiated planning explains the purpose of the first tools used by the DAs and allows us to appreciate the coherence of the two approaches.

At the same time, with negotiated planning, it became evident that the local approach required the establishment of a new kind of governance and specific tools to bring together different levels of public administration and public with private bodies. These tools must have the sole purpose of identifying specific solutions to sustain strategic projects supporting endogenous development and employment. In the wake of these measures, the Law of Orientation in agriculture introduced two distinct definitions for rural and quality agro-food districts, while maintaining a strong link with the basic concept of Local Production Systems (LPS), defined as *"the homogeneous productive contexts characterised by both a high concentration of industrial enterprises and the specialisation of business systems"*. These LPS *"characterised by homogeneous identity from a historical and territorial point of view, arising from integration between agricultural activities and other local activities, as well as the production of goods or services of particular specificity, consistent with traditions and natural and territorial vocations"* are qualified as Rural Districts (DIRU). The definition of Quality Agro-food Districts (DAQ), on the other hand, is given to those LPS *"even interregional, characterised by significant economic presence and production interrelationship, and by interdependence of farms and agri-food enterprises, and by one or more certified or protected products in compliance with applicable Community or national regulations, or by traditional or typical products"*. Without contesting Musotti (2001b), one may observe that these definitions more or less mirror the main terms of the scientific debate.

In the second period (2002–2008) Italian law continued to mature intentions and purposes. The first decade of the adoption of districts and negotiated programming tools had responded to the objective of increasing employment by supporting endogenous local development processes. In the next period, attention reverted to focussing primarily on support for districtual firms in their processes of innovation, rationalisation of production, supply chain and internationalisation. So in the 2006 finance law the definition of 'district' was revised and reworked in terms of "free association of undertakings*".

This new approach opened the way to a season of measures aimed at expanding the range of administrative, fiscal and credit incentives and benefits in favour of district enterprises, finally assimilating and equalising all the different types (and also industrial and agricultural districts) in this new notion. It is worth underlining that this is the sole, albeit important, new feature. In moving from the former to the latter definition of district, the government changed political approach, placing more emphasis on the role of entrepreneurial and private initiative than on that of the State. Nonetheless, political discussion about districts changed much more than did the legal approach. In fact the legal definition of production district included – albeit unnecessarily - explicit reference to the two principles of subsidiarity, thereby reconfirming the reciprocity as well as the cooperation necessary between private and public initiatives. In other words, the introduction of this new definition was justified more by general political aims than by a new economic policy stance.

Moving forward in the wake of the political approach above, the third period - in 2008 - saw the first regulation of business networks, which were defined as free aggregations of "single production centres" having the network contract as a specific tool. The network is an important element of the wider category of "institution" and is a basic node in the district conceptual framework very significant for the policy approach (Borràs and Tsagdis 2008), from a scientific point of view. It may therefore be considered a more focused tool for strengthening the formation of the business network, spontaneously created by firms, thanks to face-to-face interaction, or informal deals or private contracts. Instead, in the political approach, this intervention is deemed the right way of meeting the need for a simpler and faster tool than the district, recognising the changing requirements of mature industrial districts that have significantly changed their relationship with their territory and the local dimension, because companies have adopted delocalisation strategies to better compete in the global market. Rather, DAs and networks should be considered different but integrable instruments.

Over time the government has reformulated several different notions of district in its successive acts without repealing the preceding ones. The Regions are thus able to choose from these notions or mix them according to their own aims and approaches and nowadays they have become a tangled stratification of multiple definitions.

In fact, each Region has regulated the DAs differently (Toccaceli 2012: 58), often creating a path towards the formula deemed best for its objectives through subsequent interventions designed to respond to multiple pressures - either the changing economic and political scene, or precise requests from the territories concerned to create a specific district or in order to respond swiftly to the flurry of initiatives from the government. For example, following the 2006 Finance Law, ten Regions have enacted laws on DAs, either introducing them from scratch or adapting existing rules to create the conditions for access to new benefits introduced, so as to ensure maximum support to businesses in the Region. Moreover the Regions, far beyond the simple "implementation" of national provisions, have generated a lot of new types of district, e.g. the "supply chain districts" introduced by Lombardy.

Consequently, the overall picture is a heterogeneous image, changing over time, which well reflects the constant and fervent work of the Regions, unceasing and still ongoing, so it is not easy to order such a miscellany.

Therefore, in order to compare regional patterns, it was first of all necessary to identify a truly common concept of DA, highlighting the economic mechanisms actually involved, but taking a different approach than the political and juridical ones.

Method

Reconstructing the conceptual framework from a historical, economic and legal standpoint was essential in order to correctly fix the baseline upon which to found a survey on a national scale.

The scope of the survey was limited to the DAs set up in those Regions with a law in force and desk research focused only on the DAs recognised by them.

The methodology is articulated in two phases. The first phase deals with the desk research into the DAs to define their main common features on a national scale, to show how the basic economic mechanisms of the notion of 'district' were designed by the Regions. The second phase deals with an analysis of the possible integrated role of the DAs in the coming programming period.

The desk research is sub-articled, referring to the different objectives: on one hand, to create an updated map of the recognised DAs, on the other hand, to carry out a critical analysis of the complex of regional laws to clarify the very notions of 'district' used by Regions. The latter objective aims to better highlight how the basic economic mechanisms are shaped by the regions and whether their functioning is effectively ensured. Final considerations on these outcomes guided us to the methodological approach, thanks to which a common notion of DA is proposed.

As information about DAs in Italy is not collected, nor immediately accessible, the data used to build Figure 1 came from the official regional sources of information and from examination of the main bibliography on the subject.

All regional regulation since 1990 was collated and critically examined in order to learn the essential elements for developing a scheme usable for conducting comparative analysis of such complex regulatory frameworks (Toccaceli 2012: Appendix). Following this, we were able to highlight the few key elements of each law: the juridical definition of district elaborated or used (derived from those offered by the national laws) and the typologies covered; the purposes attributed to the law; the requirements for identifying DAs and the procedures for their recognition; the organisational models developed to regulate the relations between the different actors involved in districtual dynamics; how projects are planned and financed; how the interventions are monitored.

Critical reflection has suggested considering the juridical issue of clarifying and reducing to a strict number the possible DA typologies, separately from the economic question.

In order to illustrate the connection between the basic economic mechanisms of the district notion and how the Regions have shaped and implemented this mechanism, the overall conceptual framework has been split into its basic components. These are the spatial and value chain dimension, which together represent the economic engine and network making and governance, which together represent the socio-economic engine. All components, of course, work together producing external economies and business environment and generating growth both for businesses and for the Regions where they are located. Finally, there is the political dimension. In this analysis, any external economies generated are the outcome –and even the aim - of a good regulation, rather than an element directly susceptible of regulation.

The spatial and the value chain dimensions in regional legislation are shaped through the choice of how to identify DAs, which relate to localisation and concentration of activities, as well as to their specialisation and interconnection in an organised value chain.

The manifold nature of institutions involved in the district process, as well their relationships, are again outlined through the governance requirements for identifying DAs, as well as in the rules that prefigure a large number of organisation patterns. It is worth underlining that mostly the local network is regulated, perhaps intending a too 'localistic' conception of the district, as better analysed below.

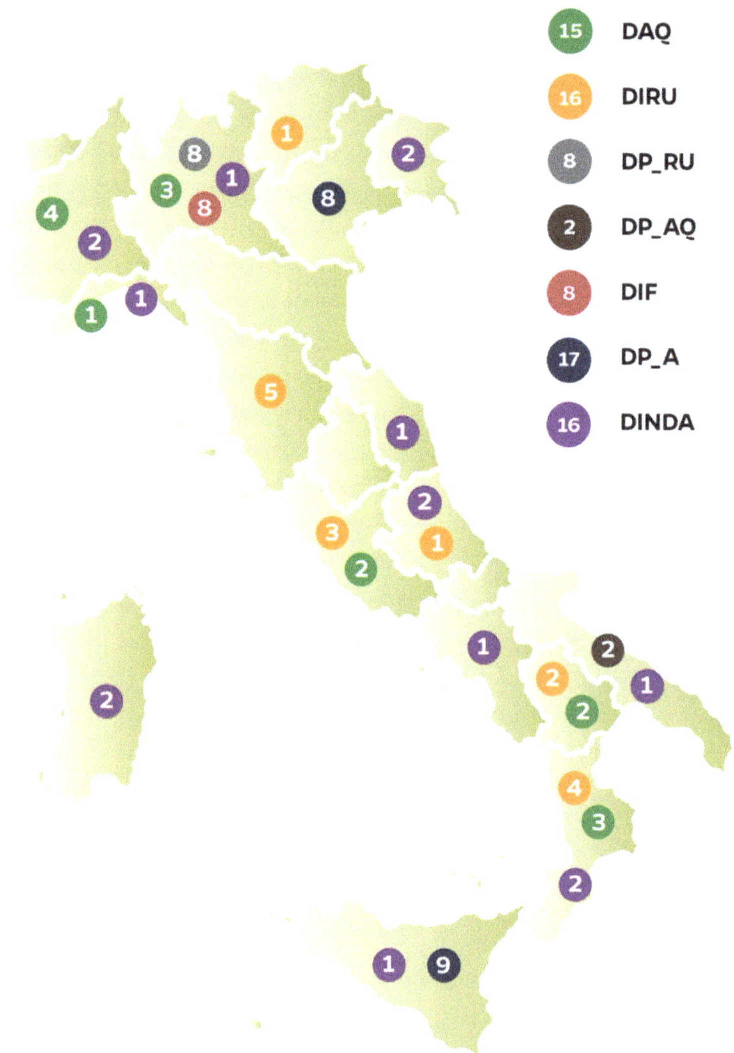

Figure 1 The DAs recognised by the Regions by typology (Source: own elaboration on Regions' Official Journal and official web sites, updated to April 2013).

The political component, albeit often deemed external to the conceptual economic framework, in this kind of design is considered essential, because it is by law that life in the districts is regulated. So this component is considered in the sense of policy to be implemented on a local scale and is expressed through the norms for regulating what kind of DA project may be accepted and how they may be supported by public finance. From this analysis derived our methodological approach and the 'methodological' common definition of DAs, which is the milestone for some general reflections about the possible future use of DAs.

In the second phase, this more general concept of DAs is compared with a critical survey of the structural policies for 2014–2020, having regard to the extension of LEADER methodology to all European and Structural Investment (ESI) Funds[k.], as well as to the increased attention paid by rural development policy to cooperation among farms, including in the form of networks and clusters.

Results and discussion

Findings from desk research into the DAs

- From the map to the rationalisation types of DAs created by the Regions

Under the legislation discussed above, over time sixteen Regions have recognised 82 districts[1.] (Figure 1) in accordance with the restricted number of typologies examined below (Table 2). The recognised DAs are either specialised, hence focused on a specific agricultural or agro-industrial production often with a designation of origin, or non-specialised (Table 3). For reasons given above, there is not a univocal correspondence between typologies of districts and typology of specialisations. For example, some milk and dairy product districts have been recognised as quality agro-food districts, others as districts of the supply chain, others as production or industrial districts. More than half of the DAs (44) specialise in the following sectors: fruit and vegetables (9) plant nurseries (7), animal husbandry (6) wine (5), milk and dairy products (5), cereals, (3) fish (2), agro-energy (2) food biotechnology (2) cork (1) coffee (1) and confectionery (1).

The other non-specialist DAs (38) focus (23) on a *basket of integrated products and services of the territory* (Pecqueur 2001; Pacciani 2003; Belletti and Marescotti 2004) because of the awareness that the district -especially the rural district - determines an overall enhancement of the territory (Iacoponi 2002; Pacciani 2003). Typical products with designation of origin (Arfini et al. 2010) are usually at the core of that basket. This classification does not coincide with that (9) named *multi-product,* which includes those agro-industrial districts less concentrated on integrated action of territorial development, even if focused on several typical products. Among the non-specialised you see also some particular uses of DAs: reinforcing the rural economy of peri-urban areas and getting better governance for managing their relationship with the city (e.g. the rural district of Milan and the other three around the Milan metropolitan area); enhancing the environment and focusing on forest heritage in the Pistoia Mountains in Tuscany. Non-specialised DAs are mostly rural districts, signifying correct use of this tool.

Hitherto, the Regions have recognised a huge number of designations and types of DA for very different economic contexts and rural environments. Unfortunately it is not always understandable from regional laws or DA designations to which kind of juridical notion they refer. So a deeper analysis is needed for correctly re-organising the types of DAs created by the Regions, finding a first criterion to make the multifaceted universe of agricultural districts more homogeneous. The more logical choice seemed to combine the two criteria derived from legal bases: on the one hand, the basic subsequent notions of local productive systems (LPS) and production districts (DP) to which regional laws refer; on the other hand, the basic types of agricultural or industrial district. So all the district types named by the Regions (summarised in Table 4) are reduced to basic typologies on which we can now focus.

Table 2 DA typologies related to the main legal basis (Source: Toccaceli D. 2012:54)

	Local Productive System – (LPS)	Productive districts - (DP)
Rural and Quality Agro-Food districts (DRAQ)	DIRU DAQ	DP_RU DP_AQ DIF
Industrial Districts (DIND)	DINDA	DP_A

Table 3 DAs recognised by the Regions by main production sectors (Source: own elaboration)

		DAQ	DIRU	DP_RU	DP_AQ	DIF	DP_A	DINDA	Total
Non specialized	Multi-product						1	8	9
	Basket products-territory	6	11	4	2				23
	Urban–rural		1	4					5
	Forestry rural		1						1
	Subtotal	6	13	8	2	0	1	8	38
Specialized	Agro-energy		1			1			2
	Food biotechnologies							2	2
	Coffe							1	1
	Cereals	1					1	1	3
	Confectionery						1		1
	Plant Nurseries	2	2			2	1		7
	Fish						2		2
	Milk and milk dairy products	1				1	2	1	5
	Fruit and vegetables	3				1	4	1	9
	Cork							1	1
	Wine	2					2	1	5
	Animal husbandry					3	3		6
	Subtotal	9	3	0	0	8	16	8	44
	Total	15	16	8	2	8	17	16	82

Moreover, it should also be noted how the methodological link between LEADER and rural districts (Belletti and Marescotti 2007) found expression in real life. Because LEADER-eligible areas have the same general characteristic as those where rural districts are recognisable, sometimes some LAGs planned to evolve into rural districts and did so. This happened in two ways, either achieving recognition from their Region (e.g. the LAG-rural district Serre Calabresi or the LAG Golem in Lombardy which led to recognition of DIRU Franciacorta-Sebino-Valtrompia); or, where there were no regional laws, defining

Table 4 Regional criteria for identifying and recognising DAs (source: own elaboration in Toccaceli 2012)

Indicators	**LPS**			**DP**
	DINDA	**DAQ**	**DIRU**	**All types**
Spatial (concentration)	Statistical indicators less restrictive than those for Industrial district		Homogeneous rural identity of the territory	Statistical indicators
	Territorial contiguity	Territorial contiguity	Territorial contiguity	Wider areas also without contiguity
Business (specialization)	Specialised and integrated supply chains	Quality of agro-food production		Economic importance of the sector
		Integrated supply chains	Multi-sector integration	Integrated supply chains
		Territorial integration	Territorial integration	
		Innovation	-	Innovation
	Governance	Governance	Governance	Governance

themselves or their planning target as a "rural district". Examples of this are the LAG "L'altraromagna" which in 2007/13 developed a local action plan for "the rural district of biodiversity", or several LAGs in the Sardinia Region[m]. Another example comes from the Marche Region where the LAG Colli Esini designed the Rural District Colli Esini as an innovative project to promote its territory. Reading their local action planning documents, the idea emerges that the DIRU are deemed able to boost the image and attractiveness of the area and re-create conditions of economic value creation in a more integrated and stronger way. In some other cases District and LAG paths are distinct but synergistically intertwined (e.g. the Quality Agro-food district of Castelli Romani and Monti Prenestini).

- The DAs' economic mechanisms shaped by the Regions

The spatial and business dimensions

Bearing in mind the evolutionary path along which the legal notion of DAs has evolved over the years, modifying its relation to the scientific concept, this analysis shows how the economic function varies for the different typologies.

The Regions have shaped these dimensions using a long list of criteria to identify and legally recognise the DAs (Table 4). We discuss the four essential types of DA identified, in chronological order and distinguishing the spatial indicators of proximity from the business indicators of businesses' specialisation and concentration.

The first definition of industrial district, later discarded, based on the spatial dimension of "local territorial area", hinged on the concept of local labour system based on commuting elaborated by Sforzi (1987). It was identifiable thanks to a set of statistical indicators signifying the concentration of enterprises, entrepreneurs and workers in a very specialised sector which was really important for a certain area.

Held to be too restrictive, after 1999, for the spatial identification of local production systems (LPS), and consequently for agro-industrial districts (DINDA), some discretional and qualitative criteria were added, in order to be able to include many more districts that did not correspond to some of the statistical requirements.

For the rural and agro-quality food districts (DRAQ), the Regions elaborated a rich array of criteria which strictly corresponded to the main points of their legal definition. First, they specified two differentiated requirements: for the rural districts (DIRU), a homogeneous rural identity of the territory as well as integration between agricultural and other local activities; for the quality agro-food districts (DAQ), the presence of quality or typical agro-food products with certified origin as well as interconnection among farms and agro-food enterprises. They also specified two common requirements: the integration of the district's enterprises with the territory, expressed by the production of typical goods, representative of the territory and a well-organised mechanism of governance requested by each Region. Again, there are some links between LEADER and DAs. For example, the Piedmont region has defined its requirements for rural districts as above, but said that they are basically similar to the LEADER programme and cover the same rural zones, so DAs are not recognised in that region to avoid ineffective overlapping.

Like industrial districts, production districts (DP), to be recognised, must demonstrate by a set of statistical indicators their economic relevance and vertical integration of the supply chains. Also the capacity for innovation is required.

Without exception, all the Regions require evidence of governance requirements, that is, a governance process and an effective partnership agreement in place.

Institutions, organisation and governance patterns (socioeconomic dimension)

The Regions have taken different political approaches (not part of this analysis, as mentioned above) in dealing with the wide category of institutions which may be public, semi-public and private actors, also non-local, and their possible networks, which may be formal, semi-formal or informal. The Regions have produced a broad range of possibilities. To cite an example, Tuscany has a network approach, and its regulation is very confident about the local network's effectiveness, so there is little legislation about governance. Conversely, a greater group of Regions that have an implicit top-down approach have much more regulated and formalised partnership arrangements. This group includes those Regions that have shaped DAs as effectively implementing rural development policy, also conferring upon them funding.

Apart from this rough distinction, it is very hard to recognise any more detailed pattern by which to organise the district's bodies. More or less formalised, those bodies are conceived for working on two basic functions: governance at local level, to identify aims and draw up the project; and management, to implement the district project and strategy. Considering these basic functions only, we can distinguish (Table 5) either a *single pattern* for those Regions which envisage only one body (usually named District Committee) to carry out both functions for the district (in some cases the president of the district is the representative); or a *dual pattern,* when two distinct bodies are planned, one for each function. Under the 2006 Finance Law, some Regions have requested a Representative of the district to be identified, charged with interfacing with the public administration at various levels. In the dual pattern, this may be the person charged with management functions, in the single pattern, when it occurs, it may be the district committee or an addition to it. A different pattern is followed by Lombardy whose DAs are composed entirely of enterprises, so the DAs are organised just by a District Company. A legal form to constitute those bodies is established in some Regions with various more specific requirements.

The very articulated, differentiated and complex procedures for promoting, constituting and sometimes formalising the governance process and partnership-forming often require proof of the bottom-up character of this process, even when an implicit desire for control is easily detected. This demonstrates a shared acknowledgement of the importance of trusting in involvement of local and private actors for a successful district.

The policy implementation dimension

Almost paradoxically, at least in relation to the conceptual framework, the Veneto and Sicilia Regions consider the development of the district project one of the most important requirements for recognising a district. For them, the district practically coincides with or is embedded in its project, as confirmed by the unique procedures for financing

Table 5 DA governance patterns: the bodies and their functions - Own elaboration of regional laws in force

Patterns	Regions	Governance functions	Management functions	Representation	Legal form
Dual	**Veneto (DP)**	Every subject involved in the district (informal network)	Pact Representative		Representation mandate
		Regional Board of Districts and meta-districts	—		—
	Sicilia (DP)	—	Pact Representative		Individual form
		Regional Board of Districts	—		-
	Piemonte (DAQ)	District Desk	Company of District		Company form or associative
	Calabria (DRAQ)	District committee	Company of District		Company form
	Lazio (DRAQ)	District Committee	Company of District		Joint-stock company also consortium
	A.P. of Trento (DIRU)	- Valley Community -Committee of participation	Agency	—	Not required
Single	**Friuli VG (DINDA)**	ASDI			Consortium company
	Abruzzo (DAQ)	Company of District			Capital company
	Basilicata (DRAQ)	District Committee		—	Not required
	Liguria (DAQ)	District Committee		District President	Not required
	Puglia (DP)	District Committee		District President	Not required
Company	**Lombardia (DP)**	—	Company of District		Company form
Network	**Toscana (DIRU)**	Coordinator	—	—	Not required

districts established by Veneto, described below. It is just an example, but one that says something about the conception and the concrete expectations about this policy tool.

This brings us into the operational dimension, as the project (also called pact or plan or programme) always has clearly expressed rules, for achieving the real aim of mobilising local resources towards innovative actions along medium-term pathways. In fact "district project" invariably means a group of individual or collective projects, always private enterprises, sometimes public too, together coherent with the main objective of the strategy. Usually the project has to be 'ready to be implemented' so it has to include feasibility studies and, when specific finance is provided, an economic-financial plan. Some regional regulations require projects to be subscribed by all the participants in the district, to assure their commitment to working together, to investing and (where needed) to co-financing as agreed. In those cases the project becomes the object of a binding contract.

Even if the procedures for elaborating the project are different, its elements (Table 6) are always the same: a socio-economic analysis of the districtual area, the chosen objectives and the strategy for achieving them, the partnership composition, a plan of action and the financial plan. Monitoring, always prescribed, is not much stressed. The project is evaluated and, if approved, financed by the Region.

So the project has three main features. It is always *integrated,* either in a vertical way in the case of districts mainly focused on products and supply chain, or in a horizontal way in the case of districts mainly focused on local development. It is *strategic,* so in the specified time of three years (the same in all Regions) it focuses only on the chosen strategic goal. It is *programmatic* in all those Regions which finance, directly or not, the districts' projects. In this case the districts are a reference point for regional programming action. Finally, the project has usually (but not always) *bottom-up* promotion, again indicating that generally the Regions recognise that a good project - and its chance of success – requires good dialogue among the actors. As a matter of fact, a number of Regions support the preparatory actions for communicating the aims of the district and involving most of the actors.

It is observed that the more or less implicit desire for control by the Regions over the district's actions is strictly related to the acknowledgement of a specific line of credit to the DA project. The case of Veneto is emblematic, which provides a specific three-year credit line, so that the DA's actions, realisations and financial reports are always under regional control.

For financing, too, it is very hard to identify any pattern (Table 7), as every Region has followed its own path regarding the procedures, juridical tools and the aspect to be supported.

Regarding financing of projects, there are three specific ways. One is to use negotiated planning tools. In other cases, a few Regions provide a specific line of credit within the regional budget, so that recognising the district, approving the project and granting funding may be contextual. Some other Regions reserve a share of the regional rural development plan for DAs.

Sometimes DA projects are included in regional planning as specific actions to implement one or more policies, in an integrated way and following multiple aims (e.g. rural development and environmental sustainability or enhancing firms' competitiveness and internationalisation and innovation processes).

Table 6 A comparison between DA project elements and CLLD strategy elements – own elaboration

Elements	DAs	CLLD (Reg. UE 1303/2013)
Analysis	Ex-ante analysis of situation in district (with help of SWOT analysis), socio-economic situation and situation of main productive sectors.	(a) definition of area and population covered by strategy; (b) analysis of development needs and potential of area, including an analysis of strengths, weaknesses, opportunities and threats;
Strategy	Definition of strategic objectives and targets. Identification of main lines of action and innovative actions. Indication of expected results.	(c) description of strategy and its objectives, description of integrated and innovative features of strategy and hierarchy of objectives, including measurable targets for outputs or results. In relation to results, targets may be expressed in quantitative or qualitative terms. The strategy shall be consistent with relevant programmes of all ESI Funds concerned;
Partnership	Identification of public and private stakeholders involved in project and indication of mode of their involvement (usually required to form a legal entity)	(d) description of community involvement process in development of strategy;
Planning	Presentation of single initiatives proposed, sometimes including initiatives of public nature, such as infrastructure of service to the district. Indication of feasibility of initiatives and reali sation times	(e) action plan demonstrating how objectives are translated into actions;
Monitoring	Usually an annual report is requested.	(f) description of management and monitoring arrangements of strategy, demonstrating capacity of local action group to implement strategy and description of specific arrangements for evaluation;
Financial plan	Development of financial plan detailing share of contribution required from Region (where it provides direct financing with regional resources) and financial commitments made by each participant and sources of public (EU, national and regional) funding identified.	(g) financial plan for strategy, including planned allocation from each ESI Fund concerned.

Table 7 Regional choices regarding financing DA projects – own elaboration based on regional laws in force

Regional support only to preparatory actions	Negotiated planning tools	Financial line of the regional budget	Reserved share of regional RDP
	Friuli V.G.	Friuli V.G.	
		Veneto	
Lombardia	Lombardia		
	Puglia		Puglia
Sicilia		Sicilia	
	A.P. Trento	P.A. Trento	
Lazio	Lazio	Lazio	
	Abruzzo		Abruzzo
Liguria			
Piemonte			
	Calabria	Calabria	
Basilicata	Basilicata		Basilicata
Toscana	Toscana		

In all events, most Regions consider coordinated use of existing policies to implement the district project as the culmination of the district's path.

The methodological approach to DAs

Having completed the desk research, we can summarise the main findings and attempt to identify some logical consequences.

It may be generally stated that the Regions see DAs as supporting the competitiveness of agricultural enterprises and their innovation processes, strengthening the agricultural and agro-food supply chains and opening them to international markets, developing rural areas, protecting the environment and managing the relationship between city and countryside in peri-urban areas.

More specifically it may be observed that those aims are pursued in a differentiated way, so that there is a range of variation in the understanding of DAs' functions and applications. These range from the totally horizontal, conceived for aims of rural development and environmental sustainability, to the totally vertical focused almost exclusively on the competitiveness of farms and firms in the agricultural and agro-food supply chains and their innovation processes. It is self-evident that this presumed assignation makes more political than scientific sense. This misconception may have thus influenced the correct use of this policy tool.

We have reduced the array of DA designations made by the Regions to four categories derived from the main legal basis, in order to obtain an initial criterion to proceed with our analysis. Although useful, this criterion was not enough to read such a complex phenomenon, e.g. for univocally linking any typology to a function or a way of financing DAs.

Without exception, the Regions have established such a set of criteria as to allow the development of the economic process, selecting districts with requirements of proximity, concentration and specialisation, coherently differentiating them in function of their main orientation towards the horizontal or the vertical dimension.

We note a variety of approaches to network, ranging from the informal involvement of multiple local actors. At the other extreme, the district coincides with the local partnership agreement, despite the fact that the operational solutions adopted for recognising and making operative the network are generally very complex and time-consuming. This variety of approaches was our primary reason for attempting a methodological approach for finding a common basic definition of Italian DAs.

This impression is reinforced by the fact that a prerequisite for the recognition of DAs and the basis of their successful functioning in all regional laws is the implementation of broad governance processes, mostly at local level.

To concretise the strategy, the highest aim of the DAs is developing and implementing a medium-term project that is strategic and integrated. In most cases, it is also programmatic, in the sense that, in several Regions, DA projects become part of the process of regional planning policy, so they are also considered as a source of input information for the purposes of policy planning, management and land use planning.

Finally, we note that all the Regions introduced monitoring of the application of their own laws, although specific indicators of effectiveness are almost never mentioned. As a result no data - no comparable data - are available to enable objective analysis and deduction.

This homogeneous method adopted by the Regions, despite differences and changes of political vision and leadership, was possible because, as has been observed (Toccaceli 2012), the methodological approach defined by the law on industrial districts has remained at the heart of all subsequent national legislation on DAs. Considering the subsequent assimilation of all kinds of district, it has been pointed out that the government has implicitly followed a methodological framework which remained constant and neutral in regard to various economic sectors of application and economic objectives, for which districts were prepared and used over time.

As a consequence, a methodological definition (see Figure 2) is proposed that qualifies DAs as a method of intervention at local level, based on a multi-level governance dynamic (Commission of the European Communities 2001; Committee of the Regions 2009). This comprises the local partnership – established among districtual firms and private and public local stakeholders and operating by a participatory method – which formulates a strategy and develops innovative projects for enhancing the districtual productive system. There is also a district representative, to dialogue with the Public Administration at various levels, even if informally identified among the local actors of the district. Through a specific juridical tool (in the form of a contract, e.g. the district pact) the projects are included in higher-level programming tools, in application of the principles of vertical and horizontal subsidiarity, so they are financed with public and private sources and implemented. This definition focuses on the complex system of relationships linking the different actors at local and also at higher level, where the privileged interlocutor is the Region, through which policies are passing as a result of the national devolution process and application of the principle of horizontal and vertical subsidiarity.

The district thus defined, because of the principles that animate it, is consistent with post-2013 rural development policy, both for its participatory approach and for the

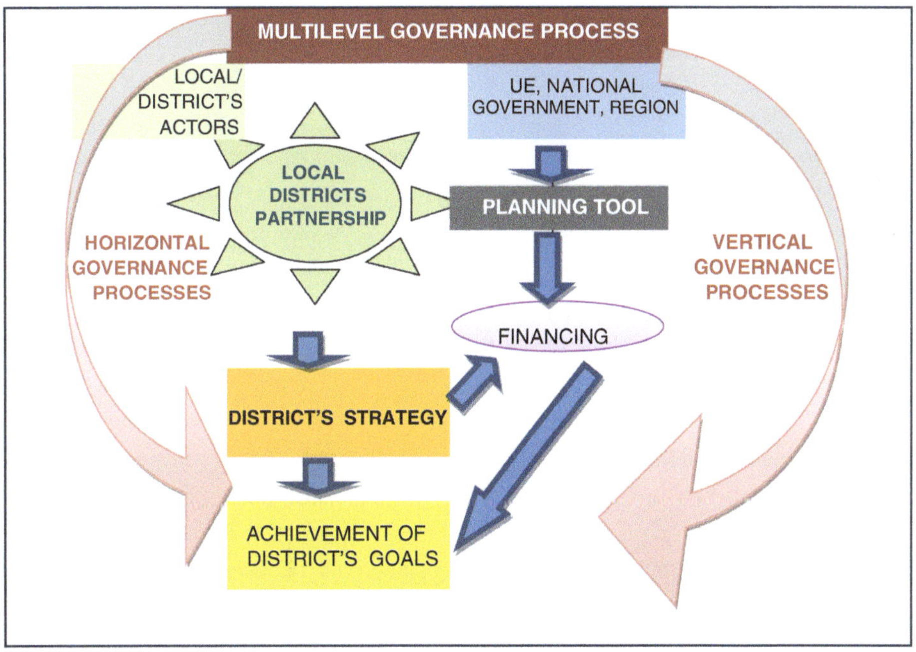

Figure 2 The general methodological scheme of DAs (Source: Toccaceli D. 2012:46).

local scale of intervention which is significant to implement European Structural and Investment Funds, more so than in the past.

The DAs in the perspective of the 2020 European goals for agriculture and rural development

At the time of Agenda 2000, rural and quality agro-food districts were established within the context of a significant set of tools to modernise and strengthen Italian agriculture, in the face of new scenarios of increased exposure of firms to global competition. Two programming cycles later, Italian agriculture and rural territories still need to be strengthened, including through new organisational solutions (Canali 2013:79) to enhance their competitiveness.

More in general, in the period of Europe 2020 strategy, old and new challenges have to be faced by means of the CAP and the rural policy, as well as the other ESI Funds (De Castro 2010; De Castro et al. 2011; De Filippis 2012). These include food security and food quality concerns generating great demands on food systems, environmental concerns linked to the loss of biodiversity, landscape deterioration and the demand for renewable sources of energy.

The EU has reinforced its strategic approach, linking all ESI funds under common provisions and a set of guiding principles, the Common Strategic Framework (CSF), as well as establishing eleven Thematic Objectives on which ESI Funds support should be concentrated. The strategy is based on (among others) the principle of multilevel governance and partnership. So, in utilising ESI Funds, every Member State is requested to prepare a Partnership Agreement reached with the Commission and a broad national partnership.

In the light of this new orientation of European policies and considering the comprehensive methodological definition of DAs, we can attempt to give a detailed answer to our main research question. We asked ourselves whether and how DAs may be still useful in new programming policies, with reference to structural policies, more in particular to rural development policy and with special attention to the relation between DAs and LEADER method and programmes.

More so than in the past, the local level is key in implementing policy measures. Indeed a specific chapter of the CSF Regulation is dedicated to community-led local development[n.] (CLLD) which supports local strategies led by a Local Action Group (LAG) and thanks to multi-fund financing. The device established by the EU is modelled on LEADER methodology so, analysing point by point the main features of CLLD, it is not surprising to find great similarity between CLLD methodology and DA principles. The new possibility of carrying out integrated and multi-sectorial area-based development strategies and of supporting them by multi-fund resources is the most significant similarity and a new opportunity for DAs too (especially for Rural Districts). In fact this new tool becomes more important given the different types of rural territory, especially considering the wide-ranging 'ruralities' from remote to peri-urban rural areas. In the new context, there should not be a dichotomy between urban and rural, rather a more proactive relationship. Starting anew from their own endogenous potential, rural areas are called on to develop "smart specialisation" based both on their territorial capital and local actors' capability to act together to put in place successful strategies.

Focussing on rural development policy, the European agricultural fund for rural development (EAFRD) contributes to achieving specific objectives[o.] (agriculture competitiveness, environment and rural territories development) following detailed priorities[p.] in accordance with the CAP framework. So this Fund has established precise measures which may be considered useful for DAs, or which may be more effectively implemented where there is a "district atmosphere" and thanks to DAs. Within the EAFRD framework, strong attention is paid to the role played by co-operation[q.], so the related measure, reinforced and enlarged in comparison with the past and intended to stimulate innovation, is very complex and multi-purpose. In fact, the concept of innovation is very broad and the possible forms of co-operation are highly articulated with respect to the possible subjects involved, in order to respond to all the strategic objectives of the fund: environmental, competitive and territorial balance (Table 8). For this paper's purpose, we note that co-operation can involve different actors from the food chain, agriculture or forestry sectors, as well as producer groups, cooperatives and interbranch organisations. Co-operation may also be intended to "create clusters or networks[r.] ", even if they have to be "newly formed" and "commencing an activity that is new to them", in order to pursue the aim of innovation.

Support may also be given to the "implementation, in particular by groups of public and private partners of local development strategies"[s.] adding it to the CLLD and limiting it to EAFRD support. So it seems that many types of DA and DA project might be covered by this measure, if Regional planning allows and provides for this.

Furthermore, a broad typology of cost is eligible and "where a business plan or an environmental plan or a forest management plan or equivalent or a development strategy is implemented, Member States may grant the aid either as a global amount covering the costs of co-operation and the costs of the projects implemented or cover only the costs of the co-operation and use funds from other measures or other Union Funds for project implementation"[t.], obviously avoiding any over-compensation. It should be underlined that many different contexts and purposes can be adapted into the bundle of this detailed and flexible measure. Also, DAs might strengthen their partnership or renew their project for innovating, whether they are quality agro-food, or rural, or aiming to manage relationships between rural and urban areas, or focused only on a single product or food supply chain, considering that these tools depend on the good functioning of multi-level governance. Generally speaking, it might be more effective in those cases where governance pathways are already set out.

Finally, the multiple contact points between DAs and LEADER suggest two different points.

On one hand, "district thinking" may suggest some indications about LEADER Programmes newly financed by EAFRD. The European Court of Auditors (2010), among other more specific observations, pointed out that local action groups' (LAGs) actions were not well focused on local strategic objectives, and that the added value due to the LEADER approach was not seen to be achieved. Even if those recommendations have been taken into account by the European Commission and Member States to formulate better rules to ensure those gaps will be filled, in practice, "district thinking" may be useful in better concentrating local strategies, and resources as well, on the most active value chains that may activate interactions among enterprises of all sectors or belonging to a same value chain.

Table 8 The 'bundle' measure of co-operation in EAFRD, Reg. (UE) 1305/2013 Article 35 (2), referred to EAFRD general objectives Article 4 – (source: Toccaceli 2012:101)

Eafrd objectives article 4	Co-operation actions supported By Eafrd Article 35
(a) fostering the competitiveness of agriculture	(a) pilot projects;
	(b) the development of new products, practices, processes and technologies in the agriculture, food and forestry sectors;
	(c) co-operation among small operators in organising joint work processes and sharing facilities and resources and for the development and/or marketing of tourism services relating to rural tourism;
	(d) horizontal and vertical co-operation among supply chain actors for the establishment and the development of short supply chains and local markets;
	(e) promotion activities in a local context relating to the development of short supply chains and local markets;
(b) ensuring the sustainable management of natural resources, and climate action	(f) joint action undertaken with a view to mitigating or adapting to climate change;
	(g) joint approaches to environmental projects and ongoing environmental practices, including efficient water management, the use of renewable energy and the preservation of agricultural landscapes;
	(h) horizontal and vertical co-operation among supply chain actors in the sustainable provision of biomass for use in food and energy production and industrial processes;
	(j) drawing up of forest management plans or equivalent instruments;
(c) achieving a balanced territorial development of rural economies and communities including the creation and maintenance of employment	(i) implementation, in particular by groups of public and private partners other than those defined in point (b) of Article 32(2) of Regulation (EU) No 1303/2013, of local development strategies other than those defined in Article 2(19) of Regulation (EU) 1303/2013 addressing one or more of the Union priorities for rural development;
	(k) diversification of farming activities into activities concerning health care, social integration, community-supported agriculture and education about the environment and food.

On the other hand, "LEADER thinking" may help to point out some strong and weak points in DAs. Our research shows that DAs shaped by Regions are always conceived as competitive development tools to enhance agricultural and agro-food assets and to reinforce the local rural system, Thanks to our new-found methodological approach, we can say that DAs, by design, have proved to be:

- A flexible method. Based on different and changing political inspirations, Regions have adjusted the district tool to fit many different cases, given the high degree of autonomy that now exists following extension of the LEADER method to ESI Funds.
- A homogeneous method. Though each Region was following its own purposes, together they have inspired and contributed to formulating a methodological pattern with common and homogeneous features.
- A stable structure for organising the local economy. The fact that all Regions ask DAs to make updatable three-year projects indicates that DAs are designed as a stable structure designed, in many cases, to become a constant reference point in the programming of regional policies. It is true that the complex mechanisms required to set up DAs may be so tiresome as to envisage the possibility of numerous failures or

dropouts. However, the very amount of time spent in setting up DAs may be considered a strong point if it means practising dialogue among local stakeholders. So DAs can be regarded as a flexible device for policy delivery.

Conversely, the most evident weak point of DAs is that, more than ten years after the Law of Orientation, it is still impossible to know more about the DA phenomenon because there is no constantly updated information about regional choices, regional legal sources or DAs actually operative, nor is there a regional system for monitoring DA performance.

In comparison with the LEADER method one may evaluate DAs' critical points, looking at three "genetic" weaknesses:

- DAs do not form a network among themselves. So, within their own regulation, they have neither territorial cooperation nor a place to exchange their experiences.
- Only some DAs can rely on a certain perspective of financial support, as successful implementation of medium term planning is requested for this.
- They lack a monitoring system by third parties. Every Region has made monitoring compulsory by law. But - especially in cases where the law already provides for a regional non-financial line dedicated to the DA - there are no actual evaluations of effectiveness of the method, nor even homogeneous surveys able to present more generalised elements of assessment.

So, even if still an expanding phenomenon, the DA method lacks an evaluation system, such as that of LEADER, to permit learning from mistakes and the updating itself, rather than fostering ineffective consumerism of policy tools.

Conclusions

This article aimed to create the conditions for carrying out research on national scale into DAs, starting from an unclear political, juridical and economic context. Thanks to the "reconstructive reading" of the legal bases as well as of the conceptual economic framework, and to desk research into DAs shaped by regional laws, we analysed how the basic components of the district conceptual framework were elaborated by the different Regions in such a way as to ensure the conditions for displaying the potential of the 'district process'. Even if we could not identify a specific pattern because of the different political approaches developed over time, a methodological approach was found that allowed us to outline the DAs' common points and their economic mechanisms and to collect them into a general definition.

On this basis, it was possible to develop a general reflection on the possible future use and utility of DAs, integrating them into the coming programming period. At the same time, some of these reflections on DAs may offer suggestions for other Countries where the same regulation is in force. So, the chances for the coming years are linked to different scenarios, whether the main purpose is either reinforcing weak and marginal rural systems or enhancing competitiveness of strategic supply chains. Two ways are possible:

- assimilating DAs and LAGs indicated for developing and implementing multi-fund local strategies (CLLD). This depends on the success of DAs , i.e. whether DAs emerge as an example of stable organisation of the local economy and good governance or not.
- including DAs in the 'bundle' of the co-operation measure : constituting new clusters and networks for the wide range of aims covered by the law; referring to supported implementation by groups of public and private partners of local development strategies.

These new opportunities would allow Regions to overcome some of the weaknesses highlighted and have systems in place for standardised, constant monitoring and evaluating in the medium term. Regions could also make use of the results of evaluation and monitoring, which would be available for the first time; to improve DA methods, just as the LEADER method was improved over time.

In conclusion we remark that this paper refers to the findings of the first survey of DAs, developed on national scale. Although it contributes towards classifying some early points, many questions remain open for further research.

Endnotes

[a]'Region' includes also the autonomous Provinces of Trento and Bolzano, which can legislate just like a Region.

[b]Legislative decree 18 May 2001 n. 228 « Orientation and modernization of the agricultural sector in accordance with Article 7 of Law 57 of March 5, 2001». Published on G.U. n.137 of 15/6/2001 Ord. Suppl. n.149. Art.13 «Rural and quality agro-food districts». It is also named 'Orientation Law' in analogy with the French 'Loi d'orientation agricole'.

[c]In the following "DA" is used for generically indicating each type of district in agriculture.

[d]A not exhaustive list of studies on regional case studies includes: Pacciani (2003), Regione and Agenzia Regionale Servizi Sviluppo Agricolo (1999); Saraceno (1992); Casati et al. (1999); Pacciani (2002b); Pacciani (2003); Belletti and Marescotti (2004); Conti (2005); Aimone (2006); Toccaceli (2006); CNEL (2007); Fazio and Ricciardi (2008); Gulisano et al. (2008); Hoffmann (2008); Marangon (2008a); Marangon 2008b); Baldi and Casati (2009); Belletti and Marescotti (2010); Contò et al. (2010); Petrovich and Manzoni (2010); Caré et al. (2010); Contò and La Sala (2011); Garofoli (2011); Maizza (2010); Montresor et al. (2011); Tarangioli (2008); Tarangioli (2011); Spampinato and Timpanaro (2012).

[e]Just defined in Article 5 of the Treaty on European Union, both were issued since 1992 in a declaration of Edinburgh European Council-

[f]LEADER is the acronym for: Liason Entre Actions de Développement de l'Économie Rurale, that is "Links between actions for the development of the rural economy".

[g]Three EU schemes promote and protect names of quality agricultural products and foodstuffs: Protected Designation of Origin - PDO: covers agricultural products and foodstuffs which are produced, processed and prepared in a given geographical area using recognised know-how. Protected Geographical Indication - PGI: covers agricultural products and

foodstuffs closely linked to the geographical area. At least one of the stages of production, processing or preparation takes place in the area. Traditional Speciality Guaranteed - TSG: highlights traditional character, either in the composition or means of production.

[h]Towards the end of the '80s the contributions (1987 and 1989) where Becattini systematizes the studies conducted on the industrial district of Prato and formulate the paradigm of the industrial district were published. At that time, some studies were carrying out on the contribution of agriculture to the determination of the pre-conditions for the development of an industrial district (Cecchi 1988; Cianferoni 1990; Cecchi et al. 1991).

[i]Here and below the symbol "*" indicates our translation from italian.

[j]See note b.

[k]'ESI (European Structural and Investment) Funds' are the European Regional Development Fund (ERDF), the European Social Fund (ESF), the Cohesion Fund, the European Agricultural Fund for Rural Development (EAFRD) and the European Maritime and Fisheries Fund (EMFF), which operate under a common framework and work together in a coordinated way according to common provision established by Reg. (EU) 1303/2013.

[l]Updated to April 2013.

[m]E.g. The LAG "Rural district Barbagia-Mandrolisai-gennargentu-Supramonte" titled its local development plan "toward the rural district" and the LAG "Shardana Lands" titled "Path toward a sustainable rural district and for quality of life".

[n]Reg. (EU) 1303/2013 Article 2(19).

[o]Reg. (UE) n. 1305/2013 Article 4.

[p]Reg. (UE) n. 1305/2013 Article 5

[q]Reg. (UE) 1305/2013, Article 35.

[r]Reg. (UE) n. 1305/2013, Article 2(1,q) «"cluster" means a grouping of independent undertakings, including start-ups, small, medium and large undertakings as well as advisory bodies and/or research organisations - designed to stimulate economic/ innovative activity by promoting intensive interactions, the sharing of facilities and the exchange of knowledge and expertise, as well as contributing effectively to knowledge transfer, networking and information dissemination among the undertakings in the cluster».

[s]Reg. (UE) n. 1305/2013, Article 35, (2,i).

[t]Reg. (UE) n. 1305/2013, Article 35, (6).

Abbreviations

CAP: Common Agricultural Policy; CLLD: Community-Led Local Development; CSF: Common Strategic Framework; DA: District in Agriculture; DAQ: Quality Agro-Food District; DIF: Supply Chain District; DIND: Industrial District; DINDA: Agro-Industrial District; DIRU: Rural District; DP: Production District; DP_AQ: Quality Agro-Food Production District; DP_RU: Rural Production District; DRAQ: Rural and Quality Agro-Food Districts; EAFRD: European Agricultural Fund for Rural Development; ESI Funds: European Structural and Investment Funds; LAG: Local Action Group; LEADER: Links between Actions for the Development of the Rural Economy; MID: MARSHALLIAN INDUSTRIAL DISTRICT.

Competing interests

The author declares that she has no competing interests.

Author information

Dr. Daniela Toccaceli Ph.D. in Economics and territory, Post-Doc Fellow Department of Science for Economics and Business, University of Florence Via delle Pandette, 9 I-50127 Florence daniela.toccaceli@unifi.it.

Acknowledgments

This paper is possible thanks to research promoted by the National Rural Network, coordinated by Mrs. Graziella Romito, within the Task Force "Integrated Planning" 2007-2013. Special thanks are due to Professor Alessandro Pacciani and Professor Ferdinando Albisinni for their precious observations during the research period which has resulted in this paper. I am very grateful to the two anonymous referees for their stimulating considerations. My thanks, too, to Julia Campbell Hamilton for proofreading the paper. Any errors remain, of course, my responsibility alone.

References

Adornato F (Ed) (2005) *Sviluppo locale e contrattazione programmata* Franco Angeli Milano
Aimone S., (Ed) (2006) Programmazione integrata e sviluppo rurale del Piemonte in: *Quaderni di ricerca* Istituto di Ricerche Economico e Sociali del Piemonte (IRES),Torino
Albisinni F (2010) Distretti e sviluppo rurale: elementi per una lettura delle regole di diritto, in: Agriregionieuropa, Anno 6n.20, pp.21-25 ISSN: 1828–5880
Albisinni F (2011a) Distretti e contratti di programma in agricoltura in Costato L. Germanò A. Rook Basile E. (Eds) Trattato di diritto agrario, Volume Primo:409–461 UTET, Torino
Albisinni F (2011b) Prodotti agro-alimentari e distretti in agricoltura In: Goldoni M, Sirsi E. (eds) Il ruolo del diritto nella valorizzazione e nella promozione dei prodotti agro-alimentari: Giuffrè Milano pp103–148 ISBN: 88-14-17236-6
Angeli L (2000) L'atmosfera di Artimino e noi economisti agrari. La Questione Agraria n.3:111–121 ISSN: 1593-8441
Arfini F, Belletti G, Marescotti A (2010) Prodotti Tipici e Denominazioni Geografiche: Strumenti di Tutela e di Valorizzazione. Quaderni, Gruppo 2013, Roma
Baldi L., Casati D (2009) Un distretto della quarta gamma? Il comparto che vende tempo libero, in: Agriregionieuropa, Anno5, n. 16, pp. 50-52, ISSN: 1828-5880
Becattini G (1987) Mercato e Forze Locali: il Distretto Industriale. Il Mulino, Bologna, ISBN: 88-15-01420-9
Becattini G (1989) Riflessioni sul distretto industriale marshalliano come concetto socio-economico. Stato e mercato 25:111–128
Becattini G (2000) "Distrettualità" fra industria e agricoltura in La Questione Agraria n.2, pp: 11–24 ISSN: 1593-8441
Becattini G (2001) Distretti e ruralità: sfide al riduzionismo economico. Una replica in: La Questione Agraria n.1pp. 119-127. Franco Angeli, Milano, ISSN: 1593-8441
Becattini G, Bellandi M, Lisa DP (eds) (2009) A Handbook of Industrial Districts. Edward Elgar, Cheltenham, UK – Northampton, MA, USA, ISBN: 978-1-84720-267-3
Bellandi M (1987) La formulazione originaria in Becattini G., Mercato e forze locali: il distretto industriale: Il Mulino, Bologna
Belletti G, Marescotti A (2004) Il Distretto Rurale: Metafora Organizzativa e Strumento di Governance. In: Hoffmann A (ed) Esperienze di Programmazione Dello Sviluppo Locale: il Caso del Parco dei Nebrodi. Franco Angeli, Milano
Belletti G, Marescotti A (2007) Il Distretto Come Strumento di Governance per lo Sviluppo Rurale. In: Rete Leader, Leader e distretti rurali: sinergie e complementarietà. Biemmegraf, Macerata
Belletti G., Marescotti A (2010), Il distretto rurale, In Pacciani A., Toccaceli D. (a cura di) Le frontiere dello sviluppo rurale. L'economia grossetana tra filiere e territorio, Franco Angeli, Milano, ISBN: 978-88-568-3268-6
Borràs S, Tsagdis D (2008) Cluster Policies in Europe. Edward Elgar, Cheltenam, Uk – Northampton, MA, USA, ISBN: 978-1-84542-758-0
Canali G (2013) L'OCM Unica e le Misure di Mercato. In: De Filippis F (ed) La Nuova PAC 2014–2020. Un'analisi Delle Proposte Della Commissione. Quaderno, Gruppo 2013, Roma
Caré V, Reda E, D'Angelillo E (2010) L'esperienza dei distretti rurali ed agroalimentari in Calabria. Evoluzione normativa ed iniziative proposte per un nuovo modello di governance e di sviluppo delle aree rurali. Rete Rurale Nazionale, Roma
Casati D (1988) Come cambia l'agricoltura nel sistema agro-industriale in Strategie e adattamento nel sistema agro-industriale INEA, Roma
Casati D, Banterle A, Baldi L (1999) Il distretto agro-industriale del riso. Franco Angeli, Milano
Cecchi C (1988) distretto industriale: l'agricoltura dalla complementarità alla dissociazione. La questione agraria n.32, pp 91-123, Franco Angeli, Milano, ISSN: 1593-8441
Cecchi C. (1992) Per una definizione di distretto agricolo e distretto agroindustriale La Questione Agraria, n. 46, pp 81–107, Franco Angeli, Milano, ISSN: 1593-8441
Cecchi C (2000) "E se facessimo tanti bei campi da golf?": ovvero Becattini e la campagna La Questione Agraria, n.3, pp 123–132, Franco Angeli, Milano, ISSN: 1593-8441
Cecchi C, Cianferoni R, Pacciani A (1991) Economia e politica dell'agricoltura e dell'ambiente. CEDAM, Padova. ISBN 9788813173692
Cianferoni R (1990) L'agricoltura e l'ambiente nel distretto industriale di Prato. Accademia dei Georgofili, Firenze
Cnel (1999) I protagonisti e le esperienze locali di programmazione negoziata. CNEL, Roma
CNEL (2007) Distretti rurali ed aroalimentari di di qualità: il manifesto di Matera alla luce delle nuove politiche di sviluppo rurale. Osservazioni e proposte Assemblea CNEL del 20 Marzo, CNEL, Roma
Collignon P (2014) Le Syndicate intercommunal pour la promotion du canton de Clerveux "SYCLER" au service du développementglobal d'un territoire rural In: Pacciani A., Toccaceli D. (Eds) Percorsi di governance per la valorizzazione delle aree rurali nella prospettiva di riforma delle politiche europee, pp61-65 I Georgofili Quaderni 2013-II Società Editrice Fiornetina, Firenze ISBN: 978-88-6032-294-4
Commission of the European Communities (1988) The future of rural society, COM(88/501) Bruxelles: Bullettin of the European communities, Supplement 4/88
Commission of the European Communities (2001), European governance – A white paper COM(2001) 428 def., Bruxelles: Official Journal of the European Communities, C287/1, October, 10
Committee of the Regions (2009) The Committee of the Regions'White Paper on Multilevel Governance Own-Initiative Opinion, Bruxelles: 80th plenary session, 17 and 18 June
Conti L. (2005) Le politiche distrettuali della Regione Piemonte in: Agriregionieuropa Anno1, n. 2, pp. 39-41, ISSN: 1828 - 5880
Contò F, La Sala P (2011) Approccio Territoriale e Sviluppo Locale. Il Programma di Sviluppo del Distretto Agroalimentare di Qualità del Metaponto. Franco Angeli, Milano

Contò F, La Sala P, Papapietro P (2010) Identità, Qualità e Competitività dell'area del Pollino: un Caso di Filiera "Atipica" Agro-Territoriale. Atti della XXXI Conferenza scientifica annuale AISRE, Aosta

Corò G, Grandinetti R (1999) Strategie di delocalizzazione e processi evolutivi nei distretti industriali italiani. L'industria AnnoXX 4:897–924

Courlet C., Pecqueur B (1992) Les systèmes industriels localisés en France: un nouveau modèle de développement In: Benko G., Liepitz A. (éds), Les régions qui gagnent. Districts et résaux: les nouveaux paradigmes dela géographie économique, PUF, Paris, pp.81–102

De Benedictis M (2000) Economia agraria e distrettualità in La Questione Agraria n.2, Franco Angeli, Milano, ISSN: 1593-8441 pp. 25–34

De Castro P (2010) L'agricoltura Europea e le Nuove Sfide Globali. Donzelli Editore, Roma

De Castro P., La Via G., Adinolfi F., Marangon F (2011) Il processo di riforma della Politica Agricola Comunitaria Stato dell'arte e prospettive. In: International Agricultural Policy, n. 1, Edizioni L'informatore Agrario. Verona, ISSN: 1722-4365.

De Filippis F (2000) Distrettualità e politica agraria in La Questione Agraria n.2, Franco Angeli, Milano, ISSN: 1593-8441

De Filippis F (Ed) (2012) La nuova PAC 2014–2020. Un'analisi delle proposte della Commissione Quaderni, Gruppo 2013, Roma

De Meo G., Nardone G (2002) Programmazione negoziata e sviluppo rurale in Basile E., Romano D., Sviluppo rurale: società, territorio, impresa, Franco Angeli, Milano

De Rita G., Bonomi A., (Eds)(1998) Manifesto per lo sviluppo locale Bollati Boringhieri, Torini

De Rosa M, de Vincenzo D (2002) Stato e Dinamica dei Sistemi Produttivi Territoriali nel Rapporto Locale/Globale. In: De Rosa M, de Vincenzo D (eds) Tra globalizzazione e localismo. Liguori, Napoli

European Union (2013) *Official Journal of the European Union, L 347*, Volume 56, 20 december 2013. ISSN: 1977–0677

European Commission's Directorates General AGRI, EMPL, MARE and REGIO (2013) Common Guidance on Community-led Local Development in European Structural and Investment Funds., http://ec.europa.eu/regional_policy/informing/dialog/pdf/clld_guidance_2013_04_29.pdf (last access 15/04/2014)

European Court of Auditors (2010) Implementation of the Leader approach for Rural Development Luxembourg: Publication Office ISBN: 978-92-9207-883-6

Fabiani G (2000) Distretti o Sistemi Agricoli Locali? In: La Questione Agraria, vol 2. Franco Angeli, Milano, pp 33–41, ISSN: 1593-8441

Fanfani R, Montresor E (1998) Istituzioni e imprese nel percorso di sviluppo dei sistemi locali di produzione agroalimentare. La questione agraria n.69 pp.87-108 Franco Angeli, Milano ISSN: 1593-8441

Favia F (1992) L'Agricoltura nei sistemi produttivi territoriali. La questione agraria n.46 pp.109-136 Franco Angeli, Milano ISSN: 1593-8441

Favia F. (2000) Economia agraria e distrettualità: variazioni sul tema *La Questione Agraria*, n.3, pp.133-145 Franco Angeli, Milano, ISSN: 1593-8441

Fazio V, Ricciardi A (eds) (2008) Il distretto della pesca di Mazara del Vallo. Una buona pratica di cooperazione tra aziende internazionali. Franco Angeli, Milano

Feser E., Isserman A (2009) The Rural Role in National Value Chains In: *Regional Studies*, 2009, vol. 43, issue 1, pp. 89-109, doi:10.1080/00343400801968429

Forleo M (2000) I Sistemi Agricoli Territoriali. In: CNEL, Secondo rapporto sull'agricoltura. L'agricoltura tra Locale e Globale. Distretti e Filiere vol N.23.Collana Documenti. CNEL, Roma, pp 295–324

Gallego Moreno F.J. (2014) La estructuraciòn y dinamizaciòn social para una mejor goberannza de las comunidades rurales: Caso de studio en Cuenca, Espana In: Pacciani A., Toccaceli D. (Eds) Percorsi di governance per la valorizzazione delle aree rurali nella prospettiva di riforma delle politiche europee,pp29-49, I Georgofili Quaderni 2013-II Società Editrice Fiornetina, Firenze ISBN: 978-88-6032-294-4

Garofoli G (2011) La "Governance" Dello Sviluppo Locale Nelle Langhe, in: Seminario Nazionale su Politiche, Istituzioni e Sviluppo Rurale:Come Migliorare i Processi di Governance. Rete Rurale Nazionale, Roma

Gulisano G, Marcianò C, a cura di (2008) I distretti rurali in Calabria Università degli Studi Mediterranea di Regio. Dipartimento di Scienze e tecnologie Agro-forestali e Ambientali, Calabria

Hoffmann A (2008) I distretti rurali in Sicilia Università Di Palermo, Dipartimento di Economia dei Sistemi Agro-forestali. Fotograf, Palermo

Iacoponi L (1990) Distretto industriale marshalliano e forme di organizzazione delle imprese in agricoltura in *Rivista di Economia Agraria* Anno XLV, n.4 pp 712-734 n. 4 , Franco Angeli, Milano. ISSN 0035-6190

Iacoponi L (2001) Distrettualità agricola: una difficile (e breve) navigazione tra opposti paradigmi economici in *La Questione Agraria* n. 4:111–118, Franco Angeli, Milano, ISSN: 1593-8441

Iacoponi L (2002) Dal distretto agricolo al distretto rurale in Valorosi F. (a cura di), Lo sviluppo del sistema agricolo nell'economia post-industriale: Franco Angeli, Milano

Lee J., Arnason A., Nightingale A., Shucksmith M (2005) Networking: social capital and identities in European rural development. *Sociologia ruralis*, 45 (4):269-283 ISSN: 1467-9523

Maizza A (2010) I Distretti Agroalimentari nel Contesto Globale. Un'analisi sul Territorio Pugliese. Franco Angeli, Milano, ISBN: 9788856836462

Marangon F (2008a) Aspetti economici per l'individuazione di un Distretto Rurale nel territorio della Pianura Friulana, CCIAA di Udine, Convegno Ipotesi di un distretto per lo sviluppo rurale della Bassa Pianura friulana, Agriest Udine, 28 gennaio

Marangon F (2008b) Lo sviluppo rurale sostenibile in Friuli Venezia Giulia: tra distretti rurali e distretti di economia solidale, Seminario L'economia locale per lo sviluppo del territorio. Università degli Studi di Scienze Gastronomiche, Pollenzo (Cn)

Mollard A (2001) Qualité et développement territorial: une grille d'analyse théorique à partir de la rente. In: *Économie rurale*. N°263, 2001. pp. 16–34. DOI: 10.3406/ecoru.2001.5240, http://www.persee.fr/web/revues/home/prescript/issue/ecoru_0013-0559_2001_num_263_1

Montresor E (2000) I sistemi locali di produzione agroalimentare CNEL, *Secondo rapporto sull'agricoltura. L'agricoltura tra locale e globale. Distretti e filiere.* Collana Documenti CNEL, II serie. N.23,Roma

Montresor E., Pecci F., Pontarollo N. (2011) *Distretti agroalimentari di qualità e processi di governance locale: due regioni a confronto* in Agriregionieuropa, Anno 7, n. 24 pp 57-61. ISSN: 1828 - 5880

Murdoch J (2000) Networks – a new Paradigm for Rural Development? J Rural Studies 16(4):407–419, http://dx.doi.org/10.1016/S0743-0167(00)00022-X

Musotti F (2001a) Il territorio: da sempre nell'analisi economico-agraria italiana *La Questione Agraria,* n. 4:119–131, Franco Angeli, Milano, ISSN: 1593-8441

Musotti F (2001b) Il "nuovo" orientamento dell'agricoltura italiana: riscoperta del territorio, ruralità, distrettualizzazioni *Sviluppo locale,* VII, 17:79–91 , Rosenberg & Sellier, Torino. ISSN:1974-2193

Oxford Research AS (2008) *Cluster policy in Europe* Oxford research AS, January http://www.clusterobservatory.eu/system/modules/com.gridnine.opencms.modules.eco/providers/getpdf.jsp?uid=100146

Pacciani A (1997) Il progetto "Maremma Distretto Rurale d'Europa" Atti dell'Accademia dei Georgofili, Anno 1997, vol XLIV. Industrie Grafiche Fiorentine, Firenze

Pacciani A (2002a) Il Distretto rurale della Maremma: dalla proposta alla realizzazione Atti dell'Accademia dei Georgofili, Anno 2002, settima serie, vol XLIX. Industrie Grafiche Fiorentine, Firenze

Pacciani A (2002b) Funzioni sociali dell'agricoltura e nuovi strumenti di intervento pubblico XXXIX Convegno SIDEA. "Nuove tipologie dell'impresa nell'agricoltura italiana", 12–13 settembre, Firenze

Pacciani A (2003) *La Maremma Distretto Rurale: un nuovo modello di sviluppo nella consapevolezza della propria identità.* Editrice Il mio Amico. Grosseto

Pacciani A Toccaceli D (2010) Territorio, imprese e istituzioni nella PAC oltre il 2013: l'agricoltura grossetana di fronte a nuovi scenari, In:Pacciani A. Toccaceli D. (Eds), *Le frontiere dello sviluppo rurale. L'economia grossetana tra filiere e territorio* Milano: Franco Angeli ISBN: 978-88-568-3268-6

Pacciani A., Toccaceli D (Eds) (2014) *Percorsi di governance per la valorizzazione delle aree rurali nella prospettiva di riforma delle politiche europee* I Georgofili Quaderni 2013-II Società Editrice Fiornetina, Firenze ISBN: 978-88-6032-294-4

Pecqueur B (2001) Qualité et développement territorial: l'hypothèse du panier de biens et de services territorialisès. *Économie Rurale.* 261: 37-49. doi: 10.346/ecoru.2001.5217, http://www.persee.fr/web/revues/home/prescript/article/ecoru_0013-0559_2001_num_261_1_5217

Perrier-cornet P., Sylvander B (2000) Firmes, coordinations et territorialité. Une lecture économique de la diversité des filières d'appelation d'origine In: Economie Rurale 258:78-89, pp78-89. DOI: 10.346/ecoru.2000.5195. http://www.persee.fr/web/revues/home/prescript/article/ecoru_0013-0559_2000_num_258_1_5195

Petrovich B, Manzoni P (2010) I distretti nella regione Lombardia. Rete Rurale Nazionale, Roma

Porter M. E. (1998) Clusters and the new economics of competition In: *Harvard business review* 76(6):77–90. ISSN 0017–8012

Porter E (2003) The economic performance of Regions, In: *Regional Studies,* Vol. 37.6&7, pp549-578, August-October DOI: 10.108/0034340032000108688

Porter M., Ketels C (2009) *Clusters and industrial districts: common roots, different perspectives* in Becattini G., Bellandi M., De Propris L. (Eds) *A handbook of industrial districts,* Edward Elgar, Cheltenam UK, Northampton, MA, USA:172–183 ISBN : 978-1-84720-267-3

Porter M.E., Ketels C., Miller K., Bryden R (2004) *Competitivness in rural U.S. Regions: Learnings and research agenda,* U.S. Economic Development Administration, February 25, 2004 http://www.hbs.edu/faculty/Pages/item.aspx?num=46871

Pyke F., Becattini G., Sengenberger W. (eds) (1990) *Industrial districts and inter-firm Co-operation in Italy,* IILS, ILO, Geneva

Ray C (1998) Culture, Intellectual Property and Territorial Rural Development. In: Sociologia Ruralis, vol 38, issue 1. doi: 10.1111/1467-9523.00060

Regione A bruzzo Agenzia Regionale Servizi Sviluppo Agricolo (1999) Il distretto agroindustriale della Marsica. Grafiche Di Censo, Avezzano

Romano D (2000) I sistemi locali di sviluppo rurale CNEL, *Secondo rapporto sull'agricoltura. L'agricoltura tra locale e globale. Distretti e filiere.* Collana Documenti CNEL, II serie. N.23, Roma

Rouger M.P. (2014) Pays de fougeres ECOBATYS: De l'observatoire à la demarche de prospective In: Pacciani A., Toccaceli D (Eds) *Percorsi di governance per la valorizzazione delle aree rurali nella prospettiva di riforma delle politiche europee,* pp51-60, I Georgofili Quaderni 2013-II Società Editrice Fiornetina, Firenze ISBN: 978-88-6032-294-4

Saraceno E. (1992) Il distretto delle barbatelle. *La Questione Agraria,* n. 46, pp.209-231 La questione agraria 46:209-231. Franco Angeli, Milano, ISSN: 1593-8441

Sforzi F (1987) L'identificazione spaziale In: Becattini G. (ed.) *Mercato e forze locali: il distretto industgriale,* Il Mulino, Bologna.

Sforzi F., Mancini M. C. (2012) A reinterpretation of the agri-food system and its spatial dynamics through the industrial district In: Arfini F., Mancini M.C., Donati M., *Local agri-food systems in a global world: market, social and environmental challenges,* pp 9–27, Cambridge Scholars Publishing, Newcastle upon Tyne, UK ISBN (10): 1-4438-3664-8, ISBN (13): 978-1-4438-3664-7

Spampinato D., Timpanaro G (2012) *La competitività dei sistemi produttivi territoriali in Sicilia: alcuni casi studio per un'analisi dei distretti agroalimentari* in *Economia e Diritto Agroalimentare* n.1 Economia & diritto agroalimentare XVII: 55-82. University Press, Firenze

Tarangioli S (2008), *Rural and quality food districts as governance instruments: the Italian experience.* WP presentato alla Conferenza Europea *"Europe's rural areas in action. Facing the challanges of tomorrow.* Cipro, 16–17 ottobre

Tarangioli S (2011) La progettazione integrata nelle colline del Medio Friuli, in: Seminario Nazionale su Politiche, istituzioni e sviluppo rurale:come migliorare i processi di governance. Rete Rurale Nazionale, Roma

Tinacci Mossello M (2002) Rischi di Crisi e Nuove Opportunità di Sviluppo per i Mercati Locali nel Quadro Della Globalizzazione. In: De Rosa M, de Vincenzo D (eds) Tra globalizzazione e localismo. Liguori Editore, Napoli, pp 33–50

Toccaceli D (2006) *Il Distretto rurale della Maremma: 1996–2006. Come si forma un Distretto rurale* in *Agriregionieuropa,* Anno 2, n. 6 pp. 54-56, ISSN: 1828 - 5880

Toccaceli D (2012) Dai distretti alle reti? I distretti in agricoltura nell'interpretazione delle Regioni e le prospettive verso il 2020. Rete Rurale Nazionale, Roma, ISBN: 978-88-8145-235-4

Toccaceli D (2014) Sustainable Development in Rural Areas: A Comparison on Governance Among Some European Experiences. In: De Gennaro BC, Nardone G (eds) Sustainability of the Agri-Food System: Strategies and Performances, Proceedings of the 50th SIDEA Conference, Lecce, Chiostro dei Domenicani. Universitas Studiorum, Mantova, pp 307–321, ISBN 978-88-97683-60-5

Tregear A., Arfini F., Belletti G., Marescotti A (2007) Regional foods and rural development: The role of product qualification In: *Journal of Rural Studies*, Vol. 23, Issue 1, Pages 12-22, Elsevier Ltd, doi:10.1016/j.jrurstud.2006.09.10

14

Understanding linkages between common agricultural policy and High Nature Value (HNV) farmland provision: an empirical analysis in Tuscany Region

Fabio Bartolini*† and Gianluca Brunori†

* Correspondence:
fabio.bartolini@unipi.it
†Equal contributors
Department of Agriculture, Food and Environment, University of Pisa, Via del Borghetto 80, 56124 Pisa, Italy

Abstract

The connection between agriculture and environment is a central topic to the agricultural policy agenda and it has been the core of policy reforms. These linkages have been strengthened by the 2003 Common Agricultural Policy (CAP) Reform that has introduced a set of environmental rules as a condition to receive Single Farm Payments (SFP). Consequently, the provision of environmental goods is addressed by both CAP pillars I and II, respectively, with cross-compliance belonging to pillar I while the second axis Rural Development Program (RDP) measures belong to the pillar II. The aim of this paper is to analyse the provision of environmental goods provided by agricultural activities and the impact the Common Agricultural Policy has on such provision. This aim is pursued firstly by quantifying an environmental quality indicator based on the concept of High Nature Value (HNV) farmland and, secondly by testing Exploratory Spatial Data Analysis (ESDA) techniques and spatial econometric models as means to identify provision determinants that improve causality in spatially non-stationary data. The analysis highlights the relevance of spatial models in improving the quality of the estimation, while model results show that rural development program measures and income support payments have opposite effects. Results confirm the relevance of productive factor use and entrepreneurship in explaining differences on environmental quality provision.

Keywords: Common agricultural policy; High nature value; Policy impacts; Spatial econometrics; Exploratory Spatial Data Analysis (ESDA); Agri-environmental schemes (AES)

Introduction

The linkage between agriculture and the environment is central to the agricultural policy agenda (Bureau et al. 2012). Since the introduction of voluntary measures in Regulations 2078/92 and 2080/92, the protection of natural resources and the quality of the environment have been addressed in all agricultural policy reforms and now form the core of the Rural Development Program (RDP) of the European Union. These linkages were strengthened by the 2003 Common Agricultural Policy (CAP) Reform. The provision of environmental goods is addressed by both CAP pillars, cross-compliance belonging to the first pillar and the second axis of RDP measures belonging to the pillar II. Cross-compliance is based on a set of environmental rules as a condition for receiving Single

Farm Payments (SFPs). Since environmental concerns represent the main justification to maintaining the payments (Hodge 2012), increasing focus has been given to the effectiveness of the environmental measures of the CAP.

One of the main issues is a joint design of agri-environmental schemes (AES) in order to improve the provision of environmental goods (Bartolini et al. 2012). In fact, cross-compliance rules have been set as the baseline for the implementation and quantification of agri-environmental payments, which in turn require additional environmental recommendations (Scheele 2008). As a result, cross-compliance provides an overall minimum environmental quality standard and prevents the fragmentary and uneven provision of environmental goods by AES (Juntti 2012).

Since Agenda 2000, a greater amount of the RDP budget has been allocated to AES in order to encourage farmers to adopt environmentally friendly practices. However, in spite of a relatively high AES budget allocation (about 2.5 billion Euros of EU funds per year), both the scientific literature and policy evaluation reports have reported weaknesses in the design, implementation and evaluation phases of AES (Oréade-Brèche 2005; Purvis et al. 2009). Recently the European Court of Auditors (2011) criticised 1) the low level of targeting of AES, 2) the low linkage between incentive systems and the environmental benefits provided, and 3) the weaknesses of the current evaluation and monitoring systems in measuring environmental benefits and in the additionality measuring.

The objective of this paper is to analyse the provision of environmental goods through agricultural activities and investigate the determinants of this provision, highlighting the relevance of the Common Agricultural Policy. This aim is pursued firstly by quantifying an environmental quality indicator based on the concept of High Nature Value (HNV) farmland and, secondly by testing Exploratory Spatial Data Analysis (ESDA) techniques and spatial econometric models.

Where an HNV farmland indicator is used as a proxy of changes in the maintenance of natural and semi-natural systems, spatial econometric models lead to a better formulation of causality when distribution is affected by spatial heterogeneities and associations (Brady and Irwin 2011). We made an empirical analysis of Tuscany region. The Region has a heterogeneous distribution of habitats (Mücher et al., 2009) and agricultural systems, due to the varying orography, climatic and market conditions. The main novelties of this paper are represented by the HNV index developed for the ex-post analysis of Tuscany and the correction of the results by adding a spatial pattern.

The remaining paper is structured as follows. In the next section the concept of HNV and the indicators applied are reviewed, then the methodology applied is presented and the results are discussed. Finally, in the last section the conclusions of the research are presented.

Background

Assessing policy impact on the environment

The impacts of agricultural activities and land-use changes on the environment have been widely studied in the scientific literature. Several authors have documented the agricultural pressure on the loss of biodiversity (Mouysset et al., 2012), nutrient run-off and water quality (Petit and Elbersen, 2006; Leip et al., 2011), soil conservation (Auerswald et al., 2003), and on the landscape (Lefebvre et al., 2014).

Generally, many studies have shown that low-intensive agriculture is the main source of environmental quality (Foley et al. 2005; Caballero and Gil, 2009) and at the same time is the most vulnerable farming system (Caballero et al., 2008). Thus a set of agricultural policy instruments has been implemented to maintain low-intensive farming and to encourage the shift towards less impacting systems. Policy instruments belonging to both CAP pillars have thus been implemented, consisting mainly of cross-compliance measures and agri-environmental schemes.

A large body of literature deals with the agricultural policy impacts on the environment focusing on both ex-ante and ex-post policy analysis. While the former mainly apply normative or positive mathematical programming models (Louhichi et al., 2010), in the latter environmental impacts are estimated using econometric or statistical models (see Buysse et al., 2007 for a review). These methods quantify the policy changes on the environment using environmental indicators that represent a proxy of impact on the environment (Primdahl et al., 2010; Chiron et al., 2013) or by developing bio-economic models that jointly simulate the agents in an analysis of the impact on the environment and on space (Louhichi et al., 2010).

Application of HNV indicator

An increasing number of authors have focused on measuring policy impacts by developing indicators that quantify these impacts on multiple dimensions, rather than providing a quantification of a single environmental dimension (Purvis et al., 2009; Primdahl et al., 2010). The current Common Monitoring and Evaluation Framework (CMEF) is based on the quantification of environmental impacts on a set of common indicators. The aim of these indicators is to track policy changes on the main RDP environmental objectives: 1) measuring biodiversity, 2) preserving natural and semi-natural ecosystems, 3) water quality, and 4) combating climate change.

One of these indicators is High Nature Value (HNV) farming and forestry, which quantifies changes in natural and semi-natural ecosystems through a composed index. However, although the HNV farmland index is useful for policy evaluation purposes, it has some weaknesses in accounting for causality between policy intervention and policy effects, in providing evidence of spatial spillover, and in tracking the direct and indirect policy effects highlighted by the scientific literature (Finn et al. 2009).

HNV is used as a proxy indicator to measure biodiversity and high environmental quality areas across the EU (Paracchini and Britz 2010). The reason for exploiting HNV for monitoring and evaluation, is that it is easier to collect information and to interpret causality between changes in agricultural practices and environmental quality than to use indicators based on the observation of species (Andersen et al. 2003). However, while HNV is a clearly understandable and usable tool for a policy maker to justify measures to promote sustainable agriculture, scientists have been debating about its definition and quantification.

Since the 1990s, the concept of HNV has been used to relate certain farm activities to "a natural value" (Baldock et al. 1993). HNV is a broad concept and includes both HNV farmland areas and HNV farming systems. According to Cooper et al. (2009: pp 4) HNV farmland "*comprises those areas in Europe where agriculture is a major (usually the dominant) land use and where that agriculture supports or is associated with either a high species and habitat diversity, or the presence of species of European, and/or national, and/ or regional conservation concern, or both*". A distinction between HNV farmland areas and HNV forestry is often provided, which is mainly the consequence of the expected benefits from the two HNVs.

HNV farmland has been measured by aggregating three elements (see the European Evaluation Network for Rural Development 2009):

- Farmland with a high proportion of semi-natural vegetation (type 1);
- Farmland with a mosaic of low intensity agriculture and natural and structural elements (type 2);
- A land mosaic or farmland supporting rare species of a high proportion of European or world populations (type 3).

A review of the different ways to measure HNV farmland indicators for monitoring and evaluation purposes can be found in Peppiette (2011). By comparing 24 case studies, Peppiette (2011) shows very heterogeneous definitions and measurements of these indicators as well as a lack of measurements of type 2 and type 3 indicators.

The agricultural and environmental economics literature has identified indicators (or proxies) against which to express or measure a farm contribution to HNV. Table 1 summarises the taxonomy of the HNV indicators and their measurements.

Table 1 shows the indicators used to quantify HNV farmland values. These indicators can be grouped into three approaches; land cover (type 1), farming system (type 2), and species (type 3). For a discussion of these methods, see the European Evaluation Network for Rural Development (2009).

Table 1 Indicators used to measure HNV farmland

Component	Indicator	Unit of measure	Sources
Type 1	Farmland area with native and old species and varieties; Forest Edges	(% of UAA)	Pointereau, et al., (2010);
Land cover approach		(%)	Pointereau, et al., (2010);
Type 2	Input per hectare (nitrogen, water, input costs)	(kgsq m/€ per ha)	Paracchini and Britz, (2010); Pointereau, et al., (2010); Signorotti et al. (2013); Pointereau, et al., (2010); Schwaiger (2012)
	Farmland mosaic	Plot dimension and ratio with natural and semi-natural elements (hedgerow)	
	Livestock density -	(livestock units per hectare of meadows and pastures)	Paracchini and Britz, 2010; Signorotti et al. (2013); Paracchini and Britz, (2010)
	Presence of extensive grassland, fallow and pasture	Hectares	Pointereau, et al., (2010);
Type 3	Shannon index or further index elaborations	Index	Paracchini and Britz, (2010), Pointereau, et al., (2010); Paracchini et al., 2008
	Presence of specific crops	(ha of area allocated to. olive or to wet grassland)	Pointereau, et al., (2010); Patterson (2012)
	Species data	(# of lower plants, vascular plants, butterflies, mammals and birds)	

Policy impacts on HNV farmland changes

While the literature regarding the definition of indicators that support RDP monitoring and evaluation is well developed, to the best of our knowledge only a few research papers have addressed how the policy impacts on the whole concept of HNV. While several papers quantify policy impacts on specific elements of HNV (e.g. land-use changes, land mosaic, managing intensity, livestock units per hectare), few of them have investigated the policy impacts on the HNV farmland concept as a whole.

Trisorio and Borlizzi (2011) provide a classification of the number of HNV farming systems in Italy, by combining individual data collected by Farm Structure Surveys (FSS) and the Farm Accountancy Data Network (FADN). They identify structural differences between HNV and non-HNV farms; HNV farms are characterised by a smaller size and lower efficiency in production. They further show differences regarding the amount of subsidies received from CAP and observed a higher dependency of HNV farms on RDP payments, while non-HNV farms rely more on pillar I payments. They concluded that subsidies are necessary in order to maintain the economic viability of HNV farms.

Through a spatial Durbin model, Reinhard and Linderhof (2013) estimate the policy effects (mainly payments to AES) on the HNV farmland indicator, and from an analysis at Nomenclature Units Territorial Systems (NUTS) 2 level, they found positive effects of AES on HNV. Signorotti et al. (2013) obtained the same result when applying spatial econometric models to estimate the policy effects on HNV at a municipality level. They found different effects of participation in several AESs and, in particular, involvement in organic farming and landscape measures showed a positive and significant effect, increasing the HNV indicator value, while other AESs implemented were not significant.

Paracchini and Britz (2010) applied an ex-ante analysis of the policy impacts on HNV, and by measuring impacts on alternative CAP policy designs on HNV, they found positive effects of pillar I policy on land-use changes and as a consequence on the quantity and quality of the HNV farmland indicator.

Methods

Overview

This section is organized into two parts. The first part develops the quantification of the HNV farmland indicator, while in the second part ESDA and spatial econometrics are used to estimate the determinants of the HNV farmland indicator. We used the approach developed by Paracchini and Britz, (2010), and applied by Reinhard and Linderhof (2013) to study the determinants of the HNV farmland indicator at a regional level, while Signorotti et al., (2013) adopted it for analysis at a municipality level. As previously discussed, the HNV farmland indicator (see Table 1) can be divided into three components, which measure 1) the crop diversity of non-grassland (DCI), 2) the management intensity of non-grassland (MII), and 3) the livestock intensity of grassland (LII)[a]. Formally, the HNV farmland indicator for a generic municipality i is calculated as follows:

$$HNV_i = 100\left(\sqrt{DCI_i * MII_i}\left(\frac{\sum_{s \notin G} x_s}{\sum_{s} x_s}\right) + LII_i\frac{\sum_{s \in G} x_s}{\sum_{s} x_s}\right)$$

where:

$$DCI_i = \min\left(1,\ -\sum_{c\forall s\geq 0.1}^{S} a_n * \log a_s\right)$$

$$MII_i = \left\{\begin{matrix} 1 & \text{when } n_i < P_5(n_i) \\ 2.25\text{–}0.97\left(\log_{10} n_i\right) & \text{when } P_5(n_{ii}) \geq nit_i \geq P_{95}(n_i) \\ 0 & \text{when } n_i > P_{95}(n_i) \end{matrix}\right\}$$

$$LII_i = \max\left(0,\ 1\text{–}0.708\sqrt{d_i}\right)$$

$\sum_{s} x_s$ Usable Agricultural Area (UAA) of a generic municipality i

$\sum_{s \in G} x_s$ = portion of UAA allocated to grassland or fodder crops (set G)

$\sum_{s \notin G} x_s$ = portion of the UAA not allocated to grassland or fodder crops

a_s = percentage of UAA allocated to n crops.

n_i = nitrogen surplus in a generic municipality i

$P_5(c_i)$ =5° percentile of nitrogen surplus

$P_{95}(c_i)$ =95° percentile of nitrogen surplus

d_i = livestock units per ha of grassland

The above-mentioned components quantify three different dimensions that together can be considered as a proxy of the HNV system. The three indexes are used to separately evaluate non-grassland crops and grassland crops, due to their different environmental impacts.

Compared with grassland, non-grassland areas (arable areas and orchards) are not considered as a biodiversity source (Paracchini and Britz, 2010). However, when non-grassland areas are managed with low inputs and present semi-natural vegetation or crop diversification, they can impact positively on biodiversity (Paracchini and Britz, 2010; Chiron et al., 2013). On the other hand, when grazing and livestock system are not managed with a low-input use, or there is high livestock intensity on the grassland, there is a reduction in biodiversity and possible negative externalities such as groundwater pollution (Caballero et al., 2007).

Within the first components (*DCI*), a modified version of the Shannon Index is used to quantify the crop-diversity on non-grassland land. This modified version of the Shannon index returns a non-linear score between zero and one (Paracchini and Britz, 2010) for each municipality.

The second index (MII) measures the intensity of nitrogen used and the index is calculated using a step function. Following Signorotti et al., (2013), for those municipalities within the 95th percentile (higher intensive), the high value of zero is assumed, while for those municipalities within the 5th percentile, a value of one is assumed. For the other municipalities, a non-linear transformation is assumed to return a value close to one when nitrogen is less than 20 kg per hectare, and close to zero when the value is higher than 190 kg per hectare. These two threshold values are compatible with the current amount of nitrogen use and are included into the range of restrictions for nitrogen sensitive areas (Bartolini et al., 2007).

The third index (*LII*) measures the impacts of grazing and livestock systems on biodiversity, by applying a step function in relation to the livestock intensity in each municipality. The function returns a value of one if livestock intensity in one municipality is close to zero, while it returns zero if the value is higher than two livestock units per ha. The upper boundary represents a set of constraints for nitrogen sensitive areas (Bartolini et al., 2007).

Spatial analysis

Spatial analysis is a growing topic within the agricultural economics literature (Anselin 2010; Corrado and Fingleton 2012). It enables visual elements to be created within cluster spatial systems, thus permitting the effects of spatial autocorrelation to be studied (Anselin 1995).

We develop the spatial analysis in two steps. Firstly ESDA is used to calculate the distribution of the HNV farmland indicator for each municipality in Tuscany (Italy), and secondly two different spatial econometrics models are used to explain the determinants of the HNV indicator. Generally, in spatial analysis one assumes that spatial autocorrelation can be derived from neighbourhood effects or from distances between observations.

ESDA is a branch of the Exploratory Data Analysis (EDA) which is applied to spatial data (Anselin 1995). It tests spatial distributions and studies spatial stationary phenomena/spatial instability or associated spatial regimes (Anselin 1995).

We use ESDA to identify global and local indicators of spatial association in order to test the existence of spatial autocorrelation between municipalities. These methods are common in regional economics and are being increasingly exploited in agricultural economics (Bell and Dalton 2007; Breustedt and Habermann 2011). Formally a global indicator of spatial correlation is calculated using the global Moran's *I* (Anselin 1995). The quantification of Moran's *I* index has the following form:

$$I = \frac{n}{\sum_{i=1}^{n}\sum_{j=1}^{n} w_{ij}} \frac{\sum_{i=1}^{n}\sum_{j=1}^{n} w_{ij}(x_i-\bar{x})(x_j-\bar{x})}{\sum_{i=1}^{n}(x_i-\bar{x})}$$

Where:

w_{ij}: Spatial weight, between municipalities i and j;

x_i, x_j: Observed values of variable HNV farmland indicator for municipalities i and j;

$\bar{x}$: Average HNV value;

n: Number of observations (municipalities).

As pointed out by Corrado and Fingleton (2012), any assumption about the spatial weight matrix is fundamental to the specification of spatial phenomena, when measuring the strength and direction of interactions between spatial units (see Mur 2013). Interactions among spatial units are generally hypothesised first by neighbourhood and second by distance. In the first approach, weight matrices are assumed as those spatial units that share part of the border, while in the second approach spatial units are related to each other in terms of distance. We tested the effects of two alternative spatial weight matrices: the order contiguity matrix (queen contiguity) and the inverse distance matrix.

The order contiguity matrix is a binary weight matrix with zeroes on the main diagonal and rows that contain zeroes when spatial units are non-contiguous, while they contain values of one between neighbouring units (Mur 2013). The inverse distance matrix is based on the Euclidean distance between centroids of each municipality. Consequently the elements of the matrix measure distances between centroids of the polygons containing municipalities (Mur 2013). Both weight matrices are row normalised, in order to get a value of one as a row sum. Consequently, the two alternative weight assumptions treat spatial association by taking into account neighbourhood or closeness among spatial units, respectively.

Following Anselin (1995), Moran's *I* is used to calculate a global indicator of spatial association, which, when significant, detects spatial instability and determines unrealistic assumptions regarding spatial stationary data. By decomposing Moran's *I*, Anselin (1995) developed Local Indicators of Spatial Association (LISA) in order to identify spatial clusters with similar values and to detect spatial outliers with dissimilar values (Anselin, 1995).

The spatial econometrics literature provides several models to control spatial effects in micro-level studies (e.g. Bell and Dalton 2007) and many models are now available (Mur, 2013). Thus, the estimation method chosen is mainly a consequence of data availability and of the expectations regarding the spatial effects of dependent variables (Mur, 2013).

According to LeSage and Pace (2009), the spatial effects can be modelled using a standard linear regression model:

$$y_i = \rho W_i y_i + X_n \beta + u$$

$$u = \lambda W_i u + \varepsilon$$

$$E\left[\varepsilon_i^2\right] = \sigma^2 h(z_i)$$

$$E[\varepsilon_i \varepsilon_j] = 0 \text{ with } i \neq j;$$

Where:

y is $i \times 1$ vector of observations HNV farmland indicator;

X is the $i \times k$ matrix of the k determinants of the HNV farmland indicator;

β is the vector of regression parameters (to be estimated);

ε is the error term;

W_i is the $i \times i$ matrix of spatial weights;

ρ and λ are the spatial lag parameter and the spatial error coefficient, respectively.

The spatial lag and the spatial error coefficients account for different spatial patterns. The spatial lag quantifies the spatial spillover within the dependent variable; while the spatial error coefficient corrects heteroskedasticity due to spatial dependence in the error term.

According to Breustedt and Habermann (2011), the above set of equations returns different models, based on spatial lag and spatial error coefficient values. Indeed, when $\rho = \lambda = 0$ the set of equations equals a standard regression model, otherwise when $\rho \neq 0$ and $\lambda = 0$ the equation equals a spatial autoregressive model (SAR) with spatial lag as an explanatory variable of HNV, while when $\rho = 0$ and $\lambda \neq 0$ the set of equation equals a spatial error model (SEM) with spatial components as an error term.

The investigation of spatial dependency and spatial heterogeneity has an economic relevance as the HNV farmland indicator is non-stationary over space and due to the spatial agglomeration of stakeholder participation of in agricultural policies. Indeed, natural and semi-natural ecosystems show edge effects (Fernàndez et al., 2002; Conceptión et al., 2012), due to the spatial heterogeneity in farming systems (Dick et al., 2014) and due to the spatial heterogeneity of climatic and market conditions (Mücher et al. 2009; Chiron et al., 2013).

Other authors (e.g. Schmidtner et al., 2012) have emphasised the spatial spillover effects of participation in agricultural policies, especially when they are based on voluntary mechanisms such as RDP measures. Thus environmental goods provision is spatially dependent on differences in farm performance or the quality of local institutions and the effects of policy targeting by policy makers (Yang et al., 2014).

Data

Data from the Italian Agriculture Census 2010 and the ARTEA[b] database of pillar I and II CAP payments from the current and previous RDP programs were used in the analysis. Table 2 shows the descriptive statistics of both dependent and explanatory variables.

The first two variables (Table 2) quantify HNV farmland indicators for the years 2010 and 2000. The HNV farmland indicator is quantified using year 2000 Census data and applying the method presented in the previous chapters. A comparison between the two time periods shows an average increase of ca. 10% in the municipalities in Tuscany.

Explanatory variables can be classified into three categories: farm structure, policy and location. Farm structure variables study the relationships between the characteristics of municipal farming systems, farm management and provision of environmental goods. Variables describing farm structure measuring the productive factors used can affect the farming system intensification/extensification (Bartolini et al., 2011) and thus return different HNV farmland scores. In addition, the characteristics of farmers affect their attitudes towards adopting public goods (Jongeneel et al., 2008). Thus, farming systems with an intensive use of productive factors can have a lower HNV farmland value (Foley et al. 2005).

However, other authors have pointed out a higher efficiency in the marginal productivity of factors used for more intensified farming systems (for example Cohn et al. 2014). This category of variables includes the following variables from Table 2: 1) age (*d_old and d_young*), 2) the amount of household labour used on the farm (*partime*), 3) external labour (*extlab*), 4) legal status (*coltdir, condother*), 5) farmer living on farm (*live_on*), 6) average farm size (*size*), 7) quality products (*d_quality*), 8) plots (*plots*) as a proxy of farm efficiency.

The second category contains policy variables, belonging to both CAP pillar I and II. The data used were obtained by aggregating payments received since 2003 under the single payments scheme and from the two RDP programs, namely RDP 2000-2006 and RDP 2007-2013, by the farms located in each municipality. Policy has a prominent role in promoting the provision of HNV farmland (Signorotti et al. 2013; Reinhard and Linderhof 2013). The first pillar policy can negatively affect the HNV farmland score due to maintenance support and the introduction of commodity crops. A negative impact could be balanced by the adoption of cross-compliance commitments on the eligible land (Juntti 2012).

The broader objectives of second pillar measures - increasing productivity, increasing the provision of environmental goods, increasing jobs and the viability of rural areas - can impact on the HNV farmland value (e.g. measures aimed at modernisation can have a negative impact on the HNV farmland value, while increase viabilities of rural areas and payments for introduction and maintenance of environment good positive effects). Hence,

Table 2 Descriptive statistics of dependent and explanatory variables

Description	Code	Mean	sd	Min	Max
HNV at year 2010	hnv_2010	44.72	13.57	-	84.07
HNV at year 2000	hnv_2000	32.14	12.29	-	75.26
Share of farmers older than 60 years old	d_old	39.57	11.43	-	67.74
Share of farmers younger than 40 years old	d_young	11.81	7.46	-	50.00
Share of part time farms (%)	partime	33.46	14.96	-	86.05
Share of farms using external labour (%)	extlab	3.69	4.52	-	37.50
Share of individual farms (%)	coltdir	97.98	3.02	75.00	100.00
Share of farms with legal status different than society and individual farm (%)	Condoth	0.33	0.80	-	7.55
Share of farmers that live on-farm (%)	live_on	86.32	8.18	53.33	100.00
Average farm size (ha)	size	10.86	8.53	0.65	50.98
Share of farms who sell quality products (%)	d_quality	13.02	14.97	-	62.67
Share of farms with only one plot (%)	Plot	62.52	19.40	12.50	100.00
Average SFP per ha (100 €)	SFP	9.79	5.02	-	35.46
Average RPD –first axis payments per ha (€)	axis1	347.86	532.06	-	3997.56
Average RPD –third axis payments per ha (€)	axis3	315.83	4102.83	-	69244.27
Share of participating in measure 121 (% of farmers)	rdp_121	2.52	3.96	-	37.50
Sum of compensation payments (ln)	comp	4.09	5.67	-	14.63
Sum of payments for organic production (ln)	org	9.27	4.2	-	14.47
Sum of payments for integrated production (ln)	int	10.35	3.14	-	14.43
Sum of forestry payments (ln)	for	13.69	2.09	-	16.88
Location in urban area (dummy)	poli_urb	0.07	0.26	-	1
Location in intensive agricultural area (dummy)	rur_int	0.11	0.31	-	1
Location in rural areas in transition (dummy)	rur_tran	0.32	0.47	-	1
Location in declining rural area (dummy)	rur_decl	0.24	0.43	-	1
Location in rural area with developing problems (dummy)	rur_probsv	0.26	0.44	-	1
UAA classified as Natura2000 sites (ha)	Nat2000	9.27	4.2	-	14.47
UAA classified as Nitrate Vulnerable directive (ha)	zvn	9.27	4.2	-	14.47

disentangling the impact of second pillar policies can account for the expected multiple impacts on HNV.

Variables using these data were tested: the effects of the intensity of payments (per hectare payments), the logarithmic transformation of the cumulative amount of payments received until 2010. The following variables were used as covariates: SFP per hectare of Usable Agricultural Area, (*SFP*), the logarithmic transformation of the cumulative sum of payments received from first (*axis1*) and third RDP axes (*axis3*); participation in measure 121, aimed at promoting farm modernisation (*rdp121*), which was one the most relevant measures of the first axis in terms of budget.

The impacts of RDP measures belonging to the second axis were tested using several separate sets of variables by dividing the measures into four groups: 1) compensation payments (measures 211 regarding compensation payments for farms located in less favourable areas, and 212 compensation payments for farms located in mountain areas), 2) organic farming (measure 214, sub-measure A1), 3) integrated production (measure 214, sub-measure A2), and 4) forestry measures (measures 223 afforestation of non-agricultural land, 225 forest-environmental payments; 227 non-productive investment in forestry farms). For each group of measures, the logarithmic transformations of cumulative payments received (*comp, org, int, for*) were used to estimate the policy impacts on HNV.

A further category includes location variables, such as dummies for RDP zoning. This zoning was performed using inhabitant density per municipality and the level of income. The area of Tuscany was classified into five categories (see Tuscany Region 2007 for a definition of the criteria used). Municipalities located inside urban areas belong to the first category, (*poli_urb*). The second category (*rur_int*) contains municipalities located in areas whose population density is lower than 300 inhabitants per square kilometre but with very intensive production. The three remaining zones correspond to rural areas characterised by increasing socio-economic concerns, namely rural areas in transition (*rur_trans*), declining rural areas (*rur_decl*), and rural areas with development problems (*rur_probsv*). To identify these additional three categories, variables concerning the share of agricultural labour in the total labour force with respect to the EU 25 average value, were considered in order to identify a declining rural area (*rur_decl*). Finally rural areas in transition and rural areas with development problems were identified using criteria of urban density less than 150 inhabitants. An additional criterion to distinguish between these two was the amount of tourism in the municipality. Thus municipalities close to urban areas or with tourist attractions (for example in coastal areas) were classified as rural areas in transition (*rur_trans*), while others were classified as declining rural areas (*rur_decl).*

Thus different zoning explains the relation between the expectation of off-farm earnings and rural viability. The last zones (*rur_decl and rur_probsv*) represent marginal municipalities located in hilly or mountain areas and characterised by low incomes compared with the others. In these zones a higher HNV farmland value is expected due to less opportunity for commodity production and low intensive production due to structural and market barriers (Bartolini et al., 2014).

The last category includes the amount of farmland located under Nature 2000 or under nitrogen-vulnerable protection areas (*Nat200; zvn*). These two zonings are aimed at controlling the inputs for farming activities. They affect crop profitability and thus should have a positive effect on the HNV farmland value.

Results and discussion

In this section the results of the spatial distribution of HNV are presented, secondly the ESDA analysis is done, for global and local indicators of spatial association, and thirdly, the results of the spatial econometrics models are presented.

The objective of the ESDA is to detect the spatial autocorrelation of the HNV farmland indicator distribution, while the objective of the spatial econometric technique is to estimate HNV farmland indicator determinants, taking into account spatial dependences and spatial heterogeneities (LeSage and Pace, 2009). These analyses are both performed using two alternative spatial weight matrices, namely, first order contiguity and inverse distances.

Figure 1 shows the spatial distribution of HNV municipalities in Tuscany.

Figure 1 presents the HNV farmland indicator for municipalities in Tuscany for 2010 and 2000. The figure containing 2010 values shows a piecemeal distribution of the HNV

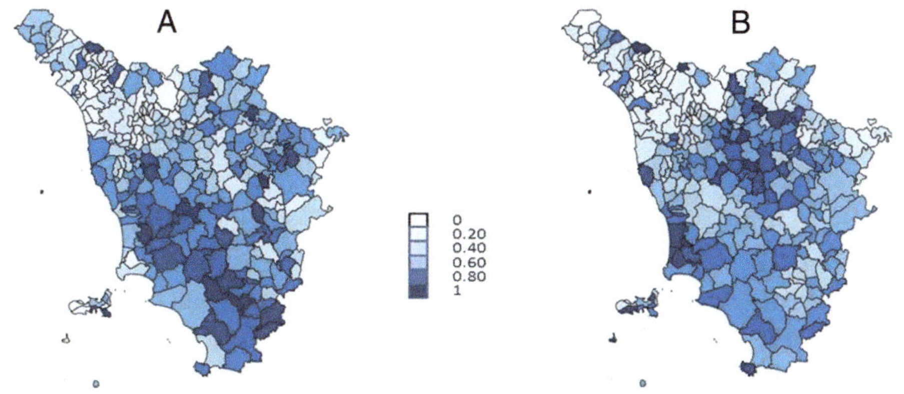

Figure 1 Spatial distribution of HNV in 2010 (A) and HNV in 2000 (B).

farmland indicator among the municipalities of Tuscany and a clear agglomeration in the south west. The municipalities with a higher HNV performance are mainly located in the provinces of Grosseto, Pisa and Arezzo[c]. High HNV indicator scores were also observed in the north east , but with an uneven distribution among municipalities. Municipalities with high HNV farmland indicators belong to different agricultural production lines with arable farming systems in the south and a combination of livestock and grazing farming systems in the north. In contrast, the intensive agricultural areas located in the provinces of Pistoia, Massa and Florence, show a very low HNV performance, due to the prevailing agricultural system, which is based on floriculture and nursery production. In addition, the map (Figure 1) shows a low HNV farmland indicator value in the mountain areas of Massa and Lucca, due to the relatively high portion of the area allocated to forestry.

The HNV score comparison between years 2010 and 2000 shows an HNV reduction in the central area, as well as in the provinces of Arezzo, Florence and Siena. The comparison highlights small changes in those municipalities with a low HNV score. These municipalities are located in the fluvial planes of the provinces of Lucca, Pisa and Florence.

The ESDA results are shown in Figure 2.

The chart on the left in Figure 2 shows Moran's scatterplot using queen contiguity (A) and the chart on the right uses inverse distances matrix (B). Moran's scatterplot is divided into four quadrants. The one at the top right represents the "hot spots" and the one on the bottom

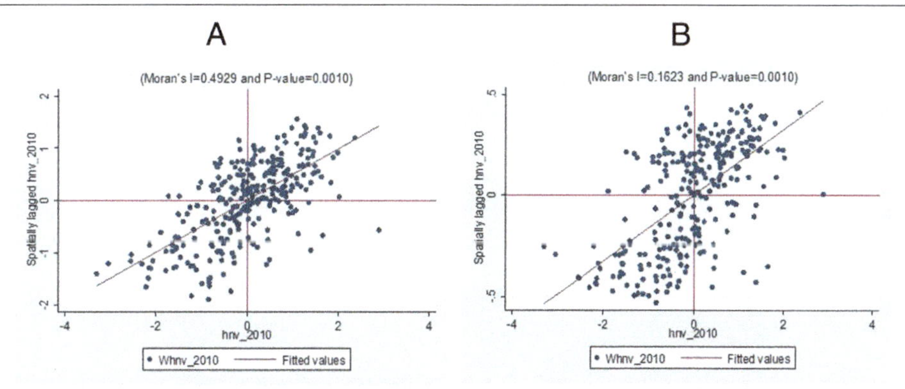

Figure 2 Moran's I indicator for the HNV farmland indicator with different spatial weight matrices. **A)** First order contiguity, **B)** Inverse distances.

left represents the "cold spots". The line represents the coefficient of Global Moran's I measuring spatial association, which assumes values between -1 and +1 (LeSage and Pace, 2009).

Both the weight matrices result in positive and significant values, which implies that the distribution of HNV is affected by spatial association. The positive and significant values mean that the hypothesis of non-spatial dependency of the observed variables can be rejected, which is the complete spatial random distribution of the HNV farmland indicator. The higher score of Moran's' I using first order contiguity compared with the inverse distance weight matrix shows a higher value of global association (0.49 versus 0.19). The results highlight that this weight matrix best describes the pattern of spatial autocorrelation of the HNV distribution. Thus the results highlight that the value of the HNV increases with proximity, but with a different magnitude. Therefore Moran's' I results allow to carry out the spatial analysis.

Figure 3 presents the spatial association (LISA) of the local indicators using alternative spatial weight matrices.

The maps in Figure 3 show the spatial clusters obtained by LISA. The map on the left assumes that spatial effects are driven by contiguity between municipalities, while the map on the right follows the assumption that municipalities are spatially related to each other by the inverse distance matrix. The coloured areas in the map (Figure 3) measure those municipalities that present a statistically significant spatial association. Both Figures 2 and 3 show a "hot spot" cluster (red municipalities), and a "cold spot" cluster (blue municipalities). Figure 3 highlights the extent of the three clusters of local spatial association. The largest cluster, in red (hot spot cluster) is located in the south of the region in the provinces of Grosseto, Siena, Pisa and Livorno. This area shows a high and significant HNV farmland score for the municipalities and for the neighbouring area or close municipalities. The second largest area with a significant local spatial association pattern is the area in the north-west of Tuscany. Here the municipalities show low HNV farmland performances. The municipalities located in the plain represent the most productive area of the region. The third cluster is located in the area between Florence and Arezzo and is characterised by a significantly high HNV farmland value.

The other municipalities highlighted represent spatial outliers, where the spatial unit has a significant and dissimilar value compared to its near-by or neighbouring municipalities. As shown in Figure 3, these spatial outliers area are located in the surrounding area of the two larger clusters.

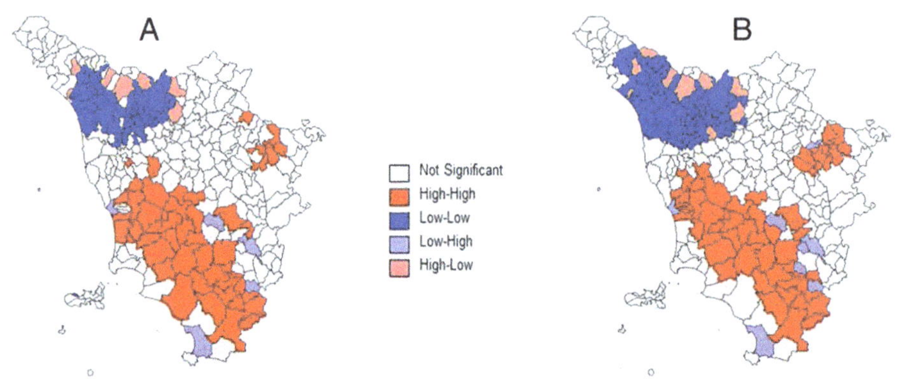

Figure 3 LISA of HNV farmland indicator assuming alternative spatial weights matrices. A) First order contiguity. **B)** Inverse distances.

The Table 3 presents the Lagrange Multiplier tests on the ordinary least squares (OLS) residual, while HNV farmland indicator determinants are estimated in Tables 4 and 5 using the two different weight matrices.

The aim of Lagrange multiplier (LM) tests (Table 3) is to compare the OLS model with alternative Spatial Autoregressive (SAR) and Spatial Error (SEM) models, (Table 4), hypothesising the absence of spatial dependence. Results show that assumptions regarding the spatial dependency of HNV are significant for each weight matrix (even with different magnitudes) and the tests suggest the application of both SAR and SEM models (Table 3).

Table 4 presents the OLS, SAR and SEM model results using a spatial weight matrix based on queen contiguity, while Table 5 presents SAR and SEM models using the inverse distance matrix.

The first column in Tables 4 and 5 presents OLS model results (model 1), the second and third columns of Table 4 present the SAR (model 2a) and SEM (model 3a) model results using a first order contiguity matrix, while the second and third columns of Table 5 present SAR (model 2b) and SEM (model 3b) model results using inverse distances matrix.

The OLS model results (Table 4) show that the HNV farmland indicator is positively correlated (0.452) by the value in 2000 (*hnv_2000*), suggesting that the provision of HNV increased in those municipalities with a high level of indicator in previous years. This would seem to indicate that environmental quality maintenance has been more effective than the conversion of a new area towards more environmentally sustainable management. Farm structure and the connection between household and external labour influences the HNV farmland indicator, changing the intensity of the productive factors (Trisorio and Borlizzi, 2011). Our results show that when external labour increases (*extlab*), there is a reduction in the HNV farmland indicator (-0.2603), due to the increased production intensity. In fact, the need to earn enough income to maintain farm household consumption leads to more intensive productive systems compared to areas with a lower level of household labour involved in agriculture. Our results also highlight that with an increasing percentage of part-time farms (*partime*), there is less efficiency in the use of natural resources and this has a negative impact (-0.0048) on the HNV farmland indicator.

The increase in the average farm size (*size*), positively affects (0.2319) the HNV farmland value due to an expected higher marginal value of land use by large farming systems.

We found that the share of farmers participating in measure 121 (rdp_121) has a negative effect (-0.6081) on the HNV farmland indicator due to the trade-off between measure 121 objective (farm modernisation) and the environmental measures of axis 2. OLS results show a positive impact (0.7216) of the organic farming payments on the HNV farmland indicator,

Table 3 Diagnostic test for spatial dependence

	First order contiguity	Inverse distances
LM (spatial error)	4.358**	0.081
Robust LM (spatial error)	2.305	2.846*
LM (spatial autoregressive)	14.33***	9.941***
Robust LM (spatial autoregressive)	12.277***	12.431***

(***significant at 0.01; **significant at 0.05; *significant at 0.1).

Table 4 Spatial regression model results using queen contiguity matrix weights

Variable			Queen contiguity	
		OLS	SAR	SEM
		Model 1	Model 2a	Model 3a
HNV at year 2000	hnv_2000	0.452***	0.3893***	0.4063***
Share of farmers older than 60 years old	d_old	0.1059	0.0722	0.0495
Share of farmers younger than 40 years old	d_young	−0.2021	−0.1957	−0.2922*
Share of farms using external labour (%)	extlab	−0.2603***	−0.2053**	−0.1874**
Share of part time farms (%)	partime	−0.0048**	−0.0374	−0.0611
Share of individual farms (%)	coltdir	0.1971	0.1857	0.0905
Share of farms with legal status different than society and individual farm (%)	Condoth	1.1146**	1.0299**	1.228***
Share of farmers that live on-farm (%)	live_on	−0.0155	−0.0121	0.021
Average farm size (ha)	size	0.2319***	0.1498*	0.1332
Share of farms with only one plot (%)	Plot	−0.0088	−0.0029	−0.0005
Share of farms who sell quality products (%)	d_quality	0.0131	−0.0163	−0.0067
Average SFP per ha (100 €)	SFP	−0.2513*	−0.2861*	−0.3933**
Average RPD –first axis payments per ha (€)	axis1	−0.0022	−0.0014	−0.0014
Average RPD –third axis payments per ha (€)	axis3	0.0015	0.0025	0.0035
Share of participating in measure 121 (% of farmers)	rdp_121	−0.6081*	−0.5564*	−0.6514*
Sum of payments for integrated production (ln)	Int	0.4073	0.2558	0.2625
Sum of payments for organic production (ln)	Org	0.7216***	0.5697**	0.4884**
Sum of compensation payments (ln)	Comp	0.0637	0.0701	0.0812
Sum of forestry payments (ln)	For	−0.2786	−0.2434	−0.3964
Location in urban area (dummy)	poli_urb	−3.3661	−0.5402	−0.3509
Location in intensive agricultural area (dummy)	rur_int	−6.2800***	−2.2724	−3.0717
Location in declining rural area (dummy)	rur_decl	5.7156***	4.205**	4.4842**
Location in rural area with developing problems (dummy)	rur_probsv	5.3520**	4.4356**	5.9949**
UAA classified as Natura2000 sites (ha)	nat2000	0.0010	0.0007	0.0005
UAA classified as Nitrate Vulnerable directive (ha)	nvz	−0.0017*	−0.0014	−0.0015
_cons		3.3090	−5.5711	23.8088
rho			0.3866***	
lambda				0.6804***
N		285	285	285
r2		0.510		
r2_a		0.481	0.585	0.510
aic		2116.438	2099.935	2104.245
bic		2211.403	2202.204	2206.515

(***significant at 0.01; **significant at 0.05; *significant at 0.1);

due to the extensification of organic production compared with integrated or conventional systems.

The classification of municipalities according to RDP zoning is significant when explaining the value of the HNV farmland indicator and consequently, there are quite heterogeneous effects among the zones. Rural areas with intensive agriculture (*rur_int*) negatively affect (-6.28) the performance in terms of HNV, while the location in less productive rural areas (*rur_dec* and *rur_probsv*) positively affects (5.7156 and 5.352

Table 5 Spatial regression model results using inverse distance matrix weights

Variable		Inverse distances	
		SAR	SEM
		Model 2b	Model 3b
HNV at year 2000	hnv_2000	0.4108***	0.444***
Share of farmers older than 60 years old	d_old	0.102	0.1021
Share of farmers younger than 40 years old	d_young	−0.1706	−0.201
Share of farms using external labour (%)	extlab	−0.2324***	−0.2546***
Share of part time farms (%)	partime	−0.0440	−0.0155
Share of individual farms (%)	Coltdir	0.2029	0.1885
Share of farms with legal status different than society and individual farm (%)	Condoth	1.0962***	1.126***
Share of farmers that live on-farm (%)	live_on	−0.006	−0.013
Average farm size (ha)	size	0.1643**	0.219**
Share of farms with only one plot (%)	Plot	−0.0043	−0.0086
Share of farms who sell quality products (%)	d_quality	−0.0015	0.0134
Average SFP per ha (100 €)	SFP	−0.2903*	−0.267*
Average RPD –first axis payments per ha (€)	axis1	−0.0023	−0.0024
Average RPD –third axis payments per ha (€)	axis3	0.0023	0.0018
Share of participating in measure 121 (% of farmers)	rdp_121	−0.5146*	−0.6105*
Sum of payments for integrated production (ln)	int	0.3487	0.4
Sum of payments for organic production (ln)	org	0.6510***	0.7006***
Sum of compensation payments (ln)	comp	0.1051	0.0757
Sum of forestry payments (ln)	for	−0.3774	−0.315
Location in urban area (dummy)	poli_urb	−1.4979	−3.0733
Location in intensive agricultural area (dummy)	rur_int	−3.8860*	−5.9831***
Location in declining rural area (dummy)	rur_decl	4.7889***	5.5520***
Location in rural area with developing problems (dummy)	rur_probsv	5.0807**	5.3670**
UAA classified as Natura2000 sites (ha)	nat2000	0.0009	0.0010
UAA classified as Nitrate Vulnerable directive (ha)	Nvz	−0.0018	−0.0017*
_cons		−21.9909	5.6722
rho		0.673***	
lambda			0.3935
N		285	285
r2			
r2_a		0.569	0.553
aic		2112.043	2120.174
bic		2214.313	2222.444

(***significant at 0.01; **significant at 0.05; *significant at 0.1).

respectively) the provision of HNV. Results do not show any significant effects of the location in urban areas on the HNV farmland indicator.

The OLS model reveals the negative (-0.002) and significance effect of location on a nitrogen vulnerable area (*nvz*). The results are somewhat counterintuitive due to the unexpected positive effects of the implementation of nitrogen use restriction. The lack of significance in the spatial models emphasizes the spatial distribution of the nitrogen vulnerable areas in zones with high pressure farming systems. Due to the location in plains and locations with a high water availability, there is a diffusion of high intensive production in these areas. Contrary to expectations, an increase in the share of UAA under Natura 2000 sites does not significantly affect the HNV farmland indicator.

All coefficients of determination – R^2, Akaike's Information Criterion (AIC) and Bayesian Information Criterion (BIC) – show an improvement in the estimation by adding spatial parameters so that, with the exception of model 3b, spatial dependence (ρ) and spatial heterogeneity coefficients (λ) are significant and have a positive value.

Both SAR models (model 2a in Table 4 and 2b in Table 5) show positive and significant values of ρ, which implies that the HNV farmland indicator is affected by spatial spillover. When other conditions are equal, ρ indicates an increase of 0.43% (model 2a) and 0.75% (model 2b) of the HNV farmland indicator, given an increase of one per cent (on average) of HNV among spatially related municipalities.

The SEM model results (model 3a in Table 4, and 3b in Table 5) show significant values of spatial error coefficient (λ) only in one case of contiguity among spatial units (model 2b).

Both spatial autoregressive models and spatial error models show notable differences in determinants of HNV farmland scores compared with OLS model results. The variables that are spatially distributed, such as part-time and location in a vulnerable nitrogen zone, were not significant in explaining HNV, due to the spatial dependence.

Conclusion

In this paper we investigated the linkages between agricultural activities and the provision of environmental goods by taking into account spatial interactions among observations (i.e. municipalities). Using HNV as a proxy to measure the environmental quality, we estimated the Common Agricultural Policy contribution to the introduction and maintenance of environmental quality. The results confirm the feasibility of the analyses for normative purposes given that the analysis can capture the causality among policy, farming practices and the impact on environmental quality.

The results emphasise the relevance of both spatial dependency and spatial heterogeneity and show the influence of spatial spillover in explaining the HNV farmland indicator. Thus, spatial analysis techniques improve the understanding of farmland environmental provision, including spatial spillover and correcting heteroskedasticity in the residual.

Our results show the heterogeneous contribution of agricultural activities on HNV farmland provision in municipalities in Tuscany, with the creation of a large agglomeration area characterised by a high and homogenous provision of environmental quality. We found that farm structure and farmer's entrepreneurship impact on the economic performance and are also relevant in explaining differences in environmental goods provision. Consequently environmental goods provision would seem to be a consequence of the right balance between increasing farm efficiency in the use of productive factors and the level of entrepreneurship. These results confirm previous research findings regarding differences in the efficiency of factor use among areas with varied HNV farmland performances (Trisorio and Borlizzi 2011).

Our results highlight the poor design of the environmental rules on environmentally vulnerable zones. In fact, the location of the farm both in the Natura 2000 area or in the nitrogen vulnerable zone is not significant when explaining a positive value of HNV. Such zones are mainly located in the most productive areas and hence, even after applying environmental rules, these areas show the worst HNV farmland performance.

The results claim for a coherent definition of the environmental pressure and for the environmental rules to be applied effectively. Our computation of the HNV farmland index refers mainly to the crops and livestock, without including species conservation which constitutes the basis for the design of Natura 2000. Thus more accurate indicators that take these elements into account could have a significant impact on the zoning.

The analysis highlights that the provision of HNV farmland seems to be connected more with the intensity of the activities and with the efficiency of a farm's in using resources, rather than the changing behaviour of farmers as pointed out by Wilson (2001). The results also confirm the dependence on public subsidies for maintaining a viable area with a high nature value.

Pillar I and II policies show opposing impacts on the HNV farmland indicator. Even with the introduction of cross-compliance rules, the amount of SFP received per municipality has a negative effect on the HNV indicator. Our results confirm expectations and previous findings (Signorotti et al. 2013; Reinhard and Linderhof 2013) regarding the positive effects of agri-environmental schemes in promoting the HNV farmland indicator. Despite this, our results highlight that only organic farming positively contributes to the HNV farmland value.

However, we found a controversial impact of SFP and RDP on the HNV farmland indicator. We confirmed previous findings regarding policy impacts on HNV provision. They showed a negative effect on HNV provision by income support and RDP payments within the first axis but, on the other hand, a positive effect of the agri-environmental schemes. The proposal for CAP after 2020 includes a greening measure aimed at encouraging crop diversification and the maintenance of natural and semi-natural grasslands. This measure could shift income support payments toward more environmentally sustainable productions and as a consequence reduce the negative impact of SFP on environmental quality.

We found that there is a trade-off within RDP measures when pursuing several RDP objectives. This result is somewhat surprising in this context, and reflects a trade-off between RDP objectives: increasing productivity and farm modernisation (axis one of RDP) on the one hand and improving and maintaining environmental quality (axis two of RDP) on the other. At the same time, there is a need for better coherence in the design and implementation of RDP programs among the several RDP axis and with pillar I policy. Thus, the development of a comprehensive RPD program based on overall regional and territorial strategies seems to be a priority within the new programming.

The paper has some weaknesses in terms of the selection and quantification of the HNV farmland indicator and the modelling components. A lack of data on the micro elements of environmental quality, for example the length/surface of hedgerows or landscape, and plot size, reduced the possibility to include further relevant variables to the HNV dimension (landscape, land mosaic, etc.).

Similarly the lack of data on the quality of forestry at a municipality level meant that forestry elements were excluded from the HNV, which in the context of Tuscany, involves a large portion of the total agroforestry area. Moreover, the HNV farmland indicator is quantified without including forestry farmers. The computation of the HNV farmland indicator only applies for farmers with UAA (and thus in the case of forestry farmers with a portion of forest this is excluded by the analysis). Thus, one of the limitations concerns the lack of forestry contribution to the HNV and thus a "pure" HNV indicator could not be developed.

The econometrics model was created using a static model, and adding a past HNV indicator as an explanatory variable. However, future work will involve a dynamic analysis, with the use of a dynamic model or working on the changes between the two time periods, in order to better understand the dynamics and policy effects on HNV, by developing space-temporal models.

Endnotes

[a]In this paper we focused on developing the HNV farmland index and have not pursued the quantification of forests and semi-natural areas on HNV. This exclusion is due to the lack of data a municipality level regarding the quality of forests. Thus all data refer to farms with UAA different from forests and exclusively forestry farmers were not analysed.

[b]ARTEA (Agenzia Regionale Toscana per le Erogazioni in Agricoltura) is the agency for agricultural policy payments.

[c]A map of the Tuscan Provinces is given in Additional file 1.

Additional file

Additional file 1: Map of Tuscan Provinces.

Competing interests

The authors declare that they have no competing interests.

Authors' contributions

FB wrote Methods, Results Sections; GB wrote Background; Introduction and Conclusion are common to all authors. All authors read and approved the final manuscript.

Acknowledgements

The authors wish to thank the three anonymous referees and the Associated Editor for their useful suggestions and comments on the earlier version of the paper.

References

Andersen E, Baldock D, Bennett H, Beaufoy G, Bignal E, Brouwer F, Elbersen B, Eiden G, Godeschalk F, Jones G, McCracken D, Nieuwenhuizen W, Van Eupen M, Hennekens SGZ (2003) Developing a High Nature Value Farming Area Indicator. IEEP, Copenhagen

Anselin L (1995) Local Indicators of Spatial Association—LISA. Geogr Anal 27(2):93–115, doi:10.1111/j.1538-4632.1995.tb00338.x

Anselin L (2010) Thirty years of spatial econometrics. Paper Reg Sci 89(1):3–25, doi:10.1111/j.1435-5957.2010.00279.x

Auerswald K, Kainz M, Fiener P (2003) Soil erosion potential of organic versus conventional farming evaluated by USLE modelling of cropping statistics for agricultural districts in Bavaria. Soil Use Manag 19(4):305–311, doi:10.1111/j.1475-2743.2003.tb00320.x

Baldock D, Beaufoy G, Bennett G, Clark J (1993) Nature Conservation and New Direction in The Common Agricultural Policy. Institute for European Environmental Policy. Institute for European Environmental Policy, London

Bartolini F, Gallerani V, Raggi M, Viaggi D (2007) Implementing the water framework directive: contract design and the cost of measures to reduce nitrogen pollution from agriculture. Environ Manage 40(4):567–577, doi:10.1007/s00267-005-0136-z

Bartolini F, Viaggi D, Ronchi D, Gomez Y, Paloma S, Sammeth F (2011) Assessing the Impact of Future CAP Reforms on the Demand of Production Factors, 122nd Seminar EAAE (European Association of Agricultural Economists). Ancona, Italy

Bartolini F, Gallerani V, Raggi M, Viaggi D (2012) Modelling the linkages between cross-compliance and agri-environmental schemes under asymmetric information. J Agr Econ 63(2):310–330, doi:10.1111/j.1477-9552.2012.00339.x

Bartolini F, Andreoli M, Brunori G (2014) Explaining the determinants of on-farm diversification: The case study of Tuscany Region. Bio-base Appl Econ 3(2):137-157, doi:10.13128/BAE-12994

Bell KP, Dalton TJ (2007) Spatial economic analysis in data-rich environments. J Agr Econ 58(3):487–501, doi:10.1111/j.1477-9552.2007.00123.x

Brady M, Irwin E (2011) Accounting for spatial effects in economic models of land use: recent developments and challenges ahead. Environ Res Econ 48(3):487–509, doi:10.1007/s10640-010-9446-6

Breustedt G, Habermann H (2011) The incidence of EU per-hectare payments on farmland rental rates: a spatial econometric analysis of german farm-level data. J Agr Econ 62(1):225–243

Bureau J-C, Tangermann S, Matthews A, Viaggi D, Crombez C, Knops L, Swinnen J (2012) The common agricultural policy after 2013. Intereconomics 47(6):316–342, doi:10.1007/s10272-012-0435-6

Buysse J, Van Huylenbroeck G, Lauwers L (2007) Normative, positive and econometric mathematical programming as tools for incorporation of multifunctionality in agricultural policy modelling. Agr Ecosyst Environ 120(1):70–81, http://dx.doi.org/10.1016/j.agee.2006.03.035

Caballero R, Gil A (2009) Binding constraints in Castile-La Mancha, Spain's cereal-sheep system. J Sustain Agr 33(1):3–27, doi:10.1080/10440040802394976

Caballero R, Riseth JA, Labba N, Tyran E, Musial W, Molik E, Boltshauser A, Hofstetter P, Gueydon A, Roeder N, Hoffmann H, Moreira MB, Coelho IS, Brito O, Gil A (2007) Comparative typology in six European low-intensity systems of grassland management. Adv Agron 96: doi:10.1016/s0065-2113(07)96001-0

Caballero R, Gil A, Fernández-Santos X (2008) An experts survey on sustainability across twenty-seven extensive European systems of grassland management. Environ Manage 42(2):190–199, doi:10.1007/s00267-008-9134-2

Chiron F, Princé K, Paracchini ML, Bulgheroni C, Jiguet F (2013) Forecasting the potential impacts of CAP-associated land use changes on farmland birds at the national level. Agr Ecosyst Environ 176:17–23, doi:10.1016/j.agee.2013.05.018

Cohn AS, Mosnier A, Havlík P, Valin H, Herrero M, Schmid E, O'Hare M, Obersteiner M (2014) Cattle ranching intensification in Brazil can reduce global greenhouse gas emissions by sparing land from deforestation. Proc Natl Acad Sci doi:10.1073/pnas.1307163111

Conceptión ED, Fernandez-Gonzàlez F, Dìaz M (2012) Plant diversity partitioning in Mediterranean croplands: effects of farming intensity, field edge, and landscape context. Ecol Appl 22(3):972–981, doi:10.1890/11-1471.1

Cooper T, Hart K, Baldock D (2009) The Provision of Public Goods Through Agriculture in the European Union. Institute for European Environmental Policy, London

Corrado L, Fingleton B (2012) Where is the economics in spatial econometrics? J Reg Sci 52(2):210–239, doi:10.1111/j.1467-9787.2011.00726.x

Dick J, Maes J, Smith RI, Paracchini ML, Zulian G (2014) Cross-scale analysis of ecosystem services identified and assessed at local and European level. Ecol Indicat 38:20–30, doi:10.1016/j.ecolind.2013.10.023

European Court of Auditors (2011) Is Agri-Environmental Support Well Designed and Managed? special report 7/2011. Publications Office of the European Union, Luxemburg

European Evaluation Network for Rural Development - EENRD (2009) The Application of the High Nature Value Impact Indicators. European Commission, Brussels

Fernàndez C, Acosta FJ, Abellà G, Lòpez F, Dìaz M (2002) Complex edge effect fields as additive processes in patches of ecological systems. Ecol Model 149(3):273–283, doi:10.1016/s0304-3800(01)00464-1

Finn JA, Bartolini F, Bourke D, Kurz I, Viaggi D (2009) Ex post environmental evaluation of agri-environment schemes using experts' judgements and multicriteria analysis. J Environ Plann Manag 52(5):717–737, doi:10.1080/09640560902958438

Foley JA, DeFries R, Asner GP, Barford C, Bonan G, Carpenter SR, Chapin FS, Coe MT, Daily GC, Gibbs HK, Helkowski JH, Holloway T, Howard EA, Kucharik CJ, Monfreda C, Patz JA, Prentice IC, Ramankutty N, Snyder PK (2005) Global consequences of land use. Science 309(5734):570–574, doi:10.1126/science.1111772

Hodge I (2012) Agri-environment policy in an era of lower government expenditure: CAP reform and conservation payments. J Environ Plann Manag 56(2):254–270, doi:10.1080/09640568.2012.664103

Jongeneel RA, Polman NBP, Slangen LHG (2008) Why are Dutch farmers going multifunctional? Land Use Policy 25(1):81–94

Juntti M (2012) Implementing cross compliance for agriculture in the EU: relational agency, power and action in different socio-material contexts. Sociologia Ruralis 52(3):294–310, doi:10.1111/j.1467-9523.2012.00564.x

Lefebvre M, Espinosa M, Gomez y Paloma S, Paracchini ML, Piorr A, Zasada I (2014) Agricultural landscapes as multi-scale public good and the role of the common agricultural policy. J Environ Plann Manag doi:10.1080/09640568.2014.891975

Leip A, Britz W, Weiss F, de Vries W (2011) Farm, land, and soil nitrogen budgets for agriculture in Europe calculated with CAPRI. Environ Pollut 159(11):3243–3253

LeSage JP, Pace RK (2009) Introduction to Spatial Econometrics. Introduction to Spatial Econometrics. Chapman and Hall/CRC, doi:10.1201/9781420064254.ch2

Louhichi K, Kanellopoulos A, Janssen S, Flichman G, Blanco M, Hengsdijk H, Heckelei T, Flichman G, Boisson JM (2010) Bio-economic modelling of soil erosion externalities and policy options: a Tunisian case study. J Bioeconomics 12(2):145–167

Mouysset L, Doyen L, Jiguet F (2012) Different policy scenarios to promote various targets of biodiversity. Ecol Indicat 14(1):209–221, doi:10.1016/j.ecolind.2011.08.012

Mücher CA, Hennekens SM, Bunce RGH, Schaminée JHJ, Schaepman ME (2009) Modelling the spatial distribution of Natura 2000 habitats across Europe. Landsc Urban Plann 92(2):148–159, http://dx.doi.org/10.1016/j.landurbplan.2009.04.003

Mur J (2013) Causality, uncertainty and Identification: three issues on the Spatial econometrics agenda. Italian J Reg Sci 12(1):5–28, doi:10.3280/SCRE2013-001001

Oréade-Brèche (2005) Evaluation des Measures Agro-Environnementales. Oréade-Brèche Auzeville, France

Paracchini ML, Britz W (2010) Quantifying Effects of Changed Farm Practices on Biodiversity In Policy Impact Assessment - An Application of CAPRI-Spat. Institute for Environment and Sustainability of the Joint Research Centre, European Commission

Paracchini M, Petersen J, Hoogeveen Y, Bamps C, Burfield I, van Swaay C (2008) High Nature Value Farmland in Europe. An Estimate of The Distribution Patterns on The Basis of Land Cover And Biodiversity Data, JRC, Scientific and Technical Report

Patterson G (2012) The Development of HNV Forestry Indicators for Scotland. Presentation for Forestry Commission Scotland

Peppiette Z (2011) The Challenge of Environmental Monitoring: The Example of HNV farmland, Paper presented at the 122nd Seminar of European Association of Agricultural Economists, Ancona, Italy, February 17-18, 2011

Petit S, Elbersen B (2006) Assessing the risk of impact of farming intensification on calcareous grasslands in Europe: a quantitative implementation of the MIRABEL framework. Ambio 35(6):297–303, doi:10.1579/06-r-125r.1

Pointereau P, Doxa A, Coulon F, Jiguet F, Paracchini ML (2010) Analysis of Spatial and Temporal Variations of High Nature Value Farmland and Links With Changes In Bird Populations: A Study On France. Institute for Environment and Sustainability of the Joint Research Centre - European Commission

Primdahl J, Vesterager JP, Finn JA, Vlahos G, Kristensen L, Vejre H (2010) Current use of impact models for agri-environment schemes and potential for improvements of policy design and assessment. J Environ Manage 91(6):1245–1254, doi:10.1016/j.jenvman.2009.12.012

Purvis G, Louwagie G, Northey G, Mortimer S, Park J, Mauchline A, Finn J, Primdahl J, Vejre H, Vesterager JP, Knickel K, Kasperczyk N, Balàzs K, Vlahos G, Christopoulos S, Peltola J (2009) Conceptual development of a harmonised method for tracking change and evaluating policy in the agri-environment: The Agri-environmental Footprint Index. Environ Sci Pol 12(3):321–337, http://dx.doi.org/10.1016/j.envsci.2009.01.005

Reinhard S, Linderhof V (2013) Report on Data Screening and Qualitative Identification of Causal Relationships., Deliverable D4.5 of the FP7 SPARD project

Scheele M (2008) Common Agricultural Policy: Landscapes Goods and Environmental Services For Rural Areas, Paper presented at the XII Congress of the European Association of Agricultural Economists - Session Brussels, Health Check and future perspectives of the CAP: Challenges for agriculture - A day of scientific dialogue, Brussels 28 August 2008

Schmidtner E, Lippert C, Engler B, Häring AM, Aurbacher J, Dabbert S (2012) Spatial distribution of organic farming in Germany: does neighbourhood matter? Eur Rev Agr Econ 39(4):661–683, doi:10.1093/erae/jbr047

Schwaiger E (2012) The concept of High Nature Value Farmland in Austria – Good practice, Presentation at European Evalutation Network for Rural Development Workshop "High Nature Value Farmland and Forestry". Edinburgh, 20 February 2012

Signorotti C, Marconi V, Raggi M, Viaggi D (2013) How do Agri-Environmental Schemes (AES's) Contribute to High Nature Value (HNV) Farmland: A Case Study in Emilia Romagna, Paper presented at the Second Congress of Italian Association of Agricultural and Applied Economics, Parma, Italy, June 6-7, 2013

Trisorio A, Borlizzi A (2011) Assessing the Impact of Rural Policy on Biodiversity: High Nature Value Farming In Italy, Paper presented at the 122nd Seminar of European Association of Agricultural Economists, Ancona, Italy, February 17-18, 2011

Wilson GA (2001) From productivism to post-productivism … and back again? Exploring the (un)changed natural and mental landscapes of European agriculture. Trans Inst Br Geogr 26(1):77–102, doi:10.1111/1475-5661.00007

Yang AL, Rounsevell MDA, Wilson RM, Haggett C (2014) Spatial analysis of agri-environmental policy uptake and expenditure in Scotland. J Environ Manage 133(0):104–115, http://dx.doi.org/10.1016/j.jenvman.2013.11.038

Permissions

All chapters in this book were first published in AFE, by Springer; hereby published with permission under the Creative Commons Attribution License or equivalent. Every chapter published in this book has been scrutinized by our experts. Their significance has been extensively debated. The topics covered herein carry significant findings which will fuel the growth of the discipline. They may even be implemented as practical applications or may be referred to as a beginning point for another development.

The contributors of this book come from diverse backgrounds, making this book a truly international effort. This book will bring forth new frontiers with its revolutionizing research information and detailed analysis of the nascent developments around the world.

We would like to thank all the contributing authors for lending their expertise to make the book truly unique. They have played a crucial role in the development of this book. Without their invaluable contributions this book wouldn't have been possible. They have made vital efforts to compile up to date information on the varied aspects of this subject to make this book a valuable addition to the collection of many professionals and students.

This book was conceptualized with the vision of imparting up-to-date information and advanced data in this field. To ensure the same, a matchless editorial board was set up. Every individual on the board went through rigorous rounds of assessment to prove their worth. After which they invested a large part of their time researching and compiling the most relevant data for our readers.

The editorial board has been involved in producing this book since its inception. They have spent rigorous hours researching and exploring the diverse topics which have resulted in the successful publishing of this book. They have passed on their knowledge of decades through this book. To expedite this challenging task, the publisher supported the team at every step. A small team of assistant editors was also appointed to further simplify the editing procedure and attain best results for the readers.

Apart from the editorial board, the designing team has also invested a significant amount of their time in understanding the subject and creating the most relevant covers. They scrutinized every image to scout for the most suitable representation of the subject and create an appropriate cover for the book.

The publishing team has been an ardent support to the editorial, designing and production team. Their endless efforts to recruit the best for this project, has resulted in the accomplishment of this book. They are a veteran in the field of academics and their pool of knowledge is as vast as their experience in printing. Their expertise and guidance has proved useful at every step. Their uncompromising quality standards have made this book an exceptional effort. Their encouragement from time to time has been an inspiration for everyone.

The publisher and the editorial board hope that this book will prove to be a valuable piece of knowledge for researchers, students, practitioners and scholars across the globe.

List of Contributors

Roberto Furesi
Department of Science for Nature and Environmental Resources, University of Sassari, Via Enrico de Nicola 1, Sassari 07100, Italy

Fabio A Madau
Department of Science for Nature and Environmental Resources, University of Sassari, Via Enrico de Nicola 1, Sassari 07100, Italy

Pietro Pulina
Department of Science for Nature and Environmental Resources, University of Sassari, Via Enrico de Nicola 1, Sassari 07100, Italy

Khalid M Al-Khalidi
Community Based Rangeland Rehabilitation Programme, Royal Botanic Garden, P.O. Box 99 11910- Amman, Jordan

Amani A Alassaf
Community Based Rangeland Rehabilitation Programme, Royal Botanic Garden, P.O. Box 99 11910- Amman, Jordan

Mustafa F Al-Shudiefat
Community Based Rangeland Rehabilitation Programme, Royal Botanic Garden, P.O. Box 99 11910- Amman, Jordan

Raed J Al-Tabini
Community Based Rangeland Rehabilitation Programme, Royal Botanic Garden, P.O. Box 99 11910- Amman, Jordan

Yibekal A Tessema
School of Agricultural Economics and Agribusiness, Haramaya University, Dire Dawa, Ethiopia

Chanyalew S Aweke
Department of Rural Development and Agricultural Extension, Haramaya University, Dire Dawa, Ethiopia

Getachew S Endris
Department of Rural Development and Agricultural Extension, Haramaya University, Dire Dawa, Ethiopia

Markus Larsson
Stockholm Resilience Centre, Stockholm University, Kräftriket 2B, Stockholm SE-106 91, Sweden
Mälardalens högskola, Box 883, Västerås 721 23, Sweden.
3Swedish Society for Nature Conservation, Box 4625, Stockholm 116 91, Sweden

Louise Morin
Stockholm Resilience Centre, Stockholm University, Kräftriket 2B, Stockholm SE-106 91, Sweden

Thomas Hahn
Stockholm Resilience Centre, Stockholm University, Kräftriket 2B, Stockholm SE-106 91, Sweden

Johanna Sandahl
Swedish Society for Nature Conservation, Box 4625, Stockholm 116 91, Sweden

Huynh Viet Khai
Department of Agricultural, Resource and Environmental Economics, School of Economics and Business Administration, Can Tho University, Can Tho City, Vietnam
Laboratory of Environmental Economics, Department of Agricultural and Resource Economics, Faculty of Agriculture, Kyushu University, Hakozaki 6-10-1, Higashi-ku, Fukuoka city 812-8581, Japan

Mitsuyasu Yabe
Laboratory of Environmental Economics, Department of Agricultural and Resource Economics, Faculty of Agriculture, Kyushu University, Hakozaki 6-10-1, Higashi-ku, Fukuoka city 812-8581, Japan

Adebayo M Shittu
Department of Agricultural Economics and Farm Management Federal University of Agriculture, Abeokuta, P.M.B. 2240, Abeokuta, Ogun State, Nigeria

Rupak Goswami
Integrated Rural Development and Management Faculty Centre, Ramakrishna Mission Vivekananda University, Narendrapur, Kolkata-700103, INDIA

Soumitra Chatterjee
Department of Agricultural Economics, Bidhan Chandra Krishi Viswavidyalaya, Mohanpur, Nadia-741252, West Bengal, INDIA

Binoy Prasad
Integrated Rural Development and Management Faculty Centre, Ramakrishna Mission Vivekananda University, Narendrapur, Kolkata-700103, INDIA

Diep Thanh Tung
Faculty of Economics, Laws and Foreign Languages, Tra Vinh University, 126, National Road 53, Ward 5, Tra Vinh, Vietnam
Institute of Project and Regional Planning, Justus-Liebig-University, Giessen, Germany

Edward Martey
CSIR-Savanna Agricultural Research Institute, Nyankpala, P.O. Box TL 52, Tamale, Ghana

Prince M Etwire
CSIR-Savanna Agricultural Research Institute, Nyankpala, P.O. Box TL 52, Tamale, Ghana

Alexander N Wiredu
CSIR-Savanna Agricultural Research Institute, Nyankpala, P.O. Box TL 52, Tamale, Ghana

Wilson Dogbe
CSIR-Savanna Agricultural Research Institute, Nyankpala, P.O. Box TL 52, Tamale, Ghana

Debdatta Pal
Economics and Business Environment, Indian Institute of Management Raipur, GEC Campus, Raipur 352015, India

Arnab Kumar Laha
Production and Quantitative Methods, Indian Institute of Management Ahmedabad, Vastrapur, Ahmedabad 380015, India

Luca Bartoli
Department of economics and Law, University of Cassino and Southern Lazio, Via Folcara Cassino (Fr) 03043, Italy

Marcello De Rosa
Department of economics and Law, University of Cassino and Southern Lazio, Via Folcara Cassino (Fr) 03043, Italy

Epiphane Sodjinou
Ecole Nationale des Sciences et Techniques Agronomiques de Djougou, Université de Parakou, 03 BP 412, Porto-Novo, République du Benin

Laurent C Glin
FiBL West Africa, BP 16 IER/CRRA Sikasso, République du Mali

Gian Nicolay
FiBL (Research Institute of Organic Agriculture), 5070 Frick, Switzerland

Silvère Tovignan
Faculté d'Agronomie, Université de Parakou, Parakou, République du Benin

Jonas Hinvi
Institut National des Recherches Agricoles du Benin, Natitingou, République du Benin

Daniela Toccaceli
Department of Science for Economics and Business, University of Florence Via delle Pandette, 9 I-50127Florence

Fabio Bartolini
Department of Agriculture, Food and Environment, University of Pisa, Via del Borghetto 80, 56124 Pisa, Italy

Gianluca Brunori
Department of Agriculture, Food and Environment, University of Pisa, Via del Borghetto 80, 56124 Pisa, Italy